T0180444

Foundations of Quantitative Finance

Chapman & Hall/CRC Financial Mathematics Series

Aims and scope:
The field of financial mathematics forms an ever-expanding slice of the financial sector. This series aims to capture new developments and summarize what is known over the whole spectrum of this field. It will include a broad range of textbooks, reference works and handbooks that are meant to appeal to both academics and practitioners. The inclusion of numerical code and concrete real-world examples is highly encouraged.

Series Editors:

M.A.H. Dempster
Centre for Financial Research
Department of Pure Mathematics
and Statistics
University of Cambridge, UK

Rama Cont
Department of Mathematics
Imperial College, UK

Dilip B. Madan
Robert H. Smith School of Business
University of Maryland, USA

Robert A. Jarrow
Lynch Professor of Investment Management
Johnson Graduate School of Management
Cornell University, USA

Commodities: Fundamental Theory of Futures, Forwards, and Derivatives Pricing, Second Edition
M.A.H. Dempster, Ke Tang

Foundations of Qualitative Finance
Book 1: Measure Spaces and Measurable Functions
Robert R. Reitano

Introducing Financial Mathematics: Theory, Binomial Models, and Applications
Mladen Victor Wickerhauser

Foundations of Qualitative Finance
Book II: Probability Spaces and Random Variables
Robert R. Reitano

Financial Mathematics: From Discrete to Continuous Time
Kevin J. Hastings

Financial Mathematics : A Comprehensive Treatment in Discrete Time
Giuseppe Campolieti and Roman N. Makarov

For more information about this series please visit: https://www.crcpress.com/Chapman-HallCRC-Financial-Mathematics-Series/book-series/CHFINANCMTH

Foundations of Quantitative Finance

Book I: Measure Spaces and Measurable Functions

Robert R. Reitano
Brandeis International Business School
Waltham, MA 02454

CRC Press
Taylor & Francis Group
Boca Raton London New York

CRC Press is an imprint of the
Taylor & Francis Group, an **informa** business

A CHAPMAN & HALL BOOK

First edition published 2022
by CRC Press
6000 Broken Sound Parkway NW, Suite 300, Boca Raton, FL 33487-2742

and by CRC Press
2 Park Square, Milton Park, Abingdon, Oxon, OX14 4RN

©2022 Robert R. Reitano

CRC Press is an imprint of Taylor & Francis Group, LLC

Reasonable efforts have been made to publish reliable data and information, but the author and publisher cannot assume responsibility for the validity of all materials or the consequences of their use. The authors and publishers have attempted to trace the copyright holders of all material reproduced in this publication and apologize to copyright holders if permission to publish in this form has not been obtained. If any copyright material has not been acknowledged please write and let us know so we may rectify in any future reprint.

Except as permitted under U.S. Copyright Law, no part of this book may be reprinted, reproduced, transmitted, or utilized in any form by any electronic, mechanical, or other means, now known or hereafter invented, including photocopying, microfilming, and recording, or in any information storage or retrieval system, without written permission from the publishers.

For permission to photocopy or use material electronically from this work, access www.copyright.com or contact the Copyright Clearance Center, Inc. (CCC), 222 Rosewood Drive, Danvers, MA 01923, 978-750-8400. For works that are not available on CCC please contact mpkbookspermissions@tandf.co.uk

Trademark notice: Product or corporate names may be trademarks or registered trademarks and are used only for identification and explanation without intent to infringe.

ISBN: 978-1-032-19120-1 (hbk)
ISBN: 978-1-032-19118-8 (pbk)
ISBN: 978-1-003-25774-5 (ebk)

DOI: 10.1201/9781003257745

Typeset in Palatino
by SPi Technologies India Pvt Ltd (Straive)

To Michael, David, and Jeffrey

Contents

Preface

The idea for a reference book on the mathematical foundations of quantitative finance has been with me throughout my professional and academic careers in this field, but the commitment to finally write it did not materialize until completing my first "introductory" book in 2010.

My original academic studies were in "pure" mathematics in a subfield of mathematical analysis, and neither applications generally nor finance in particular were even on my mind. But on completion of my degree, I decided to temporarily investigate a career in applied math, becoming an actuary, and in short order became enamored with mathematical applications in finance.

One of my first inquiries was into better understanding yield curve risk management, ultimately introducing the notion of partial durations and related immunization strategies. This experience led me to recognize the power of greater precision in the mathematical specification and solution of even an age-old problem. From there my commitment to mathematical finance was complete, and my temporary investigation into this field became permanent.

In my personal studies, I found that there were a great many books in finance that focused on markets, instruments, models and strategies, and that typically provided an informal acknowledgement of the background mathematics. There were also many books in mathematical finance focusing on more advanced mathematical models and methods, and they were typically written at a level of mathematical sophistication requiring a reader to have significant formal training and the time and motivation to derive omitted details.

The challenge of acquiring expertise is compounded by the fact that the field of quantitative finance utilizes advanced mathematical theories and models from a number of fields. While there are many good references on any of these topics, most are again written at a level beyond many students, practitioners and even researchers of quantitative finance. Such books develop materials with an eye toward comprehensiveness in the given subject matter, rather than with one toward efficiently curating and developing the theories needed for applications in quantitative finance.

Thus the overriding goal for this collection of books is to provide a complete and detailed development of the many foundational mathematical theories and results one finds referenced in popular resources in finance and quantitative finance. The included topics have been curated from a vast mathematics and finance literature for the express purpose of supporting applications in quantitative finance.

I originally budgeted 700 pages per book, in two volumes. It soon became obvious this was too limiting, and two volumes ultimately turned into ten. In the end, each book is dedicated to a specific area of mathematics or probability theory, with a variety of applications to finance that are relevant to the needs of financial mathematicians.

My target readers are students, practitioners and researchers in finance who are quantitatively literate, and who recognize the need for the materials and formal developments presented. My hope is that the approach taken in the books will motivate readers to navigate the details and master the materials.

Most importantly for a reference work, all ten volumes are extensively self-referenced. The reader can enter the collection at any point of interest, and then, using the references cited, work backward to prior books to fill in needed details. This approach also works for a course on a given volume's subject matter, with earlier books used for reference, and for both course-based and self-study approaches to sequential studies.

The reader will find that the developments herein are at a much greater level of detail than most advanced quantitative finance books. Such developments are of necessity typically longer, more meticulously reasoned, and therefore can be more demanding on the reader. Thus before committing to a detailed line-by-line study of a given result, it is always more efficient to first scan the derivation once or twice to better understand the overall logic flow.

I hope the additional details presented will support your journey to better understanding.

I am grateful for the support of my family: Lisa, Michael, David, and Jeffrey, as well as the support of friends and colleagues at Brandeis International Business School.

Robert R. Reitano
Brandeis International Business School

Author Bio

Robert R. Reitano is Professor of the Practice of Finance at the Brandeis International Business School where he specializes in risk management and quantitative finance, and where he previously served as MSF Program Director, and Senior Academic Director. He has a Ph.D. in Mathematics from MIT, is a Fellow of the Society of Actuaries, and a Chartered Enterprise Risk Analyst. He has taught as Visiting Professor at Wuhan University of Technology School of Economics, Reykjavik University School of Business, and as Adjunct Professor in Boston University's Masters Degree program in Mathematical Finance. Dr. Reitano consults in investment strategy and asset/liability risk management and previously had a 29-year career at John Hancock/Manulife in investment strategy and asset/liability management, advancing to Executive Vice President & Chief Investment Strategist. His research papers have appeared in a number of journals and have won an Annual Prize of the Society of Actuaries and two F.M. Redington Prizes awarded biennially by the Investment Section of the Society of the Actuaries. Dr. Reitano has served as Vice-Chair of the Board of Directors of the Professional Risk Managers International Association (PRMIA) and on the Executive Committee of the PRMIA Board, and is currently a member of the Financial Research Committee of the Society of Actuaries, and other not-for-profit boards and investment committees.

Introduction

Foundations of Quantitative Finance is structured as follows:

Book I. *Measure Spaces and Measurable Functions*

Book II. *Probability Spaces and Random Variables*

Book III. *The Integrals of Riemann, Lebesgue and (Riemann-)Stieltjes*

Book IV. *Distribution Functions and Expectations*

Book V. *General Measure and Integration Theory*

Book VI. *Densities, Transformed Distributions, and Limit Theorems*

Book VII. *Brownian Motion and Other Stochastic Processes*

Book VIII. *Itô Integration and Stochastic Calculus 1*

Book IX. *Stochastic Calculus 2 and Stochastic Differential Equations*

Book X. *Classical Models and Applications in Finance*

The series is logically sequential. Books I, III, and V develop foundational mathematical results needed for the probability theory and finance applications of Books II, IV, and VI, respectively. Then Books VII, VIII and IX develop results in the theory of stochastic processes. While these latter books introduce ideas from finance as appropriate, the final realization of the applications of these stochastic models to finance is deferred to Book X.

This Book I, *Measure Spaces and Measurable Functions*, is mathematically the most foundational book in the collection. Echoes of these results appear throughout the other volumes.

Chapters 1 to 4 develop the Lebesgue notion of measurability. This definition of measure generalizes interval length in a natural way and sets the stage for more general results.

Chapter 1 identifies how Lebesgue may have come upon the need for a theory of measurable sets and measurable functions, as he contemplated a new approach to the integral.

Chapter 2 develops Lebesgue measure and a number of its properties. This begins by explicitly defining a general measure to possess properties that are needed for it to be useful. We also require a measure to be defined on a collection of sets, a "sigma algebra," which again possesses properties that are needed to be useful. As even the most general set can be covered by a union of intervals, we attempt to then define the Lebesgue measure of a general set in terms of the minimum of the interval lengths of such covers.

That's a good start, but it will surprise the reader that this approach does not obtain a measure on all sets, and thus not all sets are Lebesgue measurable. The final solution requires identifying a useful collection of sets, a sigma algebra, on which this construction obtains a measure. Thence we obtain Lebesgue measurable sets and Lebesgue measure.

Chapter 3 turns to properties of measurable functions. These are easy to define: the inverse of such functions maps intervals to the measurable sets of Chapter 2. But then what kinds of functions have this property, and what are the useful properties of measurable functions? For example, are continuous functions measurable? Will various combinations and limits of measurable functions obtain measurable functions?

Chapter 4 consolidates and generalizes results from previous chapters into so-called Littlewood's three principles: Lebesgue measurable sets are nearly a finite union of open intervals; Lebesgue measurable functions are nearly continuous; and, convergent sequences of Lebesgue measurable functions are nearly uniformly convergent. What "nearly" means is a detail we defer to the interested reader.

In many books it is common to jump from the Lebesgue development of measure theory to the completely general framework. In this book we take the position that an intermediate (and very applicable) collection of measures is better developed first. To explain, note that one always defines the measure of a general set by way of approximations with simple sets such as intervals. In the Lebesgue theory one can be relatively informal in this approximation since open, closed and semi-closed intervals all have the same length. One the other hand, general measure theory requires far more structure for this collection of simple sets or intervals, known as a "semi-algebra." To possess this structure, intervals are defined to be right semi-closed, such as $(a, b]$.

Thus Chapter 5 develops the theory and properties of Borel measures, the fundamental measures underlying all of probability theory. A Borel measure generalizes Lebesgue measure in the sense that the length of a right semi-closed interval $(a, b]$ need not equal interval length. In general it is defined in terms of the change in a given increasing, right continuous function: $F(b) - F(a)$. The distribution functions of probability theory provide perfect examples. From this definition on this initial collection of intervals, we again generalize the measure definition to other sets much as for the Lebesgue theory.

Chapter 6 follows the Lebesgue and Borel derivations, developing a very general framework for generating measures. To illustrate these results, imagine that we have a proposal for a measure, currently defined only on a semi-algebra of sets, such as the right semi-closed intervals above. The general framework of this chapter provides results that answer a very fundamental question. What critical properties must this proposal satisfy to ensure that all the remaining details of the final derivation, as seen for Lebesgue and Borel measures, will go through without doubt? Of course, we could in every case just check all these remaining details. Yet in each of the chapters to come the reader will be relieved to know we will not need to do so.

Chapter 7 presents the first application of this general theory to finite product spaces. As an example, how does a measure on the real line $\mathbb{R}$ induce a measure on the plane $\mathbb{R}^2$? It is easy to go from the measures of intervals to the measures of rectangles, but what about the rest of the construction to general sets? More generally, how can n measures on $\mathbb{R}$ induce a measure on $\mathbb{R}^n$? These results have immediate application to Lebesgue and Borel measures on $\mathbb{R}^n$.

Chapter 8 generalizes these results, addressing general Borel measures on $\mathbb{R}^n$ and the connection between such measures and a certain class of defining functions. These results are fundamental in probability theory, where these functions are joint distribution functions. This theory also finds a finance application in the copula theory of Book II.

Finally, Chapter 9 extends the question of Chapter 7 to a countably infinite dimensional real space, denoted $\mathbb{R}^{\mathbb{N}}$. Each point in this space can be identified with an infinite sequence of real numbers. This generalization is subtle, requiring some new tools which will be further generalized in Book VII in the development of stochastic processes. This chapter's development will restrict the results from general measures to probability measures. The most critical application of this model is in Book II, which is to rigorously support the oftentimes informal notion of an infinite sample space.

I hope this book and the books to follow serve you well.

1

The Notion of Measure 0

The theory of the Riemann integration is typically developed in courses in calculus. It is introduced in the context of a continuous function defined on a bounded interval $[a, b]$, yet ultimately definable on bounded functions that are continuous **almost everywhere** on such intervals. That is to say, bounded functions that are continuous **except on a set of Lebesgue measure 0**. Recall the definition:

Definition 1.1 (Set of measure 0) *A set $E \subset \mathbb{R}$ has **measure 0**, or more formally **Lebesgue measure 0**, if for any $\epsilon > 0$ there is a finite or countable collection of open intervals $\{G_i\}$, with $G_i = (a_i, b_i)$, so that $E \subset \bigcup_i G_i$ and $\sum_i |G_i| < \epsilon$, where $|G_i| \equiv b_i - a_i$ denotes interval length.*

Example 1.2 *The rationals and any countable set of reals have measure zero. If r_i denotes the ith element, $i \geq 1$, take $(a_i, b_i) = (r_i - \epsilon/2^{i+1}, r_i + \epsilon/2^{i+1})$ and then $\sum_i |G_i| = \epsilon$.*

The Riemann theory is extendable in certain contexts to unbounded functions and/or unbounded intervals, and conventionally if not humorously labelled "improper" integrals. Real analysis is the mathematical discipline that studies properties of functions which in general are not continuous, and yet have enough "regularity" to allow the development of an integration theory.

Examples can be developed which demonstrate somewhat counterintuitive properties of the Riemann integral. First, a bounded function that is continuous except on a set of measure 0 is Riemann integrable on any bounded interval $[a, b]$. Thus in this case, the values given to the function on this set of measure 0 are irrelevant. That is, the values of the function on this set of measure 0 neither change the Riemann integrability of the function nor the value of this integral. The value of the integral is entirely determined by the function's values on its points of continuity.

On the other hand, let's begin with a function that is continuous everywhere, say, $z(x) = 0$ on the interval $[0, 1]$, and which is Riemann integrable with:

$$\int_0^1 z(x)dx = 0.$$

We cannot arbitrarily change this function's values on sets of measure 0 and preserve integrability. If this set has only finitely many points, integrability is preserved as is the value of this integral, independently of how this function is defined on these exceptional points. However, we cannot generalize from a finite set of exceptional points to an arbitrary countable infinite set.

DOI: 10.1201/9781003257745-1

The classical example of this is the **Dirichlet function**, $d(x)$, named for its discoverer, **J. P. G. Lejeune Dirichlet** (1805–1859):

$$d(x) = \begin{cases} 0, & x \in [0,1] \text{ irrational}, \\ 1, & x \in [0,1] \text{ rational}. \end{cases}$$

Then $\int_0^1 d(x)dx$ cannot be defined in the Riemann sense since every lower Riemann sum (see below) has value 0, and every upper Riemann sum has value 1.

While $z(x)$ is **everywhere continuous**, this property is completely destroyed by redefining this function on the rationals, a set of measure 0, in that $h(x)$ is now **nowhere continuous**. Ironically, one might argue that $d(x)$ should be integrable, and with integral 0, since except on a set that should not matter, $d(x) = z(x)$. But because Riemann integrability is so intrinsically connected to continuity, this integral cannot exist.

The problem that arises in this example is not always encountered when the exceptional set is infinite. An example of how the continuous function $i(x) \equiv 1$ on $[0,1]$ can be redefined on the rationals in a way that preserves continuity except on this set of measure 0 is given by:

$$f(x) = \begin{cases} 0, & x = 0, \\ 1 - \frac{1}{n}, & x = m/n, \\ 1, & x \text{ irrational}. \end{cases}$$

In this definition, it is assumed that the rational $x = m/n$ is expressed in lowest terms, meaning that n and m have no common factors.

This function modifies the function $i(x)$ only on the rationals on $[0,1]$, but in a way that preserves continuity on the irrationals, and consequently, preserves both integrability as well as the value of the Riemann integral. That is,

$$\int_0^1 f(x)dx = \int_0^1 i(x)dx = 1.$$

The function $f(x)$ is closely related to **Thomae's function** of Remark 3.19, named for **Carl Johannes Thomae** (1840–1921):

That $f(x)$ is continuous on the irrationals is proved as follows. If x_0 is irrational, then for any integer N let:

$$\delta_N = \min\left\{ |x_0 - m/n| \mid m/n \text{ is in lowest terms}, n \leq N \right\}.$$

By irrationality of x_0 it follows that $\delta_N > 0$. Thus any rational y in the interval $|x_0 - y| < \delta_N$ has denominator $n > N$ and thus $f(y) > 1 - 1/N$. In other words, $f(y) \to 1$ as $y \to x_0$.

Below we provide a brief review of the Riemann approach to integration, then turn to a short intuitive introduction to an alternative theory of Lebesgue integration. It will be seen in Book III that the Lebesgue theory preserves virtually all of the familiar properties of the Riemann counterpart, yet it eliminates all of the anomalous properties related to sets of measure 0. As will subsequently be developed in Book V, this Lebesgue theory also provides a natural framework for generalizations which are essential for probability spaces and their applications.

To begin this journey, a great deal needs to be developed relating to the notion of "measure," and it is the goal of this Book I to pursue this development.

1.1 Riemann Integrals

Let $f(x)$ be a nonnegative bounded function on $[a, b]$:

$$0 \leq m' \leq f(x) \leq M' < \infty.$$

The integration approach first introduced on a rigorous basis by **Bernhard Riemann** (1826–1866) sets out to derive the area of the planar region bounded by the graph of $f(x)$, the x-axis, and the vertical line segments $x = a$ and $x = b$. More generally, when $f(x)$ is bounded but it is both positive and negative, one derives the "signed" area, where signed means that regions above the x-axis have positive area, those below this axis have negative area, and the integral produces the net result.

The Riemann derivation begins with an arbitrary **partition** of the interval $[a, b]$ into subintervals $[x_{i-1}, x_i]$:

$$a = x_0 < x_1 < \ldots\ldots\ldots < x_{n-1} < x_n = b,$$

and arbitrary subinterval **tags** $\widetilde{x}_i \in [x_{i-1}, x_i]$. Then with $\Delta x_i = x_i - x_{i-1}$, an estimate of this area is obtained with the **Riemann sum** defined as:

$$\sum\nolimits_{i=1}^{n} f(\widetilde{x}_i) \Delta x_i.$$

A function is then said to be **Riemann integrable** if this Riemann sum converges to the same finite limit as the mesh size $\mu \to 0$, for all choices of tags. Here the **mesh size of the partition** μ is defined by:

$$\mu \equiv \max_{1 \leq i \leq n} \{x_i - x_{i-1}\}. \tag{1.1}$$

The approach introduced by **Jean-Gaston Darboux** (1842–1917) was to estimate such Riemann sums with upper and lower bounds:

$$m'(b-a) \leq \sum\nolimits_{i=1}^{n} m'_i \Delta x_i \leq \sum\nolimits_{i=1}^{n} f(\widetilde{x}_i) \Delta x_i \leq \sum\nolimits_{i=1}^{n} M'_i \Delta x_i \leq M'(b-a). \tag{1.2}$$

In this expression, m'_i denotes the **greatest lower bound** or **infimum** of $f(x)$, and M'_i the least **least upper bound** or **supremum** of $f(x)$, both defined on the subinterval $[x_{i-1}, x_i]$. In other words:

$$m'_i = \inf\{f(x) | x \in [x_{i-1}, x_i]\} \tag{1.3}$$
$$\equiv \max\{y | y \leq f(x) \text{ for } x \in [x_{i-1}, x_i]\},$$
$$M'_i = \sup\{f(x) | x \in [x_{i-1}, x_i]\} \tag{1.4}$$
$$\equiv \min\{y | y \geq f(x) \text{ for } x \in [x_{i-1}, x_i]\}.$$

In (1.2), the lower bounding summation is an example of what is referred to as a **lower Darboux sum** and sometimes a **lower Riemann sum**, while the upper sum is correspondingly referred to as an **upper Darboux sum** and sometimes an **upper Riemann sum**.

We then have the following definition of Riemann integrability:

Definition 1.3 (Riemann integrable) *A function $f(x)$ is **Riemann integrable** on a bounded interval $[a,b]$ if as $\mu \to 0$:*

$$\left[\sum_{i=1}^{n} M_i' \Delta x_i - \sum_{i=1}^{n} m_i' \Delta x_i\right] \to 0, \tag{1.5}$$

*where m_i' and M_i' are defined in (1.3) and (1.4). In this case, the **Riemann integral of** $f(x)$ **over** $[a,b]$ is defined by:*

$$(\mathcal{R}) \int_a^b f(x)dx = \lim_{\mu \to 0} \sum_{i=1}^{n} f(\tilde{x}_i) \Delta x_i, \tag{1.6}$$

*which exists and is independent of the choice of $\tilde{x}_i \in [x_{i-1}, x_i]$ by (1.2). The function $f(x)$ is then called the **integrand**, and the constants a and b the **limits of integration** of the integral.*

Traditionally, the Riemann integral is first proved to exist for continuous functions on bounded intervals $[a,b]$, then generalized. We recall the definition of continuity for single variable functions, and then for later results also note its generalizations to other spaces.

Definition 1.4 (Continuous function) *The function $f : \mathbb{R} \to \mathbb{R}$ is **continuous at** x_0 if:*

$$\lim_{x \to x_0} f(x) = f(x_0). \tag{1.7}$$

That is, given $\epsilon > 0$ there is a $\delta \equiv \delta(x_0, \epsilon) > 0$ so that:

$$\left|f(x) - f(x_0)\right| < \epsilon \text{ whenever } |x - x_0| < \delta.$$

*This function is said to be **continuous on an interval** $[a,b]$ if it is continuous at each $x_0 \in (a,b)$, and also continuous at a and b where the limit in (1.7) is understood as one-sided, meaning for $x < b$ or $x > a$. A function is said to be **continuous** if it is continuous everywhere on its domain.*

The same definition in (1.7) applies to a function $f: \mathbb{R}^n \to \mathbb{R}$, but where $|x - x_0| \equiv d(x, x_0)$ is interpreted in terms of the standard metric on $\mathbb{R}^n$:

$$d(x,y) = \left[\sum_{i=1}^{n} \left(x_i - y_i\right)^2\right]^{1/2}.$$

Even more generally this definition applies to a function $f : (X_1, d_1) \to (X_2, d_2)$, where X_j is a metric space with metric d_j. The limit in (1.7) is then defined as follows. Given $\epsilon > 0$ there is a $\delta \equiv \delta(x_0, \epsilon) > 0$, so that:

$$d_2\left(f(x), f(x_0)\right) < \epsilon \text{ whenever } d_1(x, x_0) < \delta.$$

See also Proposition 3.12.

The function $f : \mathbb{R} \to \mathbb{R}$ is said to be **uniformly continuous on a set** $E \subset \mathbb{R}$ *if given $\epsilon > 0$ there is a $\delta \equiv \delta\,(\epsilon) > 0$, so that for all $x_0 \in E$:*

$$\left| f(x) - f(x_0) \right| < \epsilon \text{ whenever } |x - x_0| < \delta.$$

In other words, δ depends on ϵ but not on x_0. The same definition applies in the other contexts.

The **Lebesgue existence theorem for the Riemann integral**, named for **Henri Lebesgue** (1875–1941) who first proved it, states that a bounded function $f(x)$ is Riemann integrable over a bounded interval $[a, b]$ if and only if it is continuous except on a set of Lebesgue measure 0, recalling Definition 1.1. It should be noted that this result does not require that the set of discontinuities be countable, only that this set has measure 0.

Proposition 1.5 (Lebesgue's Existence Theorem) *If $f(x)$ is a bounded function on the bounded interval $[a, b]$, then:*

$$(\mathcal{R}) \int_a^b f(x)dx$$

exists if and only if $f(x)$ is continuous except on a set of points, $E \equiv \{x_\alpha\}$, of measure 0.

The proof of this result can be found in Proposition 10.22 of **Reitano** (2010) and elsewhere. But it is with some embarrassment that I note the error in Chapter 10 of this reference where this result is called the **Riemann existence theorem** and attributed to **Bernhard Riemann**. There is indeed a **Riemann existence theorem**, but it is a result in a different area of mathematics and not a result on Riemann integration. Apparently, and in retrospect quite logically, Lebesgue's existence result came some time after Riemann, using his newly developed ideas on measures.

1.2 Lebesgue Integrals

The approach to integration introduced by **Henri Lebesgue** is sometimes described as having turned the definition of the Riemann integral on its head. Rather than divide the domain of $f(x)$ into intervals, Lebesgue's idea was to divide the range of $f(x)$ into intervals. For Lebesgue's approach, given a bounded $f(x)$ with $m < f(x) < M$, we introduce a partition on the range of $f(x)$:

$$m = y_0 < y_1 < \cdots < y_n = M,$$

and define the function's **level sets** by:

$$A_i = \{x | y_{i-1} < f(x) \le y_i\}, i = 1, 2 \ldots, n. \tag{1.8}$$

We could equivalently define these level sets by switching the inequalities and using $\leq$ on the lower bound. While the inequalities in the definition of A_i are somewhat arbitrary, they are defined to assure that the $\{A_i\}$ are disjoint and $\bigcup_i A_i = [a, b]$.

On each set A_i is defined the **infimum** and **supremum** of $f(x)$ on A_i as in (1.3) and (1.4):

$$m_i = \inf\{f(x)|x \in A_i\}, \qquad M_i = \sup\{f(x)|x \in A_i\}. \tag{1.9}$$

If it exists, the Lebesgue integral $(\mathcal{L}) \int_a^b f(x)dx$ is then bounded by **upper and lower "Lebesgue sums"**:

$$\sum_{i=0}^{n} m_i |A_i| \leq (\mathcal{L}) \int_a^b f(x)dx \leq \sum_{i=0}^{n} M_i |A_i|, \tag{1.10}$$

where $|A_i|$ denotes the yet to be defined "Lebesgue measure" of the set A_i.

As will be seen, if A_i is an interval, then the Lebesgue measure of this set will be defined to equal the "length" of this interval. However, even this simple introduction motivates the observation that this setup would work equally well if in fact the notion of measure was defined in a way that was quite different from that of interval length. The first example of this will be seen in Book III, and the general case developed in Book V.

The last step in this process is to define $(\mathcal{L}) \int_a^b f(x)dx$ by a "limit," if such a limit exists, as the range partition becomes increasingly fine:

$$\nu \equiv \max\{|y_{i+1} - y_i|\} \rightarrow 0.$$

Since we cannot expect in general that these summations will monotonically increase/decrease to a unique value, the notion of limit here needs to be loosened somewhat. Thus, what is meant by "if a limit exists," is that the infimum of all upper Lebesgue sums equals the supremum of all lower Lebesgue sums in (1.10).

What is immediately clear is that if this construction has any chance of working, it is necessary that all such sets A_i be "measurable," meaning that $|A_i|$ is well defined. This requires both a "measure theory," which identifies which sets are measurable and how the measure of such sets is defined, as well as a restriction on the integrand function $f(x)$ which assures that the level sets of f are indeed measurable. Functions which have this property will be called "measurable functions."

Remarkably, for positive and bounded functions, the condition that these level sets be measurable is also sufficient for the existence of this unique limit, and thus the existence of a Lebesgue integral.

Notation 1.6 *As suggested above, the notation* $(\mathcal{L}) \int_a^b f(x)dx$ *will often be used to denote a Lebesgue integral, and correspondingly* $(\mathcal{R}) \int_a^b f(x)dx$ *to denote a Riemann integral. Admittedly this is cumbersome, and in many chapters of this and other books in this collection, the context will be clear and these labels omitted.*

In Book III it will be shown that for bounded functions that are Riemann integrable on an interval $[a, b]$, the Lebesgue integral also exists and these integral values agree.

However, the converse is not true and the Lebesgue integral can exist when the Riemann integral does not. Consequently, the Lebesgue integral will in general expand the class of integrable functions without changing the values of the integrals that already exist in the Riemann sense.

Once all this is established, the need for this notational convention will diminish in that $\int_a^b f(x)dx$ will typically be used to mean "in the sense of Lebesgue," with the knowledge that if $f(x)$ is also Riemann integrable the value of this integral can also be defined "in the sense of Riemann."

To clarify the statement that the Lebesgue integral will **in general** expand the class of integrable functions, there is what initially appears to be a technicality in the definition of Lebesgue integral which creates a class of Riemann integrable functions which are not Lebesgue integrable. These functions have the property that the Riemann integral exists because of cancellation between infinite positive and infinite negative areas.

For example, the function defined as $f(x) = (-1)^{n+1}/n$ on $[n-1, n)$ is Riemann integrable on $[0, \infty)$ because

$$(\mathcal{R}) \int_0^\infty f(x)dx = \lim_{N \to \infty} \sum_{n=1}^N (-1)^{n+1}/n = \ln 2.$$

In the Lebesgue theory, the integral $(\mathcal{L}) \int f(x)dx$ is only defined for functions for which $(\mathcal{L}) \int |f(x)| dx$ is finite. Of course, that $|f(x)|$ is not integrable on $[0, \infty)$ in the sense of Riemann or Lebesgue follows from the divergence of the **harmonic series**:

$$(\mathcal{R}) \int_0^\infty |f(x)| dx = \lim_{N \to \infty} \sum_{n=1}^N 1/n = \infty.$$

Hence a Lebesgue integral will never be defined to exist purely because of cancellation between infinite positive and negative parts. It will be seen in the later development that this restriction on the Lebesgue theory is not simply a definitional technicality, but is fundamental to the usefulness of this theory.

Example 1.7 (Lebesgue integral of $d(x)$) *Let's apply the Lebesgue construction to the Dirichlet function $d(x)$ defined above, where $d(x) = 1$ on the rationals and $d(x) = 0$ on the irrationals. There are many ways to set this up, but as an example we define as subsets of $[0, 1]$:*

$$A_0 = \{x| - 1/n < h(x) \le 0\},$$
$$A_1 = \{x|0 < h(x) \le 1 - 1/n\},$$
$$A_2 = \{x|1 - 1/n < h(x) \le 1 + 1/n\}.$$

Then A_0 is the set of irrationals in $[0, 1]$, A_1 is the empty set, and A_2 is the set of rationals in $[0, 1]$. Now:

$$A_0 \bigcup A_1 \bigcup A_2 = [0, 1], \tag{*}$$

and so the union of these sets is an interval. As Lebesgue measure will be constructed to replicate interval length, this union is Lebesgue measurable and has measure 1.

As noted above, measures and measurable sets are required to satisfy certain properties. First, the empty set A_1 is always measurable with measure 0, and this applies to Lebesgue measure. Similarly, A_2 has measure 0 by Definition 1.1 and Example 1.2, and this will assure that once defined, A_2 is Lebesgue measurable with measure 0. So what about A_0?

The collection of measurable sets is constructed to be closed under unions, intersections and set complementation. With A^c denoting the complement of the set A:

$$A_0 = [0,1] \bigcap \left(A_1 \bigcup A_2 \right)^c,$$

and it follows that A_0 is Lebesgue measurable. Now ($*$) is a disjoint union, and measures are required to satisfy "finite additivity," meaning that the measure of such a union is the sum of the measures. Thus A_0 is Lebesgue measurable with measure 1.

In the above notation $m_0 = M_0 = 1$, and $m_2 = M_2 = 0$, so the value of both the upper and lower Lebesgue summation is 1 for any n. Consequently, the limit exists as $n \to \infty$ and we conclude that:

$$(\mathcal{L}) \int_0^1 h(x)dx = 1.$$

More generally, this integral exists because the infimum of all upper Lebesgue sums equals the supremum of all lower Lebesgue sums, and both equal 1.

To make this construction work in more general cases, a better understanding is needed of what kinds of sets can be "measured." For the first few chapters of this book this will mean measured in a way that is compatible with interval length. This will give rise to what is known as "Lebesgue measure." In addition, it is important to ensure that the functions that we aim to integrate will have the property that the level sets defined above are measurable sets. And finally, the question of the existence of this integral needs to be investigated and properties developed.

For example, if such an integral exists, must the function also be Riemann integrable, and if so, will the values agree? The example above of $d(x)$ makes it clear the answer to this question is, "no, there is at least one Lebesgue integrable function that is not Riemann integrable." So, perhaps we can ask the question the other way around. If a function is Riemann integrable on $[a, b]$, will it be Lebesgue integrable, and if so, will the values agree? We will see that the answer is, "yes, if $f(x)$ is bounded," and in other cases to be identified.

One payoff in making this idea work is that this notion of integral will automatically eliminate the problem noted above for the Riemann integral for $d(x)$. That is, for any measurable set A_i with $|A_i| = 0$, neither the existence of the Lebesgue integral, nor its value when it exists, can be affected by the values the function takes on this set. This follows because the associated terms in the Lebesgue sums will always contribute nothing.

The idea underlying the construction of the Lebesgue integral is also fundamental to a host of generalizations to other measures, and to applications, most notably in probability theory.

Remark 1.8 (On Probability Theory) *Although the Lebesgue theory was not developed to address this application, it was **Andrey Kolmogorov** (1903–1987) who first recognized that a* generalization of the Lebesgue approach to measure theory was the perfect theoretical framework for probability theory. More of the details will be developed in later books, but it is easy to appreciate the power of Kolmogorov's insight even with the little we have done.

In the Lebesgue theory, we are working on the space of real numbers $\mathbb{R}$, and will define measurable sets as an extension of the idea of the length of an interval. It will be seen below that measures on $\mathbb{R}$ can also be defined many other ways by effectively redefining what is meant by interval length. It is then not hard to imagine that the idea of measure can be introduced to other spaces, for example $\mathbb{R}^n$, which is n-dimensional Euclidean space, and even to an abstract space or **sample space** we denote by **S**.

Euclidean spaces are named for **Euclid of Alexandria** (ca. 325–265 BC) who studied $n = 2, 3$ in **Euclid's Elements**. In the abstract setting, Kolmogorov recognized that **S** could be identified with a **measure space**, the measurable sets identified with the **events** defined in the sample space by the probability model, and the **measures of these events** identified with the probabilities of these events. In other words, the probability of an event in a probability application can be interpreted in terms of a probability measure, say μ, defined on the measurable sets of a sample space **S**.

Measurable functions are then the "random variables" X, defined as $X : \boldsymbol{S} \to \mathbb{R}$, and the level sets of X are again identifiable with events in **S**. Finally, if we can take the Lebesgue integration theory from $\mathbb{R}$ and generalize it to an integration theory on **S**, integrals such as $\int_{\boldsymbol{S}} X d\mu$ can in theory be defined, and will give a rigorous foundation for the definition of the "expected value" of X.

But we have gotten ahead of ourselves by running before we know how to walk.

2

Lebesgue Measure on $\mathbb{R}$

2.1 Sigma Algebras and Borel Sets

Before attempting to define a "measurable" set, it is relatively easy to appreciate based on the above introduction to the Lebesgue integral that the collection of measurable sets ought to satisfy at least two simple properties, and these are captured in the notion of an **algebra of sets**.

Definition 2.1 (Algebra of sets) *Let $\mathcal{A}$ denote a collection of sets defined on a space X. Then $\mathcal{A}$ is called an **algebra of sets on** X if the following conditions are satisfied:*

1. *If $A \in \mathcal{A}$ and $B \in \mathcal{A}$, then the **union** $A \cup B \in \mathcal{A}$ where:*

$$A \cup B \equiv \{x | x \in A \text{ or } x \in B\}.$$

   *In other words, $\mathcal{A}$ is **closed** under unioning.*

2. *If $A \in \mathcal{A}$, then the **complement of** A, $\widetilde{A} \in \mathcal{A}$ where:*

$$\widetilde{A} \equiv \{x \in X | x \notin A\}.$$

   *In other words, $\mathcal{A}$ is **closed** under complementation. The complement of A is also denoted A^c.*

Exercise 2.2 *Prove **De Morgan's laws**, named for **Augustus De Morgan** (1806–1871), that if B is a set and $\{A_\alpha\}$ an arbitrarily indexed collection of sets, then:*

a. $\widetilde{\bigcup_\alpha A_\alpha} = \bigcap_\alpha \widetilde{A_\alpha}$,
b. $\widetilde{\bigcap_\alpha A_\alpha} = \bigcup_\alpha \widetilde{A_\alpha}$,
c. $B \cap \left[\bigcup_\alpha A_\alpha \right] = \bigcup_\alpha [A_\alpha \cap B]$,
d. $B \cup \left[\bigcap_\alpha A_\alpha \right] = \bigcap_\alpha [A_\alpha \cup B]$.

Remark 2.3 (Properties of an algebra of sets) *If $\mathcal{A}$ is an algebra of sets on X, then items 1 and 2 of the above definition imply that:*

3. $\emptyset \in \mathcal{A}$, *since $\emptyset = A \cap \widetilde{A}$ for any $A \in \mathcal{A}$;*

DOI: 10.1201/9781003257745-2

4. $X \in \mathcal{A}$ *since* $X = A \cup \widetilde{A}$ *for any* $A \in \mathcal{A}$.

   An application of De Morgan's laws also proves that:

5. *If* $A, B \in \mathcal{A}$, *then the* **set difference** $A - B \in \mathcal{A}$, *since:*

$$A - B \equiv A \cap \widetilde{B}.$$

6. *If* $A, B \in \mathcal{A}$, *then the* **intersection** $A \cap B \in \mathcal{A}$ *since:*

$$A \cap B = \left(\widetilde{\widetilde{A} \cup \widetilde{B}} \right).$$

7. *Finally,* $\mathcal{A}$ *is closed under all finite unions and finite intersections by induction. That is, if* $\{A_i\}_{i=1}^n \subset \mathcal{A}$, *then* $\bigcup_{i=1}^n A_i \in \mathcal{A}$ *and* $\bigcap_{i=1}^n A_i \in \mathcal{A}$.

In this chapter the focus is largely on algebras where the sets in the collection $\mathcal{A}$ are subsets of the real numbers, $\mathbb{R}$. But the terminology and many of the results below apply to a more general situation where the sets are subsets of an arbitrary set X, which could be $\mathbb{R}^n$ or a general space.

Example 2.4 (Algebra generated by right semi-closed intervals) *On* $\mathbb{R}$, *define:*

$$\mathcal{A}' = \{(a, b]\},$$

the collection of all **right semi-closed intervals**. *In this book, right semi-closed will always mean open on the left and closed on the right.*

It is explicitly assumed that one or both of $a_j = -\infty$ *or* $b_j = \infty$ *is possible. Thus the intervals* $\mathbb{R} = (-\infty, \infty)$, (a_j, ∞) *and* $(-\infty, b_j]$ *are included in* $\mathcal{A}'$, *with the notational convention that* $(a_j, \infty] \equiv (a_j, \infty)$. *Also since* $(a, a] = \emptyset$, *the empty set is included in* $\mathcal{A}'$. *We will see in Definition 6.8 that* $\mathcal{A}'$ *is a* **semi-algebra of sets on** $\mathbb{R}$.

Now let:

$$\mathcal{A} = \left\{ \bigcup_{j=1}^n (a_j, b_j] \,\middle|\, (a_j, b_j] \in \mathcal{A}' \right\}, \tag{2.1}$$

the collection of all finite unions of **right semi-closed intervals** *as defined above.*

While such finite unions need not be disjoint unions:

1. **If** $A \in \mathcal{A}$, **then** $A = \bigcup_{j=1}^m (a_j', b_j']$ **with** $\{(a_j', b_j']\}_{j=1}^m$ **disjoint.**

   The proof is constructive. The key observation for one approach is that the union of two intersecting right semi-closed intervals is a right semi-closed interval. For the second approach, the intersection of two right semi-closed intervals is empty or a right semi-closed intervals. In the latter case, this intersection can be removed from either interval and produce another semi-closed interval.

Let $A = \bigcup_{j=1}^{n} (a_j, b_j] \in \mathcal{A}$, and we begin with the collection $\{(a_j, b_j]\}_{j=1}^{n}$.

(a) *Choose two intervals:*

     i. *If $(a_j, b_j] \cap (a_k, b_k] = \emptyset$, leave these intervals in the collection;*

     ii. *If $(a_j, b_j] \cap (a_k, b_k] \neq \emptyset$, remove these two intervals from the collection and replace with $(a_j, b_j] \cup (a_k, b_k]$.*

     Alternatively:

     ii′ *If $I_{jk} \equiv (a_j, b_j] \cap (a_k, b_k] \neq \emptyset$ and neither interval is contained in the other, remove these two intervals from the collection and replace with I_{jk}, $(a_j, b_j] - I_{jk}$, and $(a_k, b_k] - I_{jk}$.*

     If $(a_j, b_j] \subset (a_k, b_k]$, then remove these two intervals from the collection and replace with I_{jk} and $(a_k, b_k] - I_{jk}$, noting that $(a_k, b_k] - I_{jk}$ will be one, or two disjoint semi-closed intervals.

(b) *Repeat step a. until all intervals are disjoint. This must happen in finitely many steps.*

Given the notation, it will not surprise the reader that:

2. **$\mathcal{A}$ is an algebra of sets on $\mathbb{R}$.**

First, $\mathcal{A}$ is closed under finite unions by definition. If $A_1 = \bigcup_{j=1}^{n}(a_j, b_j]$ and $A_2 = \bigcup_{j=1}^{m}(c_j, d_j]$, then $A_1 \cup A_2$ is apparently of the same form, containing all the intervals from both A_1 and A_2. Similarly, $\mathcal{A}$ is closed under complements since by De Morgan's law:

$$\widetilde{\bigcup_{j=1}^{n} (a_j, b_j]} = \bigcap_{j=1}^{n} \widetilde{(a_j, b_j]}.$$

Now for finite a_j and b_j:

$$\widetilde{(a_j, b_j]} = (-\infty, a_j] \cup (b_j, \infty],$$

and $\widetilde{(a_j, b_j]}$ is otherwise a single right semi-closed interval. Finite intersections of such sets are then seen to produce right semi-closed intervals.

As a final property of $\mathcal{A}$, we will see in Example 2.6 that if $\{A_j\}_{j=1}^{\infty} \subset \mathcal{A}$, then $\bigcup_{j=1}^{\infty} A_j$ need not be a member of $\mathcal{A}$. However:

3. **If $A \equiv \bigcup_{j=1}^{\infty} A_j$ with $\{A_j\}_{j=1}^{\infty} \subset \mathcal{A}$, then there exists disjoint $\{(a_k, b_k]\}_{k=1}^{\infty} \subset \mathcal{A}'$ so that:**

$$A = \bigcup_{k=1}^{\infty} (a_k, b_k].$$

By Proposition 2.20 below there exists disjoint $\{A'_j\}_{j=1}^{\infty} \subset \mathcal{A}$ so that:

$$\bigcup_{j=1}^{\infty} A_j = \bigcup_{j=1}^{\infty} A'_j.$$

Now apply the construction in 1 to each A'_j.

This algebra will be seen to be fundamental in the Chapter 5 development of Borel Measures on $\mathbb{R}$.

Despite the general properties of an algebra from Remark 2.3, for most applications one wants more structure than an algebra offers. Specifically, we will need to be sure that certain types of "limits" of unions and intersections of measurable sets will again be measurable.

Definition 2.5 (Sigma algebra) *A **sigma algebra on a space** X, also denoted σ-algebra, is an algebra on X that is closed under countable unions of sets. By De Morgan's laws, a sigma algebra is also closed under countable intersections of sets.*

Example 2.6 (Not all algebras are sigma algebras) *The algebra $\mathcal{A}$ defined in Example 2.4 is not a sigma algebra because, for example, the open interval $(0,1) \notin \mathcal{A}$, yet:*

$$(0,1) = \bigcup_{n=1}^{\infty} (0, 1 - 1/n].$$

The reader is encouraged to implement the construction of part 3 of Example 2.4 to express $(0,1)$ as a countable union of disjoint semi-closed intervals.

In this chapter we will generally focus on sigma algebras, as this is the logical structure to require for the collection of measurable sets. For example, to contemplate a Lebesgue integral requires by its definition limits of measurable sets. There are lots of different sigma algebras possible given a fixed set X, which in many situations will be the real line $\mathbb{R}$, the nonnegative real numbers, $\mathbb{R}^+ = \{x|x \geq 0\}$, or more general spaces such as $\mathbb{R}^n$ or a probability space S.

Example 2.7 (Sigma algebras) *Two examples of sigma algebras on any space X are:*

1. *The **trivial sigma algebra**, $\{\emptyset, X\}$, made up of only the empty set and the entire space and which we denote by $\sigma(\emptyset, X)$, and,*

2. *The **power sigma algebra**, which is defined as the collection of all subsets of X, and which we denote by $\sigma(P(X))$.*

Every other sigma algebra $\sigma(X)$ on X satisfies the property that:

$$\sigma(\emptyset, X) \subset \sigma(X) \subset \sigma(P(X)).$$

Also, if $\{X_i\}$ is an arbitrary but nonempty collection of subsets of the space X:

3. *The **sigma algebra generated by** $\{X_i\}$ is defined as the smallest sigma algebra that contains these sets, and is denoted $\sigma(\{X_i\})$. This sigma algebra always exists by Proposition 2.8 because $\sigma(P(X))$ is one example of a sigma algebra that contains $\{X_i\}$.*

Proposition 2.8 (Intersections of algebras or sigma algebras) *The intersection of any collection of algebras on X is an algebra. The intersection of any collection of sigma algebras is a sigma algebra.*
Neither statement is generally true for unions.

Proof. *The proof for intersections is left as an exercise. To be clear, note that by definition:*

$$\mathcal{A}_1 \cap \mathcal{A}_2 = \{X_\alpha | X_\alpha \in \mathcal{A}_1 \text{ and } X_\alpha \in \mathcal{A}_2\},$$

and similarly for sigma algebras.
For unions, the definition is:

$$\mathcal{A}_1 \bigcup \mathcal{A}_2 = \{X_\alpha | X_\alpha \in \mathcal{A}_1 \text{ or } X_\alpha \in \mathcal{A}_2\},$$

*and similarly for sigma algebras. This can fail to be an algebra. For example, let $\mathcal{A}_1$ denote the algebra of unions of right semi-closed intervals of Example 2.4, and $\mathcal{A}_2$ the algebra of unions of **left** semi-closed intervals $\left\{ \bigcup_{j=1}^{n} [a_j, b_j) \right\}$. Then $\mathcal{A}_1 \bigcup \mathcal{A}_2$ is not an algebra since while both $(0, 2]$ and $[1, 3)$ are members of this collection, neither the union nor intersection of these sets is a member.* ∎

By this proposition, we can define an algebra $\mathcal{A}$ as the the **smallest algebra** with a given property, and similarly for a sigma algebra, as long as there is at least one algebra or sigma algebra which satisfies this property, to avoid a vacuous construction. This algebra or sigma algebra is produced by taking the intersection of all algebras or sigma algebras with this property.

Example 2.9 (Borel sigma algebra) *On $\mathbb{R}$ define the smallest sigma algebra that contains all the open intervals, denoted $\mathcal{B}(\mathbb{R})$. Then:*

1. *$\mathcal{B}(\mathbb{R})$ is well defined based on Proposition 2.8 by demonstrating that the collection of sigma algebras with this property is nonempty. As $\sigma(P(\mathbb{R}))$ contains all the open intervals, this construction is valid.*

2. *$\mathcal{B}(\mathbb{R})$ can equivalently be defined as the smallest sigma algebra that contains all the closed intervals, or the smallest sigma algebra that contains all the half-open intervals. This follows by a modification of Example 2.6, that the sigma algebra defined with any such collection of intervals contains the other types of intervals.*

3. *$\mathcal{B}(\mathbb{R})$ contains every set with one point, and hence every set with a finite or countable collection of points in $\mathbb{R}$. Again an adaptation of the idea of Example 2.6 demonstrates that $\{a\} \in \mathcal{B}(\mathbb{R})$ for all $a \in \mathbb{R}$.*

This last example identifies the special and important sigma algebra named for **Émile Borel** (1871–1956), another pioneer in the early development of measure theory and its application to probability theory. It is "special" in the sense that it is defined to contain all the open intervals, the utility of which will be seen in the sections below on measurable functions. Recall that an **open set** in $\mathbb{R}$ generalizes the idea of an open interval.

For more background on open and closed sets in various contexts, see Chapter 4 of **Reitano** (2010), and also **Dugundji** (1970) or **Gemignani** (1967).

Definition 2.10 (Open sets in $\mathbb{R}$, $\mathbb{R}^n$ and metric X) *A set $E \subset \mathbb{R}$ is called **open** if for any $x \in E$ there is an open interval containing x that is also contained in E. In other words, there is an $\epsilon_1, \epsilon_2 > 0$ so that $(x - \epsilon_1, x + \epsilon_2) \subset E$, and there is no loss of generality by requiring $\epsilon_1 = \epsilon_2$.*
*A set $E \subset \mathbb{R}^n$ is called **open** if for any $x \in E$ there is an open ball about x of radius $r > 0$:*

$$B_r(x) = \left\{ y \,\middle|\, |x - y| < r \right\},$$

so that $B_r(x) \subset E$. Here $|x - y|$ denotes the standard metric on $\mathbb{R}^n$:

$$|x - y| \equiv \left[\sum_{i=1}^n (x_i - y_i)^2 \right]^{1/2}.$$

*More generally, if X is a metric space with metric d, a set $E \subset X$ is called **open** if for any $x \in E$ there is an open ball about x of radius $r > 0$:*

$$B_r(x) = \{ y \,|\, d(x, y) < r \},$$

so that $B_r(x) \subset E$.
*In either case, a set F is called **closed** if $\widetilde{F}$, the complement of F, is open.*

Exercise 2.11 *Verify that an arbitrary union of open sets produces an open set, and using De Morgan's laws, an arbitrary intersection of closed sets is a closed set.*

While finite intersections of open sets are open, countable intersections can produce open, closed, and sets that are neither open nor closed. The same result applies to unions of closed sets: finite unions are closed, while infinite unions can be open, closed, or neither. See $\mathcal{G}_\delta$ and $\mathcal{F}_\sigma$ sets below for examples.

Open sets in $\mathbb{R}$ can be nicely characterized as seen in the following result. However, closed sets cannot be so characterized other than by the definition that a closed set as the complement of an open set. Or, using the next result and De Morgan, every closed set is the countable intersection of complements of open intervals.

Proposition 2.12 (Characterization of open sets in $\mathbb{R}^n$) *If E is an open subset of $\mathbb{R}$, then $E = \bigcup_j I_j$, where $\{I_j\}$ is a countable collection of open intervals. Moreover, $\{I_j\}$ can be chosen to be disjoint.*
If E is an open subset of $\mathbb{R}^n$, then $E = \bigcup_j B_j$, where $\{B_j\}$ is a countable collection of open balls.

Proof. *Given $x \in E$ consider the collection of sets $\{(a, b)_x\}$, where $(a, b)_x$ denotes any open interval $(a, b) \subset E$ such that $x \in (a, b)$. This collection is never vacuous because E is open. Now define $I(x) = \bigcup (a, b)_x$. Then $I(x)$ is open by Exercise 2.11, and in fact an open interval. This follows because if $y \notin I(x)$ then with apparent notation, either $y \geq \sup\{b_x\}$ or $y \leq \inf\{a_x\}$*

(recall (1.3) and (1.4)). Thus $y' \notin I(x)$ if $y' > y$ in the first case, or $y' < y$ in the second, and so $I(x)$ is an interval.

Further, for any $x, y \in E$, either $I(x) = I(y)$ or $I(x) \bigcap I(y) = \emptyset$. So $E = \bigcup_x I(x)$, a disjoint union, and countability follows because each such disjoint interval contains a different rational number.

For open $E \subset \mathbb{R}^n$, let $\{x_j\} \subset E$ be the countable collection of points with all rational coordinates. Then for each x_j let $B_j(x_j)$ be defined:

$$B_j(x_j) = \bigcup_k B_{r_k}(x_j),$$

where this, union is over all balls with rational radius r_k with $B_{r_k}(x_j) \subset E$. This union is not vacuous by Definition 2.10. Also, $B_j(x_j)$ is open as a countable union of open sets, and is in fact an open ball with radius $r_j = \sup\{r_k\}$. This last statement follows from the observation that that if $|x - x_j| < r_j$ then $|x - x_j| < r_k$ for some r_k by definition of supremum, and thus $x \in B_{r_j}(x_j) \equiv B_j(x_j)$.

Now $\bigcup_j B_{r_j}(x_j) \subset E$ by construction. If there exists $x \in E - \bigcup_j B_{r_j}(x_j)$, then there exists $r > 0$ so that $B_r(x) \subset E$. Now choose x_j with all rational components so that $|x - x_j| < r/4$. This is possible since such points are dense in $\mathbb{R}^n$. Then by the triangle inequality:

$$x \in B_{r/2}(x_j) \subset B_r(x) \subset E,$$

and then by construction $B_{r/2}(x_j) \subset B_{r_j}(x_j)$. This contradiction proves that $E = \bigcup_j B_{r_j}(x_j)$. ∎

We now formalize and generalize Example 2.9.

Definition 2.13 (Borel sigma algebra) *The **Borel sigma algebra** on* $\mathbb{R}$, $\mathcal{B}(\mathbb{R})$, *is the smallest sigma algebra that contains the open intervals. The sets of this sigma algebra are called **Borel Sets**.*

*Similarly, the **Borel sigma algebra** on* $\mathbb{R}^n$, $\mathcal{B}(\mathbb{R}^n)$, *is the smallest sigma algebra that contains the open sets of* $\mathbb{R}^n$ *(Definition 2.10). More generally, if X is a topological space, the Borel sigma algebra* $\mathcal{B}(X)$ *is defined as the smallest sigma algebra on X that contains all the open sets as defined by the topology.*

Remark 2.14 *For many applications of Borel sets, the space X will be a Euclidean space* $\mathbb{R}$ *or* $\mathbb{R}^n$, *for which open sets are identified in Definition 2.10. However, for completeness we note that more general spaces can also be endowed with a so-called topology. A topology specifies the collection of open sets. This collection is required to satisfy conditions reminiscent of the properties of open sets from Exercise 2.11. Thus a topology is defined to obey the same properties as does the collection the open sets on* $\mathbb{R}$ *or* $\mathbb{R}^n$.

Definition 2.15 (Topology) *A **topology** on a set X is a collection of subsets, denoted* $\mathcal{T}$ *and called the "open sets," so that:*

1. $\emptyset, X \in \mathcal{T}$;
2. *If* $\{A_\alpha\}_{\alpha \in I} \subset \mathcal{T}$, *where the index set I is arbitrary, then* $\bigcup_{\alpha \in I} A_\alpha \in \mathcal{T}$;
3. *If* $\{A_i\}_{i=1}^n \subset \mathcal{T}$, *then* $\bigcap_{i=1}^n A_i \in \mathcal{T}$.

There are special classes of Borel sets often identified in the study of Real Analysis starting with the collection of open sets, often denoted $\mathcal{G}$, and the collection of closed sets, denoted $\mathcal{F}$. The class $\mathcal{G}_\delta$, pronounced "G-delta," is defined as the collection of countable intersections of elements of $\mathcal{G}$, whereas $\mathcal{F}_\sigma$, pronounced "F-sigma," is defined as the collection of countable unions of elements of $\mathcal{F}$. Because the elements in such unions and intersections can be identical, it is the case that $\mathcal{F} \subset \mathcal{F}_\sigma$ and $\mathcal{G} \subset \mathcal{G}_\delta$. There is no point to introducing $\mathcal{G}_\sigma$ or $\mathcal{F}_\delta$ since as noted above, unions of open sets are open, and intersections of closed sets are closed, and hence $\mathcal{G}_\sigma = \mathcal{G}$ and $\mathcal{F}_\delta = \mathcal{F}$.

Notation 2.16 (On terminology for Borel sets) *The letter G represents the standard notation for an open set in real analysis, and this apparently originated in Germany with the word for area,* **gebiet**. *Thus, it is common to use* $\mathcal{G}$ *for the collection of open sets, while the delta in* $\mathcal{G}_\delta$ *reflects the German word for average,* **durchschnitt**, *for intersections of sets.*

Similarly, the letter F represents the standard notation for a closed set in real analysis, apparently originated in France with the word for firm, **ferme**. *Thus* $\mathcal{F}$ *represents the collection of closed sets, while the sigma in* $\mathcal{F}_\sigma$ *reflects the French word for summation,* **somme**, *for unions of sets.*

Continuing, if one defined $\mathcal{F}_{\sigma\sigma}$ as countable unions of elements of $\mathcal{F}_\sigma$, nothing new would be produced since any such union could be defined directly in terms of $\mathcal{F}$, and so $\mathcal{F}_{\sigma\sigma} = \mathcal{F}_\sigma$. The same is true for $\mathcal{G}_{\delta\delta}$, that $\mathcal{G}_{\delta\delta} = \mathcal{G}_\delta$. However, $\mathcal{F}_{\sigma\delta}$ and $\mathcal{G}_{\delta\sigma}$ are well defined in terms of $\mathcal{F}_\sigma$ and $\mathcal{G}_\delta$, respectively, and produce the next levels of Borel sets. This process continues as:

$$\mathcal{F} \subset \mathcal{F}_\sigma \subset \mathcal{F}_{\sigma\delta} \subset \mathcal{F}_{\sigma\delta\sigma} \subset \ldots\ldots \subset \mathcal{B}(\mathbb{R}),$$
$$\mathcal{G} \subset \mathcal{G}_\delta \subset \mathcal{G}_{\delta\sigma} \subset \mathcal{G}_{\delta\sigma\delta} \subset \ldots\ldots \subset \mathcal{B}(\mathbb{R}).$$

Reading subscripts from right to left, each class is defined as countable unions (σ) or countable intersections (δ) of elements of the prior class. As noted above, each class contains the prior class.

Example 2.17 (Special Borel sets) *A* $\mathcal{G}_\delta$-*set can be open, closed or neither as can be demonstrated by such examples as:*

$$\{(1 - n/(n+1), 2 + n/(n+1))\}, \quad \{(-1/n, 1 + 1/n)\}, \quad \{(1 - 1/n, 1 + n/(n+1))\},$$

with respective intersections for $n \geq 1$ *of* $(0.5, 2.5)$, $[0,1]$ *and* $[1, 1.5)$.

The collection of irrational numbers is another $\mathcal{G}_\delta$-*set, is neither open nor closed, and this set equals the countable intersection of the open sets:*

$$G_q \equiv (-\infty, q) \cup (q, \infty),$$

for all $q \in \mathbb{Q}$, *the rational numbers.*

An $\mathcal{F}_\sigma$-*set can also be open, closed or neither as demonstrated by such examples as:*

$$\{[1/n, 1 - 1/n]\}, \quad \{[-1/n, 1 + 1/n]\}, \quad \{[1/n, 1 + 1/n]\},$$

with respective unions for $n \geq 1$ *of* $(0,1)$, $[-1,2]$ *and* $(0,2]$.

The collection of rational numbers is another $\mathcal{F}_\sigma$*-set, is neither open nor closed, and this set equals the union of closed sets* $\{q\}$*, for all* $q \in \mathbb{Q}$*.*

While each of these classes is a subset of Borel sets, interestingly it is known that $\mathcal{B}(\mathbb{R})$ contains sets that are not in any of these special classes. This will not be demonstrated here, as we will have no further application of this fact.

Perhaps even more interestingly, while it is obvious by definition that $\mathcal{B}(\mathbb{R}) \subset \sigma(P(\mathbb{R}))$, the power sigma algebra of all subsets of $\mathbb{R}$ of Example 2.7, it is by no means apparent that this inclusion is strict. In other words, it is not apparent that there is a subset of $\mathbb{R}$ that is not a Borel set. As it turns out, this inclusion is strict,

$$\mathcal{B}(\mathbb{R}) \subsetneqq \sigma(P(\mathbb{R})),$$

although the development of a non-Borel set is subtle and will not be pursued here as we again have no further use for this fact.

Because of De Morgan's laws and the complementary definitions of open and closed, it is the case that every one of these special classes can be defined in terms of complements of elements of another class. For example:

$$\mathcal{F} = \widetilde{\mathcal{G}} \equiv \{\widetilde{A}|A \in \mathcal{G}\},$$
$$\mathcal{G}_{\delta\sigma} = \widetilde{\mathcal{F}}_{\sigma\delta} \equiv \{\widetilde{A}|A \in \mathcal{F}_{\sigma\delta}\},$$

and so forth.

Example 2.18 (Special Borel sets in Riemann theory) *As an example of how one of these special collections of sets arise, we recall the study in Chapter 10 of* **Reitano** *(2010) related to the existence of a Riemann integral when* $f(x)$ *has infinitely many discontinuities. Given an open interval* $I = (x_{i-1}, x_i)$*, the* **oscillation of** $f(x)$ **on** I*, denoted* $\omega(x; I)$*, is defined:*

$$\omega(x; I) = M_i - m_i,$$

where M_i *and* m_i *are respectively defined as the least upper bound (l.u.b.) and greatest lower bound (g.l.b.) of* $f(x)$ *on the interval* I*. Recall (1.3) and (1.4). In addition, the* **oscillation of** $f(x)$ **at** x*, denoted* $\omega(x)$*, is defined:*

$$\omega(x) = g.l.b. \{\omega(x; I)\} \text{ for all } I \text{ with } x \in I.$$

It can then be shown that $f(x)$ *is discontinuous at* x *if and only if* $\omega(x) > 0$*.*
Now define:

$$E_N = \{x|\omega(x) \geq 1/N\}, \qquad E \equiv \bigcup_{N\geq 1} E_N = \{x|\omega(x) > 0\}.$$

Then E *is the collection of discontinuities of* $f(x)$*, and it is proved in that book's proposition 10.19 that* E_N *is closed for all* N*.*
Hence, the following result is proved.

Proposition 2.19 (Every discontinuity set is $\mathcal{F}_\sigma$) *The set of discontinuities of an arbitrary function on $\mathbb{R}$ is a countable union of closed sets, and hence a Borel set, and more specifically, an $\mathcal{F}_\sigma$-set.*

One last result on algebras and sigma algebras is needed below. This result was already utilized in part 3 of Example 2.4.

While stated in the general setting of an algebra $\mathcal{A}$, this result is true in sigma algebras by Definition 2.5. The major difference in applications is that one cannot assume that $\bigcup_{j=1}^{\infty} A_j \in \mathcal{A}$ for an algebra, while a sigma algebra is closed under countable unions by definition. The details of the proof are assigned as an exercise.

Proposition 2.20 (Algebra unions to disjoint unions) *Let $\{A_j\}_{j=1}^{\infty}$ be a collection of sets in an algebra $\mathcal{A}$. Then, there is a disjoint collection $\{A_j'\}_{j=1}^{\infty} \subset \mathcal{A}$ with:*

$$\bigcup_{j=1}^{\infty} A_j = \bigcup_{j=1}^{\infty} A_j',$$

and for each N:

$$\bigcup_{j=1}^{N} A_j = \bigcup_{j=1}^{N} A_j'.$$

Proof. *Hint: Define (recall De Morgan's laws):*

$$A_n' = A_n - \left[\bigcup_{j=1}^{n-1} A_j\right] = A_n \cap \left[\bigcap_{j=1}^{n-1} \widetilde{A}_j\right].$$

Check that $A_n' \in \mathcal{A}$ and $A_n' \cap A_m' = \emptyset$ if $n \neq m$. By definition, $\bigcup_{j=1}^{\infty} A_j' \subset \bigcup_{j=1}^{\infty} A_j$, while if $x \in \bigcup_{j=1}^{\infty} A_j$, check that $x \in A_k'$ for some k. Now check finite unions. ∎

2.2 Definition of a Lebesgue Measure

With sigma algebras defined, our next goal is to define the notion of a "measure" on the sets of a sigma algebra. There are many ways to do this; however, for the purposes of this chapter the focus is on what is known as **Lebesgue measure**, and this measure will ultimately apply to a sigma algebra that contains the Borel sigma algebra $\mathcal{B}(\mathbb{R})$.

Lebesgue measure is defined to equal interval length when applied to the intervals in this collection of Borel sets. This idea will be generalized below in Chapter 5 on Borel measures where different definitions for interval length will give rise to different measures.

As it turns out, the goal of defining a measure that generalizes the notion of interval length is more subtle and difficult than might be first expected. Indeed, one might well expect that such a measure can be defined on the power sigma algebra of all subsets of the real numbers, $\sigma(P(\mathbb{R}))$. But before beginning to develop these ideas, we consider the properties that we would like any such measure to satisfy to make it consistent with and applicable to the operations within a sigma algebra.

To this end, we start with the following definition. It should be noted in advance that defining something does not prove that it exists.

Definition 2.21 (Lebesgue measure) *Lebesgue measure is a **nonnegative set function** m, defined on a sigma algebra of sets $\sigma(\mathbb{R})$ which contains the intervals, that takes values in the **nonnegative extended real numbers** $\overline{\mathbb{R}}^{+} \equiv \mathbb{R}^{+} \cup \{\infty\}$, and which satisfies the following properties:*

1. $m(\emptyset) = 0$;

2. ***Countable Additivity***: *If $\{A_j\}$ is a countable collection of **pairwise disjoint** sets in the sigma algebra $\sigma(\mathbb{R})$, then:*

$$m\left(\bigcup_j A_j\right) = \sum_j m\left(A_j\right), \tag{2.2}$$

where pairwise disjoint means $A_j \cap A_k = \emptyset$ if $j \neq k$;

3. *For any interval I, whether open, closed or half open, $m(I) = |I|$, the length of I.*

*In such a case, the sets in $\sigma(\mathbb{R})$ are said to be **Lebesgue measurable**, and the triplet $(\mathbb{R}, \sigma(\mathbb{R}), m)$ is called a **Lebesgue measure space**.*

*Since $\sigma(\mathbb{R})$ contains the open intervals, it follows that $\mathcal{B}(\mathbb{R}) \subset \sigma(\mathbb{R})$ for the Borel sigma algebra of Definition 2.13. Defined on $\mathcal{B}(\mathbb{R})$, the Lebesgue measure m is a **Borel measure** and $(\mathbb{R}, \mathcal{B}(\mathbb{R}), m)$ is a **Borel measure space**. See Definition 5.1.*

Remark 2.22 (Finite additivity; monotonicity) *Lebesgue measure is also **finitely additive** on **pairwise disjoint** sets:*

$$m\left(\bigcup_{j=1}^{n} A_j\right) = \sum_{j=1}^{n} m\left(A_j\right),$$

since we can take all but a finite number of $A_j = \emptyset$ and apply 1 and 2.

*A consequence of this is that Lebesgue measure is also **monotonic**, meaning $m(A) \leq m(B)$ if $A \subset B$. This follows because:*

$$B = A \bigcup (B - A),$$

a disjoint union, and thus by nonnegativity of m:

$$m(B) = m(A) + m(B - A) \geq m(A).$$

Both comments also apply for a general measure μ in Definition 2.23.

It is really item 3 which characterizes Lebesgue measure in that it produces a notion of measure on $\mathbb{R}$ which is consistent with ordinary interval length. If this restriction is removed, the definition of a **general measure on a sigma algebra** is produced and is often denoted by μ, reserving $\mu \equiv m$ as the special notation for Lebesgue measure.

Definition 2.23 (Measure on a sigma algebra) *Let X be a set and $\sigma(X)$ a sigma algebra of subsets of X. A **measure on X** is a **nonnegative set function** μ defined on $\sigma(X)$, taking values in the nonnegative **extended real numbers** $\overline{\mathbb{R}}^+ \equiv \mathbb{R}^+ \cup \{\infty\}$, and which satisfies the following properties:*

1. $\mu(\emptyset) = 0$;

2. ***Countable Additivity:*** *If $\{A_j\}$ is a countable collection of **pairwise disjoint** sets in the sigma algebra $\sigma(X)$, then (2.2) is satisfied with μ replacing m.*

   *In such a case, the sets in $\sigma(X)$ are said to be **measurable**, and sometimes μ-**measurable**, and the triplet $(X, \sigma(X), \mu)$ is called a **measure space**.*

It should by no means be apparent at this point that a Lebesgue measure actually exists. Certainly items 1 and 3 are not a problem to implement. Indeed, we simply define $m(\emptyset) = 0$ and $m(I) = b - a$ for any open interval $I = (a, b)$. Then any singleton set $\{a\} \in \sigma(\mathbb{R})$ because $\{a\} = \bigcap_n (a - 1/n, a + 1/n)$, and since (2.2) implies that $m(a) < m((a - 1/n, a + 1/n)) = 2/n$, it follows that $m(a) = 0$. This conclusion is independent of how $\{a\}$ is constructed as an intersection of open intervals.

Extending the definition of m to a finite union of disjoint intervals would appear to be straightforward, as a sum of interval lengths, and this is then how m can be extended to any interval. For example:

$$[a, b] = (a, b) \cup \{a\} \cup \{b\},$$

and since $m(a) = 0$ for all a, closed and right semi-closed versions of such intervals have the same measure as open intervals. Extending m to countable disjoint unions of intervals will be a little trickier, since we will need to demonstrate that this can be done in a well-defined way.

But then we will encounter two nearly insurmountable challenges to continuing with this construction:

1. How can the "interval length" definition be extended in a consistent way to the wide variety of sets that can be anticipated to exist in the sigma algebra $\sigma(\mathbb{R})$, which contains $\mathcal{B}(\mathbb{R})$ as noted above? Also, will any such extension be uniquely defined?

2. Though a little tricky, it seems feasible as noted to simply define the measure of a disjoint union of intervals to equal the sum of the interval lengths. Given such an extension of m, how can we demonstrate that countable additivity continues to be valid for the more general sets in $\sigma(\mathbb{R})$?

Clearly there is some work to be done.

Lebesgue measure, and more generally any measure, is explicitly defined in anticipation that a set A may have $m(A) = \infty$. This is not a surprise in the Lebesgue context, since for example $m(I) = \infty$ for all intervals of the form $(-\infty, a)$ or (b, ∞). More generally, while measures μ are allowed to assume the value ∞, they need not.

Example 2.24 (A simple finite measure) *Define a **rationals counting measure** $\mu_{\mathbb{Q}}$ on the power sigma algebra $\sigma(P(\mathbb{R}))$ as follows. Enumerate the rationals $\{r_j\}_{j=1}^{\infty} = \mathbb{Q} \subset \mathbb{R}$ arbitrarily, and define*

$$\mu_{\mathbb{Q}}(r_j) = 2^{-j}. \tag{2.3}$$

For any set $A \subset \mathbb{R}$, define $\mu(A) = \sum_{r_j \in A} 2^{-j}$, and $\mu(\emptyset) = 0$.

By construction, $\mu_{\mathbb{Q}}$ is countably additive and is hence a measure on the power sigma algebra $\sigma(P(\mathbb{R}))$. The technical detail needed for this conclusion is that an absolutely convergent series has the same sum independent of the order of the summation (Proposition 6.15, **Reitano** (2010)). In fact $\mu_{\mathbb{Q}}$ is a **finite measure** on $\mathbb{R}$, since $\mu_{\mathbb{Q}}(\mathbb{R}) < \infty$.

But it is also the case that $\mu_{\mathbb{Q}}$ has other unusual properties compared to Lebesgue measure. For example:

1. $\mu_{\mathbb{Q}}(\mathbb{R}) = 1$.
2. If $\{I_j\}$ is a sequence of nested intervals with $I_{j+1} \subset I_j$ and $\bigcap_j I_j = x$, then $\mu_{\mathbb{Q}}(I_j)$ may converge to 0 (x irrational) or not (x rational).
3. $\mu_{\mathbb{Q}}$ is not **translation invariant**. If I is an interval and $a \in \mathbb{R}$, it will not be the case that $\mu_{\mathbb{Q}}(I) = \mu_{\mathbb{Q}}(I + a)$ where the shifted interval $I + a$ is defined as $\{a + x | x \in I\}$.

The goal of the next sections is to formally construct Lebesgue measure on a sigma algebra that will be denoted $\mathcal{M}_L$, and which naturally contains the Borel sigma algebra, $\mathcal{B}(\mathbb{R}) \subset \mathcal{M}_L$. But it is a worthwhile adventure to first attempt a more ambitious goal of achieving a Lebesgue measure on the power sigma algebra, $\sigma(P(\mathbb{R}))$. This attempt will fail, and it will be demonstrated that there are sets that cannot be included in the sigma algebra $\sigma(\mathbb{R})$ of Definition 2.21 if m is to be countably additive. That is, there will be subsets of $\mathbb{R}$ that are not Lebesgue measurable.

2.3 Is There a Lebesgue Measure on $\sigma(P(\mathbb{R}))$?

We begin with an important construction toward the ultimate goal of defining Lebesgue measure. The logic of this construction follows from the above remark that from $m((a, b)) = b - a$ it follows that for any point a, $m(a) = 0$. This construction cannot be assumed to produce the desired Lebesgue measure, as that is a conclusion that would require proof. So we will initially give this set function a different name, **Lebesgue outer measure**.

In the following definition, note that since $\bigcup_n I_n$ is open by Exercise 2.11, that it can always be assumed that these collections of open sets are disjoint by Proposition 2.12, and we will often use this observation.

Definition 2.25 (Lebesgue outer measure) *For any $A \subset \mathbb{R}$, the **Lebesgue outer measure** of A, denoted $m^*(A)$, is defined by:*

$$m^*(A) = \inf \left\{ \sum_n |I_n| \ \Big| \ A \subset \bigcup_n I_n, \right\}, \tag{2.4}$$

where I_n is an open interval, and $|I_n|$ denotes its interval length.

In other words, the Lebesgue outer measure of a set is defined as the infimum, or greatest lower bound, of the sum of lengths of any countable collection of open intervals which "covers" the given set. Here and everywhere, countable includes finite.

Note that $m^*(A)$ is defined for any $A \in \sigma(P(\mathbb{R}))$. This follows because either:

1. $\sum_n |I_n| = \infty$ for any such collection, and then $m^*(A) = \infty$, or,
2. $\sum_n |I'_n| \le K < \infty$ for some collection. We can then modify an arbitrary collection $\{I_j\}$ to $\{I_j \cap I'_n\}$ without loss of generality. Further, all such collections now satisfy $\sum_n |I_j \cap I'_n| \le K$, and thus this collection of bounded real numbers has an infimum and $m^*(A) \le K$.

One important consequence of the above definition follows from the definition of infimum. If $m^*(A) < \infty$, then for any $\epsilon > 0$ there is a countable collection of open intervals $\{I_j\}$ so that $A \subset \cup_j I_j$ and:

$$\sum_j |I_j| \le m^*(A) + \epsilon. \tag{2.5}$$

In other words, the outer measure of any set of finite outer measure can be approximated arbitrarily well by the total interval lengths of a collection of disjoint open intervals which cover the set. We develop generalizations below.

If it could be proved that m^* is in fact a measure by Definition 2.21, this would be the best possible result as it would provide a Lebesgue measure on $\sigma(P(\mathbb{R}))$. For this investigation we begin by documenting a few interesting and desirable properties of Lebesgue outer measure. The associated proof will require a result known as the **Heine-Borel theorem**, named for **Eduard Heine** (1821–1881) and **Émile Borel** (1871–1956). Borel formalized the earlier work of Heine in an 1895 publication which applied to the notion of compactness when defined in terms of countably infinite open covers. This in turn was generalized by **Henri Lebesgue** (1875–1941) in 1898 to apply to the notion of compactness as defined in terms of an arbitrary infinite open cover.

We state this result without proof in the limited context of $\mathbb{R}^n$, but note it is valid in a far wider range of contexts. See Chapter 4 of **Reitano** (2010) for $\mathbb{R}^n$, and **Dugundji** (1970) or **Gemignani** (1967) for generalizations.

Definition 2.26 (Compact) *A set $E \subset \mathbb{R}^n$ is **compact** if given an open cover $\{G_\alpha\}$ of E, meaning that $E \subset \bigcup_\alpha G_\alpha$, where this collection may be uncountably infinite, there is a finite subcollection $\{G_j\}_{j=1}^m \subset \{G_\alpha\}$ so that:*

$$E \subset \bigcup_{j \le m} G_j.$$

Proposition 2.27 (Heine-Borel Theorem) *A set $E \subset \mathbb{R}^n$ is compact if and only if E is closed and bounded, where by bounded it is meant that $E \subset [a,b]^n$ for some finite rectangle.*

We now prove that m^* satisfies conditions 1 and 3 of Definition 2.21 to be a Lebesgue measure on the power sigma algebra $\sigma(P(\mathbb{R}))$, and also has other desirable properties.

Proposition 2.28 (Basic properties of m^*) *With $m^*(A)$ defined in (2.4):*

1. $m^*(\emptyset) = 0$;
2. $m^*(A) = 0$ *if A is a point or a countable collection of points;*
3. **Monotonicity:** *If $B \subset A$, then $m^*(B) \leq m^*(A)$;*
4. $m^*(A) = |A|$ *for any interval A, where $|A|$ denotes interval length;,*
5. m^* *is* **translation invariant**: *For any set A and real number a, $m^*(A) = m^*(a+A)$ where:*

$$a + A \equiv \{a + x | x \in A\}.$$

Proof.

1. *Because $\emptyset \subset I$ for any interval I and $\inf |I| = 0$ over all such intervals, the result follows by definition.*
2. *For $A = \{a\}$ a point, the conclusion follows with cover $I = (a - \epsilon, a + \epsilon)$ for any $\epsilon > 0$. For a countable collection, $\{a_j\}$, use the covers $\{(a_j - \epsilon/2^{j+1}, a_j + \epsilon/2^{j+1})\}$ for arbitrary $\epsilon > 0$. This collection may not be disjoint, but has total length ϵ.*
3. *This follows since every open cover of A is by definition an open cover of B.*
4. *If A is an unbounded interval, then $m^*(A) = \infty = |A|$, so assume $A = \{a, b\}$ where this notation implies that the bounded interval can be open, closed or semi-closed. Letting $I = (a - \epsilon, b + \epsilon)$, item 3 obtains that for all ϵ:*

$$m^*(A) \leq |I| \equiv b - a + 2\epsilon.$$

   Hence $m^*(A) \leq |A|$ for any such interval.

   If $A = [a, b]$ is also closed, then it is compact by the Heine-Borel theorem. So any open cover in the definition of outer measure has a finite subcover, say $\{I_j\}_{j=1}^n$, where these are disjoint and ordered in terms of the left endpoints. Thus $a \in I_1$, $b \in I_n$, and since these are open intervals there is an $\epsilon > 0$ so that $(a - \epsilon, a + \epsilon) \subset I_1$, and $(b - \epsilon, b + \epsilon) \subset I_n$. Consequently, $A \subset \cup_{j=1}^n I_j$ implies $\sum_{j=1}^n |I_j| > b - a + 2\epsilon$, and hence $m^*(A) \geq |A|$. Combining, it follows that $m^*(A) = |A|$ if A is closed.

   If A is an open or semi-closed interval, choose a closed subinterval $B \subset A$ with $|B| \geq |A| - \epsilon$ for given but arbitrary $\epsilon > 0$. For example, if $A = \{a, b\}$, choose $B = [a + \epsilon/3, b - \epsilon/3]$. Then using the result for closed intervals and item 3:

$$|A| - \epsilon \leq |B| = m^*(B) \leq m^*(A) \leq m^*(\overline{A}) = \left|\overline{A}\right| = |A|,$$

   where $\overline{A} = [a, b]$ denotes the closure of A. Thus, $m^*(A) = |A|$ in the open and semi-closed case.
5. *Translation invariance follows from item 4, since interval length is translation invariant.* ∎

The next result relates to the question of countable additivity of outer measure, the final and deepest condition in Definition 2.21. This proposition falls short, showing only that outer measure is **countably subadditive**. What is worse is that we will then show below by example that m^* satisfies neither countable nor finite additivity on the sigma algebra $\sigma(P(\mathbb{R}))$.

Proposition 2.29 (Countable subadditivity of m^*) *If $\{A_n\}_{j=1}^{\infty}$ is a countable collection of sets, then m^* is **countably subadditive**:*

$$m^* \left(\bigcup_j A_j \right) \leq \sum_j m^* \left(A_j \right). \qquad (2.6)$$

*Thus m^* is also **finitely subadditive**.*

Proof. *If there is an A_j so that $m^* \left(A_j \right) = \infty$, the result follows since then $\sum_j m^* \left(A_j \right) = \infty$. The same logic follows if $m^* \left(A_j \right) < \infty$ for all j and $\sum_j m^* \left(A_j \right) = \infty$. So assume $\sum_j m^* \left(A_j \right) < \infty$. By (2.5), for every set A_j there is a collection of open intervals, $\{I_{jk}\}_k$ so that $A_j \subset \cup_k I_{jk}$ and $\sum_k |I_{jk}| < m^*(A_j) + \epsilon/2^j$. Thus $\cup_j A_j \subset \cup_{j,k} I_{jk}$ and so $\{I_{jk}\}_{j,k}$ is a countable open cover of $\cup_j A_j$. Further:*

$$\sum_{j,k} |I_{jk}| < \sum_j m^*(A_j) + \sum_j \epsilon/2^j$$
$$= \sum_j m^*(A_j) + \epsilon.$$

So taking an infimum, for any $\epsilon > 0$:

$$m^* \left(\bigcup_j A_j \right) < \sum_j m^* \left(A_j \right) + \epsilon,$$

and (2.6) follows.

Taking all but a finite number of $A_j = \emptyset$ and applying item 1 of Proposition 2.28 obtains finite subadditivity. ∎

With the help of this proposition, the result in (2.5) can be restated:

Corollary 2.30 (Approximating outer measure) *If $m^*(A) < \infty$, then for any $\epsilon > 0$ there is an open set $I \in \mathcal{G}$ with $A \subset I$ and:*

$$m^*(I) \leq m^*(A) + \epsilon. \qquad (2.7)$$

Further, there is a set $J \in \mathcal{G}_\delta$, the class of countable intersections of open sets, so that $A \subset J$ and:

$$m^*(J) = m^*(A). \qquad (2.8)$$

Proof. *Define $I = \cup_j I_j$ where $\{I_j\}$ is given in (2.5). Then $m^*(I) \leq \sum_j m^* \left(I_j \right)$ by (2.6), and since the outer measure of any interval equals its interval length, (2.7) follows from (2.5).*

The second statement is an exercise. Hint: For $\epsilon_n = 1/n$, let I_n denote the corresponding open set given by (2.7), and consider $\bigcap I_n$. ∎

The subadditivity result of Proposition 2.29 perhaps seems inconclusive because it does not demonstrate that outer measure is not in fact countably additive. What is unclear is whether the conclusion of countable additivity is false, or if is it true but not revealed by the above proof. Indeed, the proof did not even depend on whether the sets $\{A_n\}_{j=1}^{\infty}$ were disjoint. It is perhaps compelling to postulate that if these sets were disjoint, the proof could be sharpened and the inequality in (2.6) could be improved to that of equality as needed in (2.2).

Unfortunately, this is not so if one allows the **axiom of choice** in the axiomatic structure of set theory, as is fairly common. See Chapter 4 of the **Reitano** (2010) for an introduction, and **Pinter** (1971) for a more complete discussion. Simplifying, axiomatic approaches to set theory have been developed to provide a rigorous framework that eliminates the many paradoxes that arose with less formal approaches. In such approaches, it was assumed that a set could simply be defined as a collection of elements which satisfy a given property.

A famous example of such a paradox was published in 1903 when **Bertrand Russell** (1872–1970) communicated a discovery he made in 1901, and which has come to be known as **Russell's Paradox**. Russell proposed as a "set" the following:

$$X = \{R | R \text{ is a set, and } R \notin R\}.$$

In other words, X is the "set" of all sets which are not a member of themselves. The paradox occurs in attempting to answer the question:

$$\text{Is } X \in X?$$

If $X \in X$, then by the above defining property it is a set that is **not** an element of itself; whereas, if we posit that $X \notin X$, then again by definition, X should be one of the sets R which is included in X. In summary:

$$X \in X \text{ iff } X \notin X,$$

an impossibly illogical result. In other words, a paradox.

The first approach to formalizing an axiomatic structure was introduced by **Ernst Zermelo** (1871–1953) in 1908, called the **Zermelo Axioms** and which resulted in the **Zermelo set theory**. This axiomatic structure was later modified by **Adolf Fraenkel** (1891–1965) in 1922, and produced the **Zermelo-Fraenkel axioms** and the **Zermelo-Fraenkel set theory**, or **ZF set theory**. Including the so-called **axiom of choice** to ZF set theory produces what is referred to as **ZFC set theory**. This is the approach largely used today.

The axiom of choice in effect states that from any collection of nonempty sets, $\{S_\alpha\}_{\alpha \in I}$, where I is an arbitrary index set, a collection of elements $\{x_\alpha\}_{\alpha \in I}$ can be chosen for which $x_\alpha \in S_\alpha$ for all α. Alternatively, this axiom states that from $\cup_\alpha S_\alpha$, one can choose a subset S which contains exactly one element from each of the S_α-sets. It is easy to wonder why such a statement needs to be an axiom at all and why this is not something that can be proved, or simply assumed as self-evident.

If I is a finite or a countable set, then no such axiom is needed. We enumerate the sets $\{S_j\}_{j=1}^N$ for $N \leq \infty$, and then arbitrarily choose in order one x_j from each S_j. This algorithm will miss no set since they are ordered, and $\{x_j\}_{j=1}^N$ is then obtained iteratively.

The complexity that justifies the need for this axiom occurs when I is an uncountable set. In this case there can be no hope of defining an algorithm which specifies how a given $x_\alpha \in S_\alpha$ is to be selected for all α. Any such algorithm is necessarily countable and hence will omit from its "choice" all but a countable subcollection of $\{S_\alpha\}_{\alpha \in I}$.

It turns out that this axiom is **logically independent** of the other axioms in the ZF set theory. This means that one can add this axiom or its negation to the ZF framework, and in either case produce a system of mathematics that is consistent. Because of its usefulness in the efficient proof of many deep results, most but not all mathematicians today subscribe to the ZFC framework.

It will now be demonstrated that with the help of this axiom, a countable collection of disjoint sets $\{A_j\}_{j=0}^\infty$ can be constructed for which $\bigcup_j A_j = [0,1]$, and hence by Proposition 2.28, $m^*\left(\bigcup_j A_j\right) = 1$. Yet each A_j will be shown to have the same positive outer measure, and so $\sum_j m^*\left(A_j\right) = \infty$. Thus this example provides an extreme counterexample to countable additivity.

This construction was developed by **Giuseppe Vitali** (1875–1932). The collection of sets $\{A_j\}_{j=0}^\infty$ provides an example of nonmeasurable sets. By **nonmeasurable** it is meant that if m^* is indeed a measure (and thus a Lebesgue measure) on some sigma algebra $\sigma(\mathbb{R})$ of Definition 2.21, then we cannot have $\{A_j\}_{j=0}^\infty \subset \sigma(\mathbb{R})$. This follows because if $\sigma(\mathbb{R})$ contained these sets, then m^* would not be countably additive and thus not be a measure. By part 5 of Proposition 2.28 it also follows that $\sigma(\mathbb{R})$ cannot contain any translation of this collection of sets.

Proposition 2.31 (m^* is not countably additive on $\sigma(P(\mathbb{R}))$) *There exists a countable collection of disjoints sets $\{A_j\}_{j=0}^\infty$ with $\bigcup_j A_j = [0,1]$, and:*

$$1 = m^*\left(\bigcup_j A_j\right) < \sum_j m^*\left(A_j\right) = \infty. \tag{2.9}$$

In other words, the Lebesgue outer measure is not countably additive on $\sigma(P(\mathbb{R}))$, and hence is not a Lebesgue measure on $\sigma(P(\mathbb{R}))$.

Proof. *This proof is somewhat long, so it will be split into several steps as follows.*

1. First Partition of $[0,1]$: For each real number $x \in [0,1]$ define the set $E_x = \{y \in [0,1] \mid x - y \in \mathbb{Q}\}$, where $\mathbb{Q}$ denotes the set of rational numbers. Alternatively, in the notation of Proposition 2.28:

$$E_x = \{x + \mathbb{Q}\} \cap [0,1].$$

Each E_x *contains a countable collection of reals since the rationals are countable, and any two such sets are either identical or disjoint. That is, if* $E_x \cap E_{x'} \neq \emptyset$, *then* $E_x = E_{x'}$. *This follows because if* $y \in E_x \cap E_{x'}$, *then both* $x - y$ *and* $x' - y$ *are rational, and hence so too is:*

$$(x - y) - (x' - y) = x - x'.$$

Thus $x' \in E_x$ *and* $x \in E_{x'}$.

Let $\{F_\alpha\}$ *denote the collection of disjoint sets. Because each set* F_α *contains a countable collection of reals and* $\bigcup_\alpha F_\alpha = [0, 1]$, *the collection* $\{F_\alpha\}$ *contains an uncountable number of sets, and exactly one of these sets is* $[0, 1] \cap \mathbb{Q}$.

2. The Definition of $\{A_j\}$: *By the axiom of choice, select one element from each* F_α *and define* A_0 *as the uncountable set so selected. By definition, no two elements of* A_0 *can differ by a rational number, and exactly one element of* A_0 *is rational. For specificity we assume the rational in* A_0 *is the number* $1/2$. *Now, let* $\{r_j\}_{j=0}^\infty$ *be an enumeration of the rationals in* $[0, 1)$ *with* $r_0 \equiv 0$, *and define* $\{A_j\}_{j=1}^\infty$ *by:*

$$A_j = B_j \cap [0, 1],$$

where:

$$B_j \equiv \{r_j + a | a \in A_0\} \cup \{r_j + a - 1 | a \in A_0\}.$$

Thus, for each $a \in A_0$ *we select* $r_j + a$ *if this value is in the interval* $[0, 1]$, *or* $r_j + a - 1$ *if* $r_j + a > 1$. *By definition* $A_j \subset [0, 1]$, *and* $A_j \cap A_k = \emptyset$ *if* $j \neq k$. *This follows since a nonempty intersection would imply that two elements of* A_0 *differed by a rational number, a contradiction. This is the reason for not allowing* $r_j = 1$ *above, as this would produce the same set as* $r_j = 0$, *namely,* A_0. *Also, every number* $y \in [0, 1]$ *is in some* A_j *since any such* $y \in F_\alpha$ *for some* α. *Therefore, if* $a \in F_\alpha$ *was selected as the element in* A_0, *then* $y - a = r$ *a rational, and hence* $y \in A_j$ *for* j *with* $r_j = r$. *In conclusion,* $\{A_j\}_{j=0}^\infty$ *is a countable collection of disjoint sets with* $\bigcup_j A_j = [0, 1]$.

3. $m^*(A_j) = m^*(A_0)$: *The next step is to show that* $m^*(A_j)$ *is constant for all* j *by showing that any open cover of* A_0 *can be modified to be an open cover of* A_j *without changing the sum of the lengths of the given cover's intervals. To this end, let* $\{I_k\}$ *be an open cover of* A_0, *meaning* $A_0 \subset \bigcup_k I_k$. *If* r_j *defines* A_j, *split this cover into two subcollections:*

$$\{I_k\} = \{I_k^{(1)}\} \cup \{I_k^{(2)}\},$$

where $I_k^{(1)} \equiv I_k \cap (-\infty, 1 - r_j)$, *and* $\{I_k^{(2)}\} = I_k \cap (1 - r_j, \infty)$. *We claim that for any* k, *with* $|I|$ *denoting interval length:*

$$|I_k| = \left| I_k^{(1)} \right| + \left| I_k^{(2)} \right|.$$

This is apparent if $I_k \subset (1 - r_j, \infty)$ *or* $I_k \subset (-\infty, 1 - r_j)$, *so assume that* $1 - r_j \in I_k \equiv (a, b)$. *Then* $I_k^{(1)} = (a, 1 - r_j)$ *and* $I_k^{(2)} = (1 - r_j, b)$, *so:*

$$\left| I_k^{(1)} \right| + \left| I_k^{(2)} \right| = (1 - r_j) - a + b - (1 - r_j) = b - a = |I_k|.$$

The final step is to show that $\{I_k^{(1)} + r_j\} \cup \{I_k^{(2)} + r_j - 1\}$ *is an open cover of* A_j. *First, if* $x \in A_j$ *and* $x \neq r_j$, *then either* $x = r_j + a$ *or* $x = r_j + a - 1$ *for some* $a \in A_0$. *In the first case,* $a \leq 1 - r_j$, *and in the second,* $a > 1 - r_j$. *But since* $\{I_k\}$ *is an open cover of* A_0, $a \in I_k$ *for some* k, *and hence* $a \in I_k^{(1)}$ *in the first case, or* $a \in I_k^{(2)}$, *in the second. This then implies that* $x \in I_k^{(1)} + r_j$, *or* $x \in I_k^{(2)} + r_j - 1$, *respectively. Finally, the case of* $x = r_j \in A_j$ *cannot occur, since this would require that* A_0 *contained* 0, *which it does not since we chose* $1/2$ *to be the rational element of this set.*

Note: The failure of countable additivity is actually proved at this point, since either $m^*(A_0) = 0$ *or* $m^*(A_0) > 0$, *and in either case:*

$$\sum_j m^*(A_j) \neq m^*\left(\bigcup_j A_j\right) = 1.$$

The next step simply settles how failure occurs, proving (2.9).

4. $m^*(A_0) > 0$: *If* $m^*(A_0) = 0$, *then for any* $\epsilon > 0$ *and integer* j *there would be an open cover of outer measure* $\epsilon / 2^j$ *which we could modify as in part 3 to an open cover of* A_j. *Then since* $[0, 1] = \bigcup_j A_j$, *it would follow by countable subadditivity of Proposition 2.29 that:*

$$m^*([0,1]) \leq \sum_j m^*(A_j) = \epsilon,$$

a contradiction.

5. *Combining results,* $m^*(A_j) = c > 0$ *for all* j, *and so:*

$$1 = m^*\left(\bigcup_j A_j\right) < \sum_j m^*(A_j) = \infty,$$

which is (2.9).   ∎

The key set in the above result, A_0, required the axiom of choice for its construction. This set and its translates provided a demonstration that Lebesgue outer measure is not countably additive on $\sigma(P(\mathbb{R}))$ and hence is not a Lebesgue measure on $\sigma(P(\mathbb{R}))$. Consequently, if the axiom of choice is not assumed, this "set" cannot be constructed as it was above, and it is therefore possible that there will be no such counterexample to the countable additivity of Lebesgue outer measure on $\sigma(P(\mathbb{R}))$.

In fact, **Robert M. Solovay** (1938-) proved in 1970 that if a specified weaker version of the axiom of choice is added to the **Zermelo-Fraenkel set theory**, then every set in $\sigma(P(\mathbb{R}))$ is Lebesgue measurable. This weaker version is called the **axiom of dependent choice**.

Remark 2.32 (Nonmeasurable sets) *In most texts, the construction in the above proposition is referred to as a **construction of a nonmeasurable set** A_0, whereas we have for the moment chosen to present this result as a failure of Lebesgue outer measure to be countably additive on* $\sigma(P(\mathbb{R}))$. *But as noted above, the collection of A_j sets cannot be measurable under Lebesgue outer measure, meaning this collection cannot be contained in the sigma algebra on which Lebesgue outer measure is countably additive. The same is true for any countable subcollection, since the* outer measure of the union must be bounded by 1 by monotonicity of m*, while the sum of the outer measures is unbounded as above.

The problem here is not with Lebesgue outer measure, which in fact will be the proposed candidate for Lebesgue measure. Indeed it may be hard to imagine an alternative proposal for a definition of Lebesgue measure which is both consistent with interval length, and adaptable to the measure of more general sets. Moreover, defining a measure explicitly on the intervals and then extending the definition by a limiting approximation as in (2.4) is a natural approach, and it will be used again and again in this book.

So rather than change the outer measure definition, it will be seen below that a Lebesgue measure can be produced with outer measure if we simply restrict the sigma algebra from $\sigma(P(\mathbb{R}))$ to one that excludes sets like $\{A_j\}$ that will defeat countable additivity. This restriction on allowable sets will be introduced below in (2.14).

Finally, we will see in Remark 2.37 that in addition to failing countable additivity on $\sigma(P(\mathbb{R}))$, m^* is not even finitely additive on this sigma algebra. The reader is invited to think through how the above proof provides this result. Moreover, we will see there that not only is it true that the collection of A_j sets cannot be measurable under Lebesgue outer measure, but no set of this collection can be so measurable.

Thus we will finally conclude that A_0 is a nonmeasurable set.

2.4 Lebesgue Measurable Sets: $\mathcal{M}_L(\mathbb{R}) \subsetneq \sigma(P(\mathbb{R}))$

In this section a sigma algebra will be identified on which Lebesgue outer measure is countably additive, and hence will become **Lebesgue measure** on this sigma algebra. The sigma algebra will be denoted $\mathcal{M}_L(\mathbb{R})$, and as the section heading implies, it is and by the prior section's construction must be, a proper subset of the power sigma algebra $\sigma(P(\mathbb{R}))$. It will be demonstrated that $\mathcal{M}_L(\mathbb{R})$ contains the sigma algebra of Borel sets, $\mathcal{B}(\mathbb{R})$, but it will not be demonstrated that this inclusion is again strict:

$$\mathcal{B}(\mathbb{R}) \subsetneq \mathcal{M}_L(\mathbb{R}). \tag{2.10}$$

In this section the sigma algebra $\mathcal{M}_L(\mathbb{R})$ will be defined, then in the next it will be seen that on $\mathcal{M}_L(\mathbb{R})$, Lebesgue outer measure is indeed a measure.

There are two standard approaches to developing a sigma algebra on which countable additivity of Lebesgue outer measure applies. In both approaches there is a restriction on the sets on which Lebesgue outer measure will be applied, and in both approaches the same sigma algebra $\mathcal{M}_L(\mathbb{R})$ is produced. The restricted collection of sets will be called "Lebesgue measurable" in anticipation of the fact that on this sigma algebra Lebesgue outer measure is countably additive and hence, recalling Proposition 2.28, a measure.

The standard approaches are:

1. Define **Lebesgue inner measure** of a set $A \subset \mathbb{R}$, denoted $m_*(A)$, by:

$$m_*(A) = \sup\left\{ \sum_n |J_n| \, \Big| \, \bigcup_n J_n \subset A \right\}. \tag{2.11}$$

Here $\{J_n\}$ are closed sets, and $|J_n|$ denotes the "set measure" of a closed set, defined using the interval lengths of open sets. Specifically, if I is an open set with closed $J \subset I$, then since the complement $\tilde{J}$ of J is also open, so is $I \cap \tilde{J} = I - J$. Thus we define:

$$|J| = |I| - |I - J|. \tag{2.12}$$

Since both I and $I - J$ are open and hence the union of disjoint open intervals by Proposition 2.12, the right-hand side of this definition is then defined in terms of interval lengths. To be well defined, we must show that this definition is independent of the choice of open I.

In other words it must be proved that if J is a closed set, and I and I' are open sets with $J \subset I$, $J \subset I'$, then the definition of $|J|$ in (2.12) is independent of the choice between I and I', meaning:

$$|I| - |I - J| = |I'| - |I' - J|.$$

The proof of this requires that given open sets G and G', that

$$|G| + |G'| = |G \cup G'| - |G \cap G'|.$$

This result appears "obvious" but the proof is subtle. Given this, let $G = I$ and $G' = I' - J$. Then $G = I - J$ and $G' = I'$, and so (2.12) is well defined. For details on this see **Halmos** (1950) or **Royden** (1971).

We then have:

Proposed Definition 1: A set $A \subset \mathbb{R}$ is **Lebesgue measurable** if:

$$m^*(A) = m_*(A). \tag{2.13}$$

2. The second approach was introduced by **Constantin Carathéodory** (1873–1950) in the development of the general theory of outer measures. It defines a set A to be measurable if given **any** other set E, the Lebesgue outer measure of E equals the sum of the outer measures of the subsets formed by splitting E into two disjoint subsets with A and $\tilde{A}$. Formally, this is stated:

Proposed Definition 2: A set $A \subset \mathbb{R}$ is **Lebesgue measurable** if it satisfies the **Carathéodory criterion**: For any set $E \subset \mathbb{R}$:

$$m^*(E) = m^*(A \cap E) + m^*(\tilde{A} \cap E).$$

It turns out that if $m^*(A) < \infty$, these proposed definitions are equivalent, although we will not pursue this development here. Instead, we take the second approach as the definition of Lebesgue measurability, and proceed to show that the collection of such sets forms a sigma algebra $\mathcal{M}_L(\mathbb{R})$, which contains the Borel sigma algebra $\mathcal{B}(\mathbb{R})$, and that on this sigma algebra outer measure is countably additive and hence a measure.

First, we formalize this definition:

Definition 2.33 (Lebesgue measurable set) *A set* $A \subset \mathbb{R}$ *is said to be **Lebesgue measurable** if it satisfies the **Carathéodory criterion**, that for any set* $E \subset \mathbb{R}$:

$$m^*(E) = m^*(A \cap E) + m^*(\widetilde{A} \cap E). \tag{2.14}$$

The collection of Lebesgue measurable sets is denoted $\mathcal{M}_L \equiv \mathcal{M}_L(\mathbb{R})$.

Remark 2.34 *Since Lebesgue outer measure is countably subadditive by (2.6), and hence it is always the case that:*

$$m^*(E) \le m^*(A \cap E) + m^*(\widetilde{A} \cap E),$$

we have alternatively that $A \subset \mathbb{R}$ *is **Lebesgue measurable** if for any set* $E \subset \mathbb{R}$:

$$m^*(E) \ge m^*(A \cap E) + m^*(\widetilde{A} \cap E). \tag{2.15}$$

Given this definition, the goal of the forthcoming propositions is to show that the collection of sets which satisfy (2.14), $\mathcal{M}_L$, is a sigma algebra which contains the Borel sigma algebra $\mathcal{B}(\mathbb{R})$, as well as all subsets of $\mathbb{R}$ which have outer measure 0.

The first result below demonstrates that $\mathcal{M}_L$ is an algebra that contains all sets with Lebesgue outer measure 0. Then after a technical result, we will show that $\mathcal{M}_L$ is in fact a sigma algebra that contains the open intervals, and hence contains the Borel sigma algebra.

Proposition 2.35 (Initial properties of $\mathcal{M}_L(\mathbb{R})$**)** *Let* $\mathcal{M}_L$ *denote the collection of subsets of* $\mathbb{R}$ *which are Lebesgue measurable and hence satisfy (2.14). Then* $\mathcal{M}_L$ *is an algebra of sets which includes all sets of outer measure 0. Specifically:*

1. $\emptyset, \mathbb{R} \in \mathcal{M}_L$;
2. $A \in \mathcal{M}_L$ *if and only if* $\widetilde{A} \in \mathcal{M}_L$;
3. *If* $A_j \in \mathcal{M}_L$ *for* $1 \le j \le n$, *then* $\bigcup_{j=1}^{n} A_j \in \mathcal{M}_L$;
4. *If* $A \subset \mathbb{R}$ *with* $m^*(A) = 0$, *then* $A \in \mathcal{M}_L$;
5. *If* $A \in \mathcal{M}_L$, *then* $A + x \in \mathcal{M}_L$ *for all* $x \in \mathbb{R}$, *where* $A + x \equiv \{a + x | a \in A\}$.

Proof. *Taking each item in turn:*

1. *For any* $E \subset \mathbb{R}$, $E \cap \emptyset = \emptyset$, *and so* $m^*(\emptyset \cap E) = 0$ *by Proposition 2.28. Hence, (2.15) is satisfied with equality since* $\widetilde{\emptyset} = \mathbb{R}$. *That* $\mathbb{R} \in \mathcal{M}_L$ *follows from item 2.*

2. *The definition in (2.14) is symmetric in* A *and* $\widetilde{A}$, *so both are measurable, or neither is.*

3. *This result only requires proof for* $n = 2$, *since the general result will then follow by mathematical induction. Given* $A, A' \in \mathcal{M}_L$, $\widetilde{A \cup A'} = \widetilde{A} \cap \widetilde{A'}$ *by DeMoivre's law. To prove that* $A \cup A' \in \mathcal{M}_L$, *it must be shown by (2.15) that for any* $E \subset \mathbb{R}$:

$$m^*([A \cup A'] \cap E) + m^*(\widetilde{A} \cap \widetilde{A}' \cap E) \le m^*(E).$$

First:

$$[A \cup A'] \cap E = [A \cap E] \cup [A' \cap \widetilde{A} \cap E],$$

so finite subadditivity of outer measure from Proposition 2.29 obtains:

$$m^*([A \cup A'] \cap E) \le m^*[A \cap E] + m^*[A' \cap \widetilde{A} \cap E].$$

Applying the definition of $A' \in \mathcal{M}_L$ to split the set $\widetilde{A} \cap E$:

$$m^*[A' \cap \widetilde{A} \cap E] + m^*(\widetilde{A}' \cap \widetilde{A} \cap E) = m^*[\widetilde{A} \cap E].$$

Combining:

$$m^*([A \cup A'] \cap E) + m^*(\widetilde{A} \cap \widetilde{A}' \cap E)$$
$$\le m^*[A \cap E] + m^*[\widetilde{A} \cap E]$$
$$= m^*(E).$$

4. *If $m^*(A) = 0$, then $m^*(E \cap A) = 0$ for any $E \subset \mathbb{R}$ by monotonicity since $E \cap A \subset A$. Similarly, $m^*(E \cap \widetilde{A}) \le m^*(E)$ since $E \cap \widetilde{A} \subset E$. Combining yields (2.15) and hence $A \in \mathcal{M}_L$.*

5. *We must show that for all $E \subset \mathbb{R}$:*

$$m^*(E) = m^*((A + x) \cap E) + m^*(\widetilde{A + x} \cap E).$$

Now $\widetilde{A + x} = \widetilde{A} + x$, while:

$$(A + x) \cap E = [A \cap (E - x)] + x,$$
$$\widetilde{A + x} \cap E = \left[\widetilde{A} \cap (E - x)\right] + x.$$

Thus since $A \in \mathcal{M}_L$, and using translation invariance of item 5 of Proposition 2.28:

$$m^*((A + x) \cap E) + m^*(\widetilde{A + x} \cap E)$$
$$= m^*(A \cap (E - x)) + m^*(\widetilde{A} \cap (E - x))$$
$$= m^*(E - x) = m^*(E). \qquad \blacksquare$$

The next step is to show that $\mathcal{M}_L$ is in fact a sigma algebra. To do this, we need a technical result which is stated next. In effect, this result generalizes the "splitting" assumption on measurable sets in (2.14). In addition to splitting arbitrary E with disjoint measurable A and $\tilde{A}$, this splitting assumption applies to any disjoint collection of measurable sets.

Proposition 2.36 (Finite additivity of m^* on $\mathcal{M}_L(\mathbb{R})$) *If $\{A_j\}_{j=1}^n \subset \mathcal{M}_L$ is a collection of pairwise disjoint measurable sets, then for any $E \subset \mathbb{R}$:*

$$m^* \left(E \bigcap \left[\bigcup_{j=1}^n A_j \right] \right) = \sum_{j=1}^n m^* \left(E \bigcap A_j \right). \tag{2.16}$$

*Thus m^* is **finitely additive** on $\mathcal{M}_L$:*

$$m^* \left(\bigcup_{j=1}^n A_j \right) = \sum_{j=1}^n m^*(A_j). \tag{2.17}$$

Proof. *Using mathematical induction, this result is by definition true for $n = 1$, so assume true for $n - 1$. Given measurable A_n that is pairwise disjoint from $\{A_j\}_{j=1}^{n-1}$, apply (2.14) to the set $E \bigcap \left[\bigcup_{j=1}^n A_j \right]$. Because $\{A_j\}_{j=1}^n$ is a pairwise disjoint collection, $\left[\bigcup_{j=1}^n A_j \right] \bigcap A_n = A_n$ and $\left[\bigcup_{j=1}^n A_j \right] \bigcap \tilde{A}_n = \bigcup_{j=1}^{n-1} A$, and so:*

$$m^* \left(E \bigcap \left[\bigcup_{j=1}^n A_j \right] \right) = m^* \left(E \bigcap \left[\bigcup_{j=1}^n A_j \right] \bigcap A_n \right)$$

$$+ m^* \left(E \bigcap \left[\bigcup_{j=1}^n A_j \right] \bigcap \tilde{A}_n \right)$$

$$= m^* \left(E \bigcap A_n \right) + m^* \left(E \bigcap \left[\bigcup_{j=1}^{n-1} A_j \right] \right).$$

Thus (2.16) is true for $\{A_j\}_{j=1}^n$ since it is true for $\{A_j\}_{j=1}^{n-1}$, and the proof is complete.
For finite additivity, we take $E = \mathbb{R}$. ∎

Remark 2.37 (Finite additivity of m^*; nonmeasurable sets) *Two comments:*
*1. m^* is **not** finitely additive on $\sigma(P(\mathbb{R}))$: When $\{A_j\}_{j=1}^n$ are disjoint intervals, (2.17) seems fairly transparent. But when moving from disjoint intervals to disjoint sets, care must be taken when assuming such generalizations.*

For example, finite additivity of m^* is not valid on the power sigma algebra $\sigma(P(\mathbb{R}))$. Indeed, let $\{A_j\}$ denote the collection of disjoint sets constructed in Proposition 2.31. Then for any finite subcollection $\{A_j\}_{j=1}^N$, since $\bigcup_{j=1}^N A_j \subset [0,1]$, monotonicity of Proposition 2.28 obtains:

$$m^* \left(\bigcup_{j=1}^N A_j \right) < 1.$$

If m^* is finitely additive on $\sigma(P(\mathbb{R}))$, this would obtain that for all N:

$$\sum_{j=1}^{N} m^*(A_j) = Nm^*(A_0) < 1.$$

This is an apparent contradiction since $m^*(A_0) > 0$.

This is an important and often under-emphasized property of m^*. Because m^* is apparently finitely additive on any collection of disjoint intervals, it is natural to assume that finite additivity extends to all finite collections of disjoint sets. This would then imply that the failure of Lebesgue outer measure to be a Lebesgue measure on $\sigma(P(\mathbb{R}))$ is only caused by the failure of countable additivity.

In summary, not only is m^* not countably additive on $\sigma(P(\mathbb{R}))$, it is not even finitely additive on this sigma algebra.

2. Nonmeasurable sets: *As noted in Remark 2.32, $\mathcal{M}_L$ cannot contain the full collection $\{A_j\}$ of Proposition 2.31 nor any countable subcollection. As seen above it can also not contain any finite collection $\{A_j\}_{j=1}^{N}$ with $N > 1/m^*(A_0)$. In fact, $\mathcal{M}_L$ cannot contain even one of these sets.*

For example, if $A_0 \in \mathcal{M}_L$ then $A_0 + r_j \in \mathcal{M}_L$ for any rational $r_j \in [0,1)$ by Proposition 2.28. As $\mathcal{M}_L$ is a sigma algebra by the next result, it follows that $A_j^{(1)} \equiv (A_0 + r_j) \bigcap [0,1]$ and $A_j^{(2)} \equiv (A_0 + r_j) \bigcap (1,2]$ are in $\mathcal{M}_L$, as is $A_j^{(2)} - 1$. But then, with A_j the set from Proposition 2.31:

$$A_j = A_j^{(1)} \bigcup \left[A_j^{(2)} - 1 \right] \in \mathcal{M}_L.$$

The same conclusion follows if we assume that $A_j \in \mathcal{M}_L$. We now only have to prove that this implies that $A_0 \in \mathcal{M}_L$. Details are left to the reader.

Thus no A_j-set from Proposition 2.31 is Lebesgue measurable by definition of $\mathcal{M}_L$.

The final result is to show that $\mathcal{M}_L$ is in fact a sigma algebra, and that it contains the Borel sigma algebra $\mathcal{B}(\mathbb{R})$.

Proposition 2.38 ($\mathcal{M}_L(\mathbb{R})$ is a sigma algebra and $\mathcal{B}(\mathbb{R}) \subset \mathcal{M}_L(\mathbb{R})$) *Let $\mathcal{M}_L$ denote the collection of subsets of $\mathbb{R}$ which satisfy (2.14). Then:*

1. *If $\{A_j\}_{j=1}^{\infty} \subset \mathcal{M}_L$ is a collection of measurable sets, then $\bigcup_{j=1}^{\infty} A_j \in \mathcal{M}_L$, and hence $\mathcal{M}_L$ is a sigma algebra.*

2. *For any $a \in \mathbb{R}$, the interval $(-\infty, a) \in \mathcal{M}_L$, and hence $\mathcal{M}_L$ contains the Borel sigma algebra:*

$$\mathcal{B}(\mathbb{R}) \subset \mathcal{M}_L.$$

Proof. *Taking these properties of $\mathcal{M}_L$ in turn:*

1. *If $\{A_j\}_{j=1}^{\infty} \subset \mathcal{M}_L$ is given, then by Proposition 2.20 there is a disjoint collection $\{A'_j\}_{j=1}^{\infty} \subset \sigma(P(\mathbb{R}))$ with $\bigcup_{j=1}^{\infty} A_j = \bigcup_{j=1}^{\infty} A'_j$. It is an exercise to show that $\{A'_j\}_{j=1}^{\infty} \subset \mathcal{M}_L$ using the construction in this proposition, and properties 2 and 3 of Proposition 2.35. Given this, we*

now show that $\bigcup_{j=1}^{\infty} A_j' \in \mathcal{M}_L$ *using* (2.15). *Because* $\mathcal{M}_L$ *is an algebra by Proposition* 2.35, $\bigcup_{j=1}^{n} A_j' \in \mathcal{M}_L$ *for any n and so for any* $E \subset \mathbb{R}$:

$$m^*(E) = m^* \left(E \bigcap \left[\bigcup_{j=1}^{n} A_j' \right] \right) + m^* \left(E \bigcap \left[\widetilde{\bigcup_{j=1}^{n} A_j'} \right] \right)$$

$$\geq m^* \left(E \bigcap \left[\bigcup_{j=1}^{n} A_j' \right] \right) + m^* \left(E \bigcap \left[\widetilde{\bigcup_{j=1}^{\infty} A_j'} \right] \right).$$

The inequality follows since $\bigcup_{j=1}^{n} A_j' \subset \bigcup_{j=1}^{\infty} A_j'$ *implies that* $\widetilde{\bigcup_{j=1}^{\infty} A_j'} \subset \widetilde{\bigcup_{j=1}^{n} A_j'}$. *Because* $\{A_j'\}_{j=1}^{\infty}$ *are disjoint,* (2.16) *can be applied to* $m^*(E \bigcap [\bigcup_{j=1}^{n} A_j'])$ *to obtain:*

$$m^*(E) \geq \sum_{j=1}^{n} m^*(E \bigcap A_j') + m^* \left(E \bigcap \left[\widetilde{\bigcup_{j=1}^{\infty} A_j'} \right] \right).$$

This is true for all n and hence true for $n = \infty$. *Finally, since* m^* *is subadditive:*

$$\sum_{j=1}^{\infty} m^*(E \bigcap A_j') \geq m^* \left(\bigcup_{j=1}^{\infty} [E \bigcap A_j'] \right) = m^* \left(E \bigcap \left[\bigcup_{j=1}^{\infty} A_j' \right] \right).$$

Combining the above inequalities produces (2.15), *and so* $\bigcup_{j=1}^{\infty} A_j \in \mathcal{M}_L$. *Since an algebra by Proposition* 2.35, $\mathcal{M}_L$ *is a sigma algebra.*

2. *To show* $\mathcal{B}(\mathbb{R}) \subset \mathcal{M}_L$ *it is enough by part 1 to show that* $\mathcal{M}_L$ *contains all the open intervals, and we first prove that* $A \equiv (-\infty, a) \in \mathcal{M}_L$ *for all* $a \in \mathbb{R}$. *Given* $E \subset \mathbb{R}$, *if* $m^*(E) = \infty$ *then* (2.15) *is automatically true. If* $m^*(E) < \infty$ *apply* (2.5) *with* $\epsilon > 0$. *Thus there is a collection of open intervals* $\{I_j\}$ *with* $E \subset \bigcup_{j=1}^{\infty} I_j$ *and:*

$$\sum_j |I_j| < m^*(E) + \epsilon.$$

Since $I_j \bigcap A$ *and* $I_j \bigcap \widetilde{A}$ *are also intervals or empty sets, and interval length equals interval outer measure by Proposition* 2.28, *it follows that*

$$|I_j| = m^*(I_j \bigcap A) + m^*(I_j \bigcap \widetilde{A}),$$

and so:

$$\sum_j m^*(I_j \bigcap A) + \sum_j m^*(I_j \bigcap \widetilde{A}) < m^*(E) + \epsilon.$$

Now observe that $E \subset \bigcup_{j=1}^{\infty} I_j$ *implies* $m^*(E \bigcap A) \leq \sum_j m^*(I_j \bigcap A)$ *and* $m^*(E \bigcap \widetilde{A}) \leq \sum_j m^*(I_j \bigcap \widetilde{A})$. *Combining, we conclude that for any* $\epsilon > 0$:

$$m^*(E \bigcap A) + m^*(E \bigcap \widetilde{A}) < m^*(E) + \epsilon.$$

This implies (2.15) and hence $A \equiv (-\infty, a) \in \mathcal{M}_L$ for all $a \in \mathbb{R}$.

As a sigma algebra this implies that for all $b \in \mathbb{R}$:

$$(-\infty, b] = \bigcap_{n=1}^{\infty} (-\infty, b + 1/n] \in \mathcal{M}_L.$$

Thus the complement $(b, \infty) \in \mathcal{M}_L$, as are intersections to prove that $(b, a) \in \mathcal{M}_L$. ∎

This proposition assures that the sigma algebra of Borel sets, $\mathcal{B}(\mathbb{R})$, is contained in the sigma algebra of Lebesgue measurable sets $\mathcal{M}_L(\mathbb{R})$, meaning $\mathcal{B}(\mathbb{R}) \subset \mathcal{M}_L(\mathbb{R})$. But this leaves open the question: Is it possible that $\mathcal{B}(\mathbb{R}) = \mathcal{M}_L(\mathbb{R})$? As noted in (2.10), it turns out that $\mathcal{B}(\mathbb{R}) \subsetneqq \mathcal{M}_L(\mathbb{R})$, and so there are in fact Lebesgue measurable sets that are not Borel sets. The construction of such a set is subtle and will not be pursued here. However, after the forthcoming proposition on approximating Lebesgue measurable sets, we will see that the non-Borel status of these sets is caused by sets with Lebesgue outer measure of 0.

2.5 Calculating Lebesgue Measures

The above section showed that the collection of sets **defined** to be Lebesgue measurable by Definition 2.33, and denoted $\mathcal{M}_L$, is a sigma algebra which contains the Borel sigma algebra $\mathcal{B}(\mathbb{R})$, as well as all sets of Lebesgue outer measure 0. Thus a Lebesgue measurable set A has been defined as a set A that satisfies the Carathéodory criterion in (2.7), that for all $E \subset \mathbb{R}$:

$$m^*(E) = m^*(A \bigcap E) + m^*(\tilde{A} \bigcap E).$$

What has not yet been addressed is: What is the Lebesgue measure of such a "Lebesgue measurable" set? That is, what nonnegative set function is defined on $\mathcal{M}_L$ which satisfies the requirements of Definition 2.21 to be called a Lebesgue measure on $\mathcal{M}_L$, and what is then the Lebesgue measure of sets in this sigma algebra?

We now show that restricted to $\mathcal{M}_L$, Lebesgue outer measure is in fact countably additive, and hence the Lebesgue measure we seek.

Proposition 2.39 (m^* is a Lebesgue measure on $\mathcal{M}_L(\mathbb{R})$ and $\mathcal{B}(\mathbb{R})$) *Lebesgue outer measure m^*, is a Lebesgue measure on $\mathcal{M}_L(\mathbb{R})$ and on $\mathcal{B}(\mathbb{R})$. Thus both $(\mathbb{R}, \mathcal{M}_L(\mathbb{R}), m)$ and $(\mathbb{R}, \mathcal{B}(\mathbb{R}), m)$ are **Lebesgue measure spaces**.*

Proof. *We know from the earlier analysis that $m^*(\emptyset) = 0$ and $m^*(I) = b - a$ for any interval $I = \{a, b\}$, where this interval can be open, closed or semi-open. While finite additivity of m^* on $\mathcal{M}_L$ was proved in Proposition 2.36, still to prove is that m^* is countably additive on $\mathcal{M}_L$. To this end, let $\{A_i\}$ be a countable collection of pairwise disjoint sets. Then for any n,*

$$\bigcup_{i=1}^{n} A_i \subset \bigcup_{i=1}^{\infty} A_i,$$

and by monotonicity from Proposition 2.28 and finite additivity of m^* *on* $\mathcal{M}_L$:

$$m^*\left[\bigcup_{i=1}^{\infty} A_i\right] \geq m^*\left[\bigcup_{i=1}^{n} A_i\right] = \sum_{i=1}^{n} m^*(A_i).$$

Since true for all n, it follows that

$$m^*\left[\bigcup_{i=1}^{\infty} A_i\right] \geq \sum_{i=1}^{\infty} m^*(A_i).$$

This plus countable subadditivity proves countable additivity on $\mathcal{M}_L$. *The same is then true on* $\mathcal{B}(\mathbb{R})$ *since* $\mathcal{B}(\mathbb{R}) \subset \mathcal{M}_L$. ∎

We are now ready to define **Lebesgue measure**.

Definition 2.40 (Lebesgue measure) *For* $A \in \mathcal{M}(\mathbb{R})$, *the* **Lebesgue measure** *of A, denoted* $m(A)$, *is defined by:*

$$m(A) = m^*(A). \tag{2.18}$$

Remark 2.41 *It is worth a moment to assess why the above proof does not work if applied to* $\sigma(P(\mathbb{R}))$. *What special property of* $\mathcal{M}_L$ *was used? In fact it is the seemingly innocent property of finite additivity. As was demonstrated in the prior section,* m^* *is finitely additive on* $\mathcal{M}_L$ *but not finitely additive on* $\sigma(P(\mathbb{R}))$, *and hence the above proof does not generalize.*

2.6 Approximating Lebesgue Measurable Sets

The next proposition states that we can approximate Lebesgue measurable sets arbitrarily well with open and closed sets, denoted $\mathcal{G}$-sets and $\mathcal{F}$-sets in Notation 2.16, and we can approximate these measurable sets to within an outer measure 0 with $\mathcal{G}_\delta$-sets and $\mathcal{F}_\sigma$-sets. This result sharpens somewhat the conclusions in Corollary 2.30 and will be applied in the next sections and elsewhere in this book, and further extended in Proposition 4.2.

Proposition 2.42 (Approximations with Borel sub/supersets) *Let* $A \in \mathcal{M}_L(\mathbb{R})$. *Then for any* $\epsilon > 0$ *there is an open set* $G \in \mathcal{G}$ *and closed set* $F \in \mathcal{F}$ *so that* $F \subset A \subset G$ *and:*

$$m(G - A) \leq \epsilon, \quad m(A - F) \leq \epsilon. \tag{2.19}$$

In addition, there are sets $G' \in \mathcal{G}_\delta$ *and* $F' \in \mathcal{F}_\sigma$ *so that* $F' \subset A \subset G'$ *and:*

$$m(G' - A) = m(A - F') = 0. \tag{2.20}$$

Proof. *Note that if $A \in \mathcal{M}_L$, then because $\mathcal{B}(\mathbb{R}) \subset \mathcal{M}_L$ it follows that $G - A \in \mathcal{M}_L$ and $A - F \in \mathcal{M}_L$ whether G is in $\mathcal{G}$ or $\mathcal{G}_\delta$ and whether F is in $\mathcal{F}$ or $\mathcal{F}_\sigma$. Since $m = m^*$ on $\mathcal{M}_L$ by (2.18), this proof utilizes as properties of m, previous results that were stated in terms of m^*.*

If $m(A) < \infty$ we have from (2.7) the existence of open $G \in \mathcal{G}$ so that $A \subset G$, and $m(G) \leq m(A) + \epsilon$. Now since $m(G - A) \equiv m(G \cap \tilde{A})$, it follows from (2.14) that since A is measurable:

$$m(G - A) = m(G) - m(G \cap A).$$

But $A \subset G$ implies that $G \cap A = A$ and thus $m(G \cap A) = m(A)$. This proves the first result in 2.19 when $m(A)$ is finite.

If $m(A) = \infty$, define for $n = 1, 2, 3..,$

$$I_n = [-n, -n+1) \bigcup [n-1, n),$$
$$A_n = A \bigcap I_n.$$

Then $m(A_n) < m(I_n) < \infty$, and by the prior proof there is open G_n with $A_n \subset G_n$ and $m^*(G_n - A_n) \leq \epsilon/2^n$. Now define $G = \bigcup_n G_n$. Then $A = \bigcup_n A_n \subset G$, and from De Morgan's laws:

$$G - A \equiv \left(\bigcup_n G_n \right) \bigcap \left(\widetilde{\bigcup_n A_n} \right)$$
$$= \bigcup_n \left(G_n \bigcap \left(\bigcap_n \tilde{A}_n \right) \right)$$
$$\subset \bigcup_n \left(G_n \bigcap \tilde{A}_n \right).$$

Using monotonicity and subadditivity from Propositions 2.28 and 2.29:

$$m(G - A) \leq m \left[\bigcup_n \left(G_n \bigcap \tilde{A}_n \right) \right]$$
$$\leq \sum_n m(G_n - A_n) \leq \epsilon.$$

Thus the first result in (2.19) is proved for $m(A) = \infty$.

For the closed subsets, apply the prior conclusion to $\tilde{A} \in \mathcal{M}_L$. Then there exists open G with $\tilde{A} \subset G$ and $m(G - \tilde{A}) \leq \epsilon$. Defining $F = \tilde{G}$ obtains that F is closed, $F \subset A$, and $m(A - F) \leq \epsilon$, since:

$$A - F \equiv A \bigcap \tilde{F} = \tilde{\tilde{A}} \bigcap G = G - \tilde{A}.$$

For (2.20), define G_n and F_n so that (2.19) is true with $\epsilon = 1/n$. Then with $G' \equiv \bigcap G_n$ and $F' \equiv \bigcup F_n$, it follows by definition that $G' \in \mathcal{G}_\delta$, $F' \in \mathcal{F}_\sigma$, and $F' \subset A \subset G'$. Define $G'_n \equiv \bigcap_{j \leq n} G_j$, then:

$$A \subset G'_n \subset G_n.$$

But $G' \subset G'_n \subset G_n$ implies that $G' - A \subset G'_n - A \subset G_n - A$, *and hence by monotonicity and* (2.19):

$$m(G' - A) \leq m(G_n - A) \leq 1/n.$$

This is true for all n and thus the first part of (2.20) is proved.

The same argument applies to F' with $F'_n \equiv \bigcup_{j \leq n} F_j$, since then $F_n \subset F'_n \subset F'$ and this implies that $A - F' \subset A - F'_n \subset A - F_n$. ∎

2.7 Properties of Lebesgue Measure

There are two important properties of Lebesgue measure that will be seen to be true more generally for other measures. The first important property is **regularity**, which is a property that is closely related to the approximation results of (2.19) of Proposition 2.42, but with the added feature that closed subsets can be taken to be compact. This property will be seen to be true for all Borel measures, using similar approximation results.

The second important property is **continuity**, which addresses the behavior of a measure on unions and intersections of nested sets. In short, it states that with a minor restriction, the measure of the limit set is the limit of the measures of the sets. This property is closely connected with countable additivity, and thus will be seen to be a property of all measures.

The results in this section will be stated and proved directly in terms of Lebesgue measure m, again utilizing earlier properties of m^*. This reflects the comment at the beginning of the proof of Proposition 2.42, which says that because all referenced sets are seen to be elements of $\mathcal{M}_L$, it follows that $m = m^*$.

2.7.1 Regularity

The first important property of regularity of Lebesgue measure encompasses two notions:

1. **Outer regularity**: The measure of a measurable set equals the **infimum** of the measures of **open supersets**, and,
2. **Inner regularity**: The measure of a measurable set equals the **supremum** of the measures of **compact subsets**.

A measure that is both outer and inner regular is called **regular**.

Recall that compact is defined as in Definition 2.26, but by the **Heine-Borel theorem** of Proposition 2.27, a set in $\mathbb{R}$ is compact if and only if it is closed and bounded.

Proposition 2.43 (Regularity of Lebesgue Measure) *Lebesgue measure m is regular on* $\mathcal{M}_L(\mathbb{R})$. *Specifically, if $A \in \mathcal{M}_L$:*

$$m(A) = \inf_{A \subset G} m(G), \; G \text{ open}. \tag{2.21}$$

and:

$$m(A) = \sup_{F \subset A} m(F), \; F \text{ compact.} \tag{2.22}$$

Proof. 1. *Outer Regularity*: *If $m(A) = \infty$ then $m(G) = \infty$ for all such G by monotonicity of m, and (2.21) follows in this case.*

For $m(A) < \infty$ and $\epsilon > 0$, let G_ϵ be the open set defined in terms of (2.19), meaning $m(G_\epsilon - A) \leq \epsilon$. Then since A is measurable, $A \subset G_\epsilon$ and $m(\widetilde{A} \bigcap G_\epsilon) = m(G_\epsilon - A)$:

$$m(G_\epsilon) = m(A \bigcap G_\epsilon) + m(\widetilde{A} \bigcap G_\epsilon)$$
$$\leq m(A) + \epsilon.$$

Then $m(A) \leq m(G_\epsilon)$ by monotonicity, and so:

$$m(A) \leq m(G_\epsilon) \leq m(A) + \epsilon.$$

Thus (2.21) follows with the infimum defined over all such G_ϵ. As $m(A) \leq m(G)$ for any $G \supset A$, (2.21) follows with the infimum defined over all such G.

2. *Inner Regularity*: *Given $\epsilon > 0$, let $F_\epsilon \subset A$ be the closed set in (2.19), so $m(A - F_\epsilon) \leq \epsilon$. Measurability of F_ϵ and A obtain:*

$$m(A) = m(A \bigcap F_\epsilon) + m(A \bigcap \widetilde{F}_\epsilon),$$

and this plus monotonicity of m imply:

$$m(A) - \epsilon \leq m(F_\epsilon) \leq m(A). \tag{1}$$

Thus (2.22) follows with supremum over such closed F_ϵ. This implies the same result with supremum over all closed F since $F \subset A$ assures $m(F) \leq m(A)$.

Define $F_\epsilon^{(n)} = F_\epsilon \bigcap I_n$ where $I_n \equiv [-n, -(n-1)) \bigcup [n-1, n)$. Then $F_\epsilon^{(n)} \subset F_\epsilon$ and $F_\epsilon^{(n)}$ is bounded for all n. Since $F_\epsilon = \bigcup_n F_\epsilon^{(n)}$ and $F_\epsilon^{(n)}$ are disjoint, countable additivity obtains:

$$m(F_\epsilon) = \sum_{n=1}^{\infty} m\left(F_\epsilon^{(n)} \right). \tag{2}$$

If $m(A) < \infty$ then $m(F_\epsilon) < \infty$, so there is an N with $\sum_{n=N+1}^{\infty} m\left(F_\epsilon^{(n)} \right) < \epsilon$. Again by countable additivity and monotonicity:

$$\epsilon > \sum_{n=N+1}^{\infty} m\left(F_\epsilon^{(n)} \right)$$
$$= m\left(F_\epsilon \bigcap \left\{ (-\infty, -N) \bigcup [N, \infty) \right\} \right) \tag{3}$$
$$\geq m\left(F_\epsilon \bigcap \left[(-\infty, -N) \bigcup (N, \infty) \right] \right).$$

Defining $\bar{F}_\epsilon^N = F_\epsilon \cap [-N, N]$, *then* $\bar{F}_\epsilon^N \subset F_\epsilon$ *is closed and bounded and thus compact, and* $m(F_\epsilon) - m(\bar{F}_\epsilon^N) < \epsilon$ *by (3). Combining with (1) obtains:*

$$m(A) - 2\epsilon \leq m(\bar{F}_\epsilon^N) \leq m(A),$$

and (2.22) follows with compact $\bar{F}_\epsilon^N$, *and then as above for all* $F \subset A$ *compact.*

If $m(A) = \infty$ *then* $m(F_\epsilon) = \infty$ *by (1), and thus* $\sum_{n=1}^N m\left(F_\epsilon^{(n)}\right)$ *is unbounded in* N. *But if* $\bar{F}_\epsilon^N = F_\epsilon \cap [-N, N]$ *as above:*

$$\bigcup_{n=1}^N F_\epsilon^{(n)} = F_\epsilon \cap [-N, N)$$
$$\subset \bar{F}_\epsilon^N.$$

Thus by monotonicity $m(\bar{F}_\epsilon^N)$ *is unbounded with compact* $\bar{F}_\epsilon^N$ *and (2.22) is proved in this case.* ■

2.7.2 Continuity

Another important result identifies how Lebesgue measure operates on unions and intersections of **nested** measurable sets. By nested it is meant that the collection $\{A_i\}$ satisfies:

$$A_i \subset A_{i+1}, \text{ for all } i,$$

or

$$A_{i+1} \subset A_i, \text{ for all } i.$$

In the former case, we are interested in the measure of the union, and in the latter, the measure of the intersection.

The result is stated for the Lebesgue measure space, $(\mathbb{R}, \mathcal{M}_L, m)$, but an identical proof works without modification in any measure space $(X, \sigma(X), \mu)$. This is because the proof requires only countable additivity and monotonicity of measures, and not any special properties of Lebesgue measure.

The properties identified in this proposition are often referred to in terms of the "continuity" of measures and understood in the following sense. Given a collection of measurable sets $\{B_i\}$, define A_n by:

1. $A_n = \bigcup_{i=1}^n B_i$, or,
2. $A_n = \bigcap_{i=1}^n B_i$, with $m(B_1) < \infty$.

The following result states that in both cases:

$$m\left(\lim_{n\to\infty} A_n\right) = \lim_{n\to\infty} m(A_n),$$

reminiscent of the familiar property for continuous functions. In the first case, A_n is an **increasing sequence of sets** and the result is called **continuity from below**, while in the second case the sequence is **decreasing sequence of sets** and the result is called **continuity from above**.

This proposition is stated in terms of "nested" sets, where $A_i \subset A_{i+1}$ or $A_{i+1} \subset A_i$ for all i. But these results can be applied to the limits of partial unions and partial intersections of arbitrary collections of measurable sets as noted above.

Finally, the requirement that $m(A_1) < \infty$ is necessary for (2.24), as the example $A_n \equiv [n, \infty)$ illustrates.

Proposition 2.44 (Continuity of Lebesgue Measure) *Let $\{A_i\} \subset \mathcal{M}_L(\mathbb{R})$. Then:*

1. **Continuity from Below:** *If $A_i \subset A_{i+1}$ for all i:*

$$m\left(\bigcup_{i=1}^{\infty} A_i\right) = \lim_{i \to \infty} m(A_i), \qquad (2.23)$$

   where the limit on the right may be finite or infinite.

2. **Continuity from Above:** *If $A_{i+1} \subset A_i$ for all i and $m(A_1) < \infty$:*

$$m\left(\bigcap_{i=1}^{\infty} A_i\right) = \lim_{i \to \infty} m(A_i). \qquad (2.24)$$

Proof. *To prove item 1, first note that $A_i \subset A_{i+1}$ implies that $m(A_i) \leq m(A_{i+1})$ by monotonicity of m. Define $B_1 = A_1$ and for $i \geq 2$, let $B_i = A_i - A_{i-1}$. Then $\{B_i\} \subset \mathcal{M}_L$, are disjoint sets, and*

$$\bigcup_{i=1}^{\infty} A_i = \bigcup_{i=1}^{\infty} B_i.$$

By countable additivity:

$$m\left(\bigcup_{i=1}^{\infty} A_i\right) = \sum_{i=1}^{\infty} m(B_i)$$

$$= m(A_1) + \lim_{i \to \infty} \sum_{j=2}^{i} m(A_j - A_{j-1}).$$

Since A_{j-1} and $A_j - A_{j-1}$ are disjoint with union A_j, finite additivity assures that:

$$m(A_j - A_{j-1}) = m(A_j) - m(A_{j-1}).$$

Thus by cancellation in this telescoping summation:

$$m\left(\bigcup_{i=1}^{\infty} A_i\right) = \lim_{i \to \infty} m(A_i).$$

For item 2, note that $\bigcap_{j=1}^{i} A_j = A_i$ by the nesting property, while monotonicity and the assumption that $m(A_1) < \infty$ yields for all i:

$$m\left(\bigcap_{j=1}^{i} A_j\right) = m(A_i) < \infty.$$

Again by monotonicity $\{m(A_i)\}$ *is a bounded, nonincreasing sequence, and thus has a well-defined limit as* $i \to \infty$, *which proves* (2.24). ∎

Proposition 2.45 (Continuity of all Measures) *Given the measure space* $(X, \sigma(X), \mu)$, *and* $\{A_i\} \subset \sigma(X)$:

1. **Continuity from Below:** *If* $A_i \subset A_{i+1}$ *for all* i:

$$\mu\left(\bigcup_{i=1}^{\infty} A_i\right) = \lim_{i\to\infty} \mu(A_i), \tag{2.25}$$

   where the limit on the right may be finite or infinite.
2. **Continuity from Above:** *If* $A_{i+1} \subset A_i$ *for all* i *and* $\mu(A_1) < \infty$:

$$\mu\left(\bigcap_{i=1}^{\infty} A_i\right) = \lim_{i\to\infty} \mu(A_i). \tag{2.26}$$

Proof. *The proof is identical since only the sigma algebra structure of* $\mathcal{M}_L$, *and monotonicity and countable additivity of the measure,* m, *were used in the Lebesgue proof.* ∎

2.8 Discussion of $\mathcal{B}(\mathbb{R}) \subsetneqq \mathcal{M}_L(\mathbb{R})$

Proposition 2.42 demonstrated that Lebesgue measurable sets can be approximated arbitrarily well by closed subsets or open supersets. Moreover, if we extend the approximating sets to $\mathcal{G}_\delta$-sets and $\mathcal{F}_\sigma$-sets, every Lebesgue measurable set can be approximated to within a set of Lebesgue measure 0. Thus, by (2.20), if $A \in \mathcal{M}_L$ is any measurable set, there exists $G' \in \mathcal{G}_\delta$ and $F' \in \mathcal{F}_\sigma$ so that $F' \subset A \subset G'$, and:

$$A = F'\bigcup Z_F, \qquad A\bigcup Z_G = G'. \tag{2.27}$$

Here $Z_G \equiv G' - A$ and $Z_F \equiv A - F'$ have Lebesgue measure 0 by Proposition 2.42.

   Since $\mathcal{G}_\delta$-sets and $\mathcal{F}_\sigma$-sets are subsets of the Borel sigma algebra $\mathcal{B}(\mathbb{R})$, this characterization obtains the following.

Proposition 2.46 (Non-Borel measurable sets) *Given a Lebesgue measurable A set that is non-Borel, that is:*

$$A \in \mathcal{M}_L(\mathbb{R}) - \mathcal{B}(\mathbb{R}),$$

then in any application of (2.20), $Z_G \equiv G' - A$ *and* $Z_F \equiv A - F'$ *are disjoint, non-Borel sets of outer measure 0.*

Proof. *These sets are disjoint and of measure 0 by construction. If Z_F is a Borel set, then $A = F' \bigcup Z_F$ is a Borel set. Thus if A is non-Borel, so too is Z_F.*
   For Z_G, since:

$$F' \bigcup Z_F \bigcup Z_G = G',$$

the disjointness of F' and $Z_F \bigcup Z_G$ implies that:

$$Z_F \bigcup Z_G = G' - F'.$$

Thus $Z_F \bigcup Z_G$ is always a Borel set.
   If A is non-Borel and Z_G is a Borel set, then the disjointness of Z-sets implies:

$$Z_F = \left(Z_F \bigcup Z_G\right) - Z_G,$$

and then Z_F must also be a Borel set, a contradiction.   ■

   The above section title, that $\mathcal{B}(\mathbb{R}) \subsetneqq \mathcal{M}_L(\mathbb{R})$, implies that there are Lebesgue measurable sets which have non-Borel Z components. Put another way, there exist Lebesgue measurable sets A that can only be defined in terms of a Borel set and a non-Borel set of measure 0:

$$A = F' \bigcup Z_F = G' - Z_G.$$

As we have no further need for this fact, we will not prove it.

Remark 2.47 (On $\mathcal{M}_L(\mathbb{R})$ vs. $\mathcal{B}(\mathbb{R})$) *One advantage of defining Lebesgue measure on the sigma algebra $\mathcal{M}_L$ rather than on $\mathcal{B}(\mathbb{R})$, is that this sigma algebra is **complete**, and thus Lebesgue measure is **complete** on this sigma algebra. In the standard terminology, the triplet $(\mathbb{R}, \mathcal{M}_L, m)$ is a **complete measure space**.*

Definition 2.48 (Complete measure space) *A measure space $(X, \sigma(X), \mu)$ is a **complete measure space** if given $A \in \sigma(X)$ with $\mu(A) = 0$, then for every $B \subset A$ we have that $B \in \sigma(X)$.*

   Of course, if $B \subset A$ and $B \in \sigma(X)$, the assumption that $\mu(A) = 0$ ensures that $\mu(B) = 0$ by monotonicity.
   Completeness is a property of the sigma algebra $\sigma(X)$, and not a property of the measure μ. The sigma algebra $\mathcal{M}_L$ is complete by Proposition 2.35 since it contains every set of Lebesgue outer measure 0. The Borel sigma algebra $\mathcal{B}(\mathbb{R})$ is not complete, meaning a Borel set of measure 0 can have non-Borel subsets. As we have no further use for this fact, we will not prove it.
   Depending on the context or application, the **Lebesgue measure space** $(\mathbb{R}, \mathcal{M}_L, m)$ is typically used when completeness is a desired property. In other contexts or applications, the measure space may be defined as $(\mathbb{R}, \mathcal{B}(\mathbb{R}), m)$, and then it is an example of a **Borel measure space**.

3

Measurable Functions

Given the Chapter 1 derivation of the Lebesgue measure spaces $(\mathbb{R}, \mathcal{M}_L(\mathbb{R}), m)$ and $(\mathbb{R}, \mathcal{B}(\mathbb{R}), m)$, it is logical that this chapter should address the definition and properties of "measurable" functions on these spaces. This will indeed be done, but our true goal is more ambitious.

In coming chapters we will be developing a variety of measure spaces, and the reader will no doubt have noticed that this is the only chapter in this book on measurable functions. Indeed, it will soon be appreciated that the current chapter, if focused only on $(\mathbb{R}, \mathcal{M}_L(\mathbb{R}), m)$ and $(\mathbb{R}, \mathcal{B}(\mathbb{R}), m)$, could literally be cut-and-pasted into a focused chapter after each of these measure space developments with only a change of notation. How boring this could be for the reader.

So instead this chapter serves a dual purpose. We want the focus to be on Lebesgue spaces simply because for many readers, this will be their frame of reference after Chapter 1. On the other hand, we also want to focus on the generality of these results, that the conceptual notion of a measurable function is independent of the given measure space on which this function is defined.

Thus virtually any property of a Lebesgue measurable function that relies only on the sigma algebra properties of $\mathcal{M}_L(\mathbb{R})$ or $\mathcal{B}(\mathbb{R})$ will generalize immediately to any measurable function defined on any measure space. If the given result requires completeness of the measure space, it will generalize immediately to any measurable function defined on any complete measure space. Finally, if the result requires notions of open and closed sets in the space's sigma algebra, it will generalize immediately to any measurable function defined on any measure space with a topology, and where the given sigma algebra contains the open sets. By Definition 2.13, this means any measure space where the given sigma algebra contains the Borel sigma algebra.

The reader is encouraged to keep these general ideas in mind as they work through this chapter. Every attempt will be made to make these more general connections, sometimes only implicitly by not identifying the measure space.

3.1 Extended Real-Valued Functions

In contrast with calculus, which focuses on various smoothness properties of functions, it is not uncommon in real analysis to explicitly allow functions to assume the values ∞ or $-\infty$ at points of its domain, D. Put another way, the range of such functions is defined to be the **extended real numbers**.

DOI: 10.1201/9781003257745-3

Definition 3.1 (Extended real numbers) *The **extended real numbers** are defined:*

$$\overline{\mathbb{R}} \equiv \mathbb{R} \bigcup \{\infty\} \bigcup \{-\infty\}.$$

The natural ordering on $\mathbb{R}$ is extended to $\overline{\mathbb{R}}$ by assuming that for all real x, that $-\infty < x < \infty$. Also, for such real x we define:

$$x + \infty = \infty, \quad x - \infty = -\infty,$$

and for $x > 0$:

$$x \cdot \infty = \infty, \quad x \cdot -\infty = -\infty.$$

Finally,

$$\infty + \infty = \infty,$$
$$-\infty - \infty = -\infty,$$
$$\infty \cdot (\pm \infty) = \pm \infty.$$

Note that the various numerical results "defined" are consistent with the unambiguous results that would be achieved if calculations involving $\pm\infty$ were defined in terms of limits. For example:

$$x + \infty = \lim_{y \to \infty} (x + y) = \infty.$$

Also note that expressions such as $\infty - \infty$ or $0 \cdot \infty$ cannot be unambiguously assigned a value through such a limiting definition, and care must be taken when such values are encountered.

Definition 3.2 (Extended real-valued function) *A **real-valued function** defined on a domain $D \subset \mathbb{R}$ is a function $f : D \to \mathbb{R}$. An **extended real-value function** defined on a domain D is a function $f : D \to \overline{\mathbb{R}}$. The same terminology applies to functions defined on any domain $D \subset X$, for an arbitrary space X.*

Depending on the result below, we may specify whether the function under discussion is real-valued or extended real-valued. Since extended real-valued implies real-valued, any result stated for extended real-valued functions applies to both classes of functions. Some results are stated for real-valued functions, meaning that the proof requires only finite values in the range.

3.2 Equivalent Definitions of Measurability

Given the informal introduction to the Lebesgue integral in Section 1.2, the following definition will be of no surprise. In essence, we need to ensure that any set of the form:

$$\{x | y_i < f(x) \leq y_{i+1}\}$$

is Lebesgue measurable. In addition, sets defined with other combinations of inequalities, as well as with an equality, are required to be Lebesgue measurable. In each case Lebesgue measurable means measurable within the measure space $(\mathbb{R}, \mathcal{B}(\mathbb{R}), m)$ or $(\mathbb{R}, \mathcal{M}_L, m)$, which in turn means that such sets are members of the respective sigma algebras.

Any such set can be defined as the intersection of two sets defined with one inequality, for example:

$$\{x | y_i < f(x) \leq y_{i+1}\} = \{x | f(x) > y_i\} \bigcap \{x | f(x) \leq y_{i+1}\}.$$

Since the intersection of measurable sets is measurable, the notion of "measurability" of a function can be defined in terms of these one-sided inequalities. Perhaps surprising, as long as the domain of the function $f(x)$ is itself Lebesgue measurable, all four possible one-sided definitions of measurability are equivalent.

Remark 3.3 (Role of the sigma algebra) *It is important to emphasize at the outset that the concept of measurability of a function is intimately connected to the sigma algebra used for the measure space. For example, every function on $\mathbb{R}$ is measurable with respect to the rationals counting measure space $(\mathbb{R}, \sigma(P(\mathbb{R})), \mu_{\mathbb{Q}})$ illustrated in (2.3), because the power sigma algebra includes all sets. On the other hand, for the trivial measure space, $(\mathbb{R}, \sigma(\mathbb{R}, \emptyset), \mu)$, with $\mu(\mathbb{R}) = 1$ and $\mu(\emptyset) = 0$, the only functions on $\mathbb{R}$ that are measurable are the constant functions, $f(x) = c$.*

As noted in the introduction to this chapter, many of the results below will be valid in a more general measure space $(X, \sigma(X), \mu)$. A good example of this is the next proposition for which we do not even need to identify the sigma algebra in the domain space. Only definitional properties of the sigma algebra are required in the proof.

The following result is completely general and applies to all measure spaces. Thus it is silent on the given sigma algebra or the type of measurability. If desired the reader can insert "Lebesgue" or "Borel" before "measurable" to anchor this result into the current discussion. But note that its proof only requires sigma algebra manipulations.

Proposition 3.4 (Equivalent formulations for measurability) *Let $f(x)$ be a real-valued or an extended real-valued function defined on a measure space X. The following statements are equivalent:*

1. *For every real number y, the set $\{x | f(x) < y\}$ is measurable.*
2. *For every real number y, the set $\{x | f(x) \geq y\}$ is measurable.*
3. *For every real number y, the set $\{x | f(x) > y\}$ is measurable.*
4. *For every real number y, the set $\{x | f(x) \leq y\}$ is measurable.*

If $f(x)$ is measurable by any of these statements and restricted to a measurable domain D, these statements are again equivalent relative to the restricted domain.

Proof. *First note that $1 \Leftrightarrow 2$ and $3 \Leftrightarrow 4$, where by $\Leftrightarrow$ is meant "if and only if." For example:*

$$\{x | f(x) < y\} = \{x | f(x) \geq y\}^c,$$

where $A^c = \tilde{A}$ denotes the set complement of A. So, if either set is measurable, so too the other set is measurable since sigma algebras are closed under complementation.

To link these two pairs of equivalent results, note that:

$$\{x|f(x) \geq y\} = \bigcap_n \{x|f(x) > y - 1/n\},$$

and so the sigma algebra structure of measurable sets ensures that 3 $\Rightarrow$ 2. Similarly, since:

$$\{x|f(x) > y\} = \bigcup_n \{x|f(x) \geq y + 1/n\},$$

this ensures that 2 $\Rightarrow$ 3.

Finally, if D is a measurable domain, then for example:

$$\{x \in D|f(x) < y\} = \{x|f(x) < y\} \bigcap D,$$

and thus the proof of equivalence applies to the restricted function.   ∎

This result justifies the following definition of **measurable function**. See also Definition 3.9.

Definition 3.5 (Measurable function) *An extended real-valued function $f(x)$ defined on $(\mathbb{R}, \mathcal{M}_L(\mathbb{R}), m)$ is said to be **Lebesgue measurable** if any of the conditions in the above proposition are satisfied with measurable defined relative to $\mathcal{M}_L$. Similarly, an extended real-valued function $f(x)$ defined on $(\mathbb{R}, \mathcal{B}(\mathbb{R}), m)$ is said to be **Borel measurable** if any of the conditions in the above proposition are satisfied with measurable defined relative to $\mathcal{B}(\mathbb{R})$.*

*More generally, an extended real-valued function defined on a measure space $(X, \sigma(X), \mu)$ is said to be **measurable** and sometimes $\sigma(X)$-**measurable** if any of the conditions in the above proposition are satisfied with measurable defined relative to $\sigma(X)$.*

The same definitions apply to such a function defined on a Lebesgue or Borel or $\sigma(X)$-measurable domain D.

*If $f(x)$ is measurable in any of the above meanings and real valued, it will be called a **real-valued measurable** function, qualified as appropriate with Lebesgue or Borel or $\sigma(X)$.*

Remark 3.6 (Exercise opportunities) *Throughout the current chapter on measurable functions, the focus is on the Lebesgue measure space $(\mathbb{R}, \mathcal{M}_L, m)$ and sometimes the Borel measure space $(\mathbb{R}, \mathcal{B}(\mathbb{R}), m)$. To simplify the presentation, we will often state and prove results relative to Lebesgue measurable functions and leave it as an exercise for the reader to investigate if the given results apply in the Borel measurable case. In virtually all cases where the completeness of $\mathcal{M}_L$ is not needed for the stated result, the applicability to the Borel case is almost always assured. In cases where the result depends explicitly on the sigma algebra $\mathcal{B}(\mathbb{R})$, we will explicitly identify the function as Borel measurable.*

Also, we will generally state results for measurable functions defined on $\mathbb{R}$ or X. When such functions are restricted to respectively measurable domains D, nothing new occurs as noted above. So to simplify the exposition, we will avoid continuing to call out this case unless this domain is relevant to the discussion.

Thinking ahead to Book V, it will also be productive for the reader to notice that a great many of these results would apply in the context of a general measure space $(X, \sigma(X), \mu)$ because the associated proofs depend only on the definitional properties of measures and sigma algebra manipulations. An example was already seen in Proposition 3.4.

Another definitional question to address is the relationship between measurable as defined by any one of the above four criteria, and measurability as defined by sets of the form $\{x | f(x) = y\}$. In this case, the measurability of such sets is implied by the above definition, but the implication is not reversible for what might seem to be a surprising reason.

Proposition 3.7 (Consequence of measurability) *Let $f(x)$ be a real-valued or extended real-valued function defined on a measure space X and measurable by any of the 4 equivalent criteria of Proposition 3.4. Then for all real numbers $y \in \mathbb{R}$ or extended real numbers $y \in \overline{\mathbb{R}}$, respectively, the set $\{x | f(x) = y\}$ is measurable.*

Proof. *If $f(x)$ is extended real-valued,*

$$\{x | f(x) = \infty\} = \bigcap_n \{x | f(x) > n\},$$

$$\{x | f(x) = -\infty\} = \bigcap_n \{x | f(x) < -n\},$$

and so $\{x | f(x) = y\}$ is measurable for $y = \infty$ or $y = -\infty$. For finite y:

$$\{x | f(x) = y\} = \{x | f(x) \geq y\} \bigcap \{x | f(x) \leq y\},$$

and the conclusion again follows. ∎

It seems natural to posit that the measurability of $\{x | f(x) = y\}$ for all extended real numbers y should imply the measurability of the sets defined by any one of the inequalities. For example,

$$\{x | f(x) \geq y\} = \bigcup_{z \geq y} \{x | f(x) = z\},$$

and the union of measurable sets is a measurable set. Or is it? The answer is, not necessarily. The problem here is that we are taking the union of an uncountable collection of sets, and sigma algebras are only required to be closed under countably many operations.

So knowledge that every set of the form $\{x | f(x) = y\}$ is an element of a sigma algebra and hence is measurable tells us little about whether sets defined with inequalities are measurable. Fortunately, it is rarely the case that we need the implication in this direction. In almost all cases it is the measurability of the sets like $\{x | f(x) = y\}$ that we wish to infer, and this proposition assures that these sets are always measurable for measurable functions.

Exercise 3.8 *Before delving into the next section, prove that if $f : \mathbb{R} \to \mathbb{R}$ is an increasing or decreasing function, then f is Borel and hence Lebesgue measurable. Hint: Start with $f(x) = e^x$ for example.*

3.3 Examples of Measurable Functions

The definition of "measurable" introduced in the last section is at once perfectly applicable within the theoretical context of defining the Lebesgue integral of Chapter 1, and at the same time perfectly opaque in terms of identifying what kinds of functions satisfy this definition. This is not uncommon. Other properties of functions such as continuity and differentiability are also introduced with a somewhat abstract definition which characterizes a given desirable property.

The way to get comfortable with these ideas is to begin to catalogue examples of functions with these properties. This initial listing is then expanded by investigating if various manipulations of functions with the given property, such as sums, differences, products, quotients, composites, limits, etc., then preserve this property. This will be the approach taken here. We first investigate some simple examples and then develop results that address what happens when we combine or otherwise manipulate these simple examples.

Starting with the simplest functions, it is not difficult to verify directly that functions such as $f(x) = ax^n$ with a real are Lebesgue measurable functions on $\mathbb{R}$, as are all exponential, logarithmic and general monotonic functions. For example, if $f(x) = x^2$ and $y \geq 0$:

$$\{x|f(x) \geq y\} = \{x|\,|x| \geq \sqrt{y}\}$$

$$= \{x|x \geq \sqrt{y}\} \bigcup \{x|x \leq -\sqrt{y}\}.$$

This set is Lebesgue measurable as a union of intervals. For $y < 0$, $\{x|f(x) \geq y\} = \mathbb{R}$ and again is Lebesgue measurable.

Such demonstrations quickly become difficult even for marginally more complicated functions such as polynomials $p(x)$, or rational functions. These latter functions are defined as the ratio of polynomial functions, $f(x) = p(x)/q(x)$. The domain is then usually defined by $\{x|q(x) \neq 0\}$, or $\mathbb{R}$ by defining $f = \infty$ on the roots of $q(x)$. This set of roots has Lebesgue measure 0 since finite.

This definitional exploration is a good start to our investigation because it verifies the existence of infinitely many measurable functions, and also highlights the limitation of this definitional approach. In general this limitation is caused by the need to explicitly evaluate for a given set A, the **inverse function** defined by:

$$f^{-1}(A) \equiv \{x|f(x) \in A\}. \tag{3.1}$$

For example, in part 1 of Proposition 3.4:

$$\{x|f(x) < y\} = f^{-1}((-\infty, y)).$$

Thus Definition 3.5 on measurable function can be restated in terms of f^{-1} as follows.

Definition 3.9 (Measurable function - Alternative Formulation) *The extended real-valued function $f(x)$ defined on $\mathbb{R}$ is said to be **Lebesgue (or Borel) measurable** if any of the following conditions are satisfied for every real number y:*

1. *The set $f^{-1}((-\infty, y))$ is Lebesgue (or Borel) measurable.*
2. *The set $f^{-1}([y, \infty))$ is Lebesgue (or Borel) measurable.*
3. *The set $f^{-1}((y, \infty))$ is Lebesgue (or Borel) measurable.*
4. *The set $f^{-1}((-\infty, y])$ is Lebesgue (or Borel) measurable.*

Of course, Borel measurability implies Lebesgue measurability since $\mathcal{B}(\mathbb{R}) \subset \mathcal{M}_L(\mathbb{R})$.

*More generally, an extended real-valued function on a measure space $(X, \sigma(X), \mu)$ is said to be $\sigma(X)$-**measurable** if any of the above conditions are satisfied replacing "Lebesgue (or Borel)" with "$\sigma(X)$-measurable."*

As will be seen in Book II, **random variables** are defined as measurable functions in just this way. To make this connection, let X be a function defined on some probability space $(\mathcal{S}, \sigma(\mathcal{S}), \mu)$, so $X : \mathcal{S} \to \mathbb{R}$, and by **probability space** we mean a measure space with $\mu(\mathcal{S}) = 1$. Such a measure μ is called a probability measure. The **distribution function** of X, $F(x)$, is then defined:

$$F(x) = \Pr\{X \leq x\}.$$

Now $\{X \leq x\} \subset \mathcal{S}$ and can equivalently be defined as:

$$\{X \leq x\} = X^{-1}(-\infty, x].$$

In addition, by definition of probability:

$$\Pr\{X \leq x\} \equiv \mu\left[\{X \leq x\}\right].$$

Combining:

$$F(x) = \mu\left[X^{-1}(-\infty, x]\right],$$

and this would make no sense unless $X^{-1}(-\infty, x] \in \sigma(\mathcal{S})$. That is, to make probability statements, we require such functions to be measurable, and a random variable is thus defined as a measurable function on this space.

To advance this investigation into examples of measurable functions more quickly, we investigate if Lebesgue measurability can also be determined based on various familiar properties of functions.

3.3.1 Continuous Functions

Perhaps the most familiar property of a function to investigate is continuity, and so this section investigates if continuous functions are Lebesgue or Borel measurable. If true, this would provide both a long list of examples, as well as provide a simple sufficient criterion to verify measurability. This would at best be a one-way implication. The **Dirichlet function** of Chapter 1 is continuous nowhere, and yet Lebesgue (and Borel) measurable because both the rationals and irrationals are Borel measurable sets, and specifically $\mathcal{F}_\delta$ and $\mathcal{G}_\sigma$ sets, by Example 2.17. See also Example 3.14.

Remark 3.10 (Generalizations of Proposition 3.11) *The next proposition proves that for a one-variable function $f : \mathbb{R} \to \mathbb{R}$, that continuity on $\mathbb{R}$ implies Borel measurability, and thus since $\mathcal{B}(\mathbb{R}) \subset \mathcal{M}_L$, such a function is also Lebesgue measurable.*

This result is another example of Remark 3.6 in that it also extends to a function $f : \mathbb{R}^n \to \mathbb{R}$ defined on $\mathbb{R}^n$. The proof that continuity implies Borel measurability is nearly identical to the 1-dimensional case, and requires only a notational re-interpretation. Recalling Definition 1.4, we simply replace the standard absolute value metric $d(x, x_0) = |x - x_0|$ on the domain $\mathbb{R}$ with the corresponding metric on $\mathbb{R}^n$:

$$d(x, y) = \left[\sum_{i=1}^n \left(x_i - y_i \right)^2 \right]^{1/2}.$$

Once the n-dimensional counterpart to $\mathcal{M}_L$ is derived in Chapter 7, the Lebesgue measurability result of Proposition 3.11 will again apply to such functions since this Lebesgue sigma algebra will contain $\mathcal{B}(\mathbb{R}^n)$.

This result also extends to continuous functions defined on a general metric space X with metric d, as long as the sigma algebra σ (X) contains the open sets of X, and thus by Definition 2.13 contains the Borel sigma algebra $\mathcal{B}(X)$. For this proof we replace $|x - x_0|$ with $d(x, x_0)$.

Finally, this result even applies to more general topological spaces, but to prove this requires a reformulation of continuity that does not involve a metric. See Proposition 3.13.

Proposition 3.11 (Continuous $\Rightarrow$ Borel/Lebesgue measurable) *Let $f : \mathbb{R} \to \mathbb{R}$ be a continuous function defined on $\mathbb{R}$. Then $f(x)$ is a Borel measurable function and thus also Lebesgue measurable.*

Proof. *Let real y be given and consider $\{x | f(x) < y\}$. If this set is empty, then it is Borel measurable by definition, so assume $x_1 \in \{x | f(x) < y\}$ and let $y_1 = f(x_1)$. By Definition 1.4, since $f(x)$ is continuous at x_1, for any $\epsilon > 0$ there is a $\delta > 0$ so that:*

$$\left| f(x) - y_1 \right| < \epsilon \text{ if } |x - x_1| < \delta.$$

Since $y_1 < y$ by definition, choose $\epsilon_1 = .5(y - y_1)$ and let δ_1 be defined as above. Then with $I_1 \equiv (x_1 - \delta_1, x_1 + \delta_1)$, it follows that the open interval I_1 is Borel measurable and $I_1 \subset \{x | f(x) < y\}$. This follows because $|x - x_1| < \delta$ implies $\left| f(x) - y_1 \right| < .5(y - y_1)$, and then:

$$f(x) < .5(y + y_1) < y.$$

Consider next $\{x | f(x) < y\} - I_1$. If this set is empty, we are done since then $\{x | f(x) < y\} = I_1$ is Borel measurable. Otherwise, there is $x_2 \in \{x | f(x) < y\} - I_1$. We repeat the argument with $y_2 = f(x_2) < y$, and $\epsilon_2 = .5(y - y_2)$ to generate $I_2 \equiv (x_2 - \delta_2, x_2 + \delta_2)$. Then $I_1 \bigcup I_2$ is Borel measurable and again contained in $\{x | f(x) < y\}$.

Continuing in this way at most countable many times, as each interval $(x_j - \delta_j, x_j + \delta_j)$ contains a different rational number, it can be concluded that there is a finite or countable number of Borel measurable sets $\{I_j\}$, so that:

$$\{x | f(x) < y\} = \bigcup_j I_j.$$

Hence $\{x | f(x) < y\}$ is Borel measurable. ∎

To contemplate generalizations of this result beyond metric spaces as noted in Remark 3.10, we need to introduce an equivalent characterization of continuity (Definition 1.4) that does not depend on a metric. To limit abstraction, we state and prove this next result for continuous functions $f : \mathbb{R}^n \to \mathbb{R}$, but note that it is also true for continuous functions $f : (X_1, d_1) \to (X_2, d_2)$ where X_j is a metric space with metric d_j.
For this proof, recall Definition 2.10 on open sets.

Proposition 3.12 (Generalizing the continuity criterion) *Let $f : \mathbb{R}^n \to \mathbb{R}$ be a given function. Then f is continuous if and only if for any open set $G \subset \mathbb{R}$, the set*

$$f^{-1}(G) \equiv \{x | f(x) \in G\}$$

is open in $\mathbb{R}^n$.

Proof. *Assume that f is continuous, that an open set $G \subset \mathbb{R}$ is given, and that $x_0 \in f^{-1}(G)$. If we show that there exists $r > 0$ and an open ball $B_r(x_0)$ with $B_r(x_0) \subset f^{-1}(G)$, then $f^{-1}(G)$ is open by Definition 2.10. As G is open there exists $\epsilon > 0$ so that the open ball $B_\epsilon(y_0) \subset G$, where $y_0 = f(x_0)$. Thus $|y - y_0| < \epsilon$ if $y \in B_\epsilon(y_0)$, and by definition of continuity there exists $\delta > 0$ so that $|f(x) - y_0| < \epsilon$ if $|x - x_0| < \delta$. Translating, this obtains $f(B_\delta(x_0)) \subset B_\epsilon(y_0)$, and so:*

$$B_\delta(x_0) \subset f^{-1}(B_\epsilon(y_0)) \subset f^{-1}(G).$$

Conversely, assume $f^{-1}(G)$ is open for all open $G \subset \mathbb{R}$. Let $x_0 \in \mathbb{R}^n$ be given and $y_0 = f(x_0) \in \mathbb{R}$. Choose any open set $G \subset \mathbb{R}$ that contains y_0, for example we could choose $G = \mathbb{R}$. In any case by definition of open there exists $\epsilon > 0$ so that $B_\epsilon(y_0) \subset G$. By assumption $f^{-1}(B_\epsilon(y_0))$ is open in $\mathbb{R}^n$ and contains x_0. Again by definition of open there exists $B_\delta(x_0) \subset f^{-1}(B_\epsilon(y_0))$ and thus $f(B_\delta(x_0)) \subset B_\epsilon(y_0)$. This now translates to the $\epsilon - \delta$ definition for continuity and the proof is complete. ∎

In a general topological space X of Definition 2.15, a real-valued function $f : X \to \mathbb{R}$ is **defined** to be continuous if $f^{-1}(G)$ is open for any open $G \subset \mathbb{R}$. This is equivalent to the standard definition on metric spaces by Proposition 3.12.
With this definition, we are ready for the very general result relating continuity and measurability on a topological space. With this reformulation of continuity, the proof is remarkably simple.

Proposition 3.13 (Continuous $\Rightarrow$ $\sigma(X)$-measurable) *Assume that X is a topological space and that $\sigma(X)$ contains the open sets of X and hence contains $\mathcal{B}(X)$, the Borel sigma algebra on X. If $f : X \to \mathbb{R}$ is continuous, then it is a $\sigma(X)$-measurable function.*

Proof. *Consider the open set $G \equiv (-\infty, y)$. Since continuous, $f^{-1}((-\infty, y))$ is open in X, and thus by definition*

$$f^{-1}((-\infty, y)) \in \mathcal{B}(X) \subset \sigma(X).$$

By Definition 3.9, the conclusion follows. ∎

Example 3.14 (Dirichlet function) *Define:*

$$h(x) = \begin{cases} 0, & x \text{ irrational}, \\ 1, & x \text{ rational}. \end{cases}$$

Then $d(x)$ is nowhere continuous since given any x_0 and any $\epsilon < 1$, there is no $\delta > 0$ for which $|h(x) - h(x_0)| < \epsilon$ for $|x - x_0| < \delta$. This is because in any interval about x_0, there are infinitely many x for which $|h(x) - h(x_0)| = 1$.

On the other hand, this function is Borel measurable and thus also Lebesgue measurable since:

$$h^{-1}((-\infty, y)) = \begin{cases} \varnothing, & y < 0, \\ \mathbb{R} - \mathbb{Q}, & 0 \leq y < 1, \\ \mathbb{R}, & 1 \leq y. \end{cases}$$

Each of these sets is Borel measurable by Example 2.17.

*Introduced in Chapter 1, this function is closely related to the **Dirichlet function**, $d(x)$, named for its discoverer, **J. P. G. Lejeune Dirichlet** (1805–1859) and usually defined with domain [0, 1].*

Though nowhere continuous, this function possesses a lot of regularity. If we define $g(x) = 0$, then $g(x) = h(x)$ except on the rationals $\mathbb{Q}$, a set of Lebesgue measure 0.

The following proposition generalizes this example and states that if a function is Lebesgue measurable, then it remains Lebesgue measurable even if arbitrarily redefined on a set of Lebesgue measure 0. In the proof, the completeness of $\mathcal{M}_L(\mathbb{R})$ is key. Thus, this is a good example of a result that is not valid for Borel measurable functions since $\mathcal{B}(\mathbb{R})$ is not complete.

Exercise 3.15 (Generalize Proposition 3.16) *Demonstrate that the following result remains true for a general measurable function defined on a general but complete measure space. Specifically, let $f(x)$ be a real-valued measurable function, $f : X \to \mathbb{R}$, where $(X, \sigma(X), \mu)$ is a complete measure space, and let $g(x)$ be a function with $f(x) = g(x)$ except on a set of μ-measure 0. Show that $g(x)$ is $\sigma(X)$-measurable and identify where completeness of the measure space is needed.*

Proposition 3.16 (Modifications of Lebesgue measurable functions) *Let $f(x)$ be a Lebesgue measurable function and let $g(x)$ be a function with $f(x) = g(x)$ except on a set of Lebesgue measure 0. Then $g(x)$ is Lebesgue measurable.*

Proof. *Let E be the set of measure 0 on which $f(x) \neq g(x)$. Then*

$$\{x | g(x) < y\} = \{x \in E | g(x) < y\} \bigcup \{x \notin E | g(x) < y\}$$

$$= \{x \in E | g(x) < y\} \bigcup \{x \notin E | f(x) < y\}.$$

The first set is a subset of a set of Lebesgue measure 0 and is hence Lebesgue measurable by completeness. For the second set, with $\tilde{E}$ denoting the complement of E:

$$\{x \notin E | f(x) < y\} = \tilde{E} \bigcap \{x | f(x) < y\}.$$

This is Lebesgue measurable as the intersection of Lebesgue measurable sets. ∎

Definition 3.17 (Almost everywhere (a.e.)) *The property in the above proposition, that $f(x) = g(x)$ except on a set of Lebesgue measure 0, is alternatively described as $f(x) = g(x)$ **almost everywhere**, and abbreviated as $f(x) = g(x)$ **a.e.***

*To be unambiguous in a general measure space, $(X, \sigma(X), \mu)$, one states that $f(x) = g(x)$, μ-**almost everywhere**, or $f(x) = g(x)$ μ-**a.e.** This means that $f(x) = g(x)$ except on a set of μ-measure 0.*

The usual application of Proposition 3.16 is in the situation where it is known that a function $f(x)$ is measurable, and that another function of interest $g(x)$ satisfies $g(x) = f(x)$ a.e. In addition to measurability, we would often like to be able to say something more about $g(x)$. For example, if $f(x)$ is integrable, will $g(x)$ also be integrable and if so, with the same integral as $f(x)$? Integration theory is initiated in Book III, where measurability will be seen to be fundamental.

Summary 3.18 (On the collection of measurable functions) *So far, the above propositions provide a basic collection of Lebesgue measurable functions:*

1. *All continuous functions, since the sigma algebra $\mathcal{M}_L$ contains the open sets,*
2. *All functions which equal a measurable function outside a set of measure 0, since $\mathcal{M}_L(\mathbb{R})$ is complete.*

The first conclusion applies to Borel measurable functions, since the sigma algebra $\mathcal{B}(\mathbb{R})$ contains the open sets, while the second does not, since this sigma algebra is not complete.

Remark 3.19 (Continuity and a.e. statements) *While Proposition 3.16 applies generally to measurable functions, when applied to continuous functions it states that continuous functions can be arbitrarily redefined on sets of Lebesgue measure 0 and remain Lebesgue measurable. In contrast, while continuous functions are Riemann integrable over an interval $[a, b]$, this property of Riemann integrability is not preserved for arbitrary redefinitions of such functions on sets of measure 0.*

*The most general result of this type is provided in **Lebesgue's existence theorem for the Riemann integral**, and named for **Henri Lebesgue** (1875–1941). As noted in Proposition 1.5, a function on $[a, b]$ is Riemann integrable if and only if it is continuous "outside a set of measure 0." For these results, there is no difference in the definitions of "measure 0," even though the first result explicitly refers to Lebesgue measure and the second often does not.*

For the Riemann integrability result, a set E is said to be of measure 0 if given $\epsilon > 0$ there is a countable collection of open intervals $\{I_j\}$, so that $E \subset \bigcup I_j$ and $\sum |I_j| < \epsilon$ where $|I_j|$ is interval length. So while in many developments of Riemann integration the notion of Lebesgue measure is not formally introduced, it is clear from this definition that "a set of measure 0" in the Riemann result is a set with Lebesgue outer measure 0. Thus from Proposition 2.35 this set is Lebesgue measurable and has Lebesgue measure 0.

Similarly, if a set has Lebesgue measure 0, then it has Lebesgue outer measure 0 and hence by (2.5) this set can be covered by a sequence of open intervals which satisfy the Riemann specification. So the notions of "measure 0" are identical.

The first paragraph contains two statements that involve continuity and exceptional sets of measure 0:

1. ***Riemann Integrability Result****: If continuous except on a set of measure 0.*
2. ***Lebesgue Measurability Result (Limited statement)****: If equal to a continuous function except on a set of measure 0.*

*The goal of this discussion is to demonstrate that though sounding similar, **neither notion implies the other**.*

*By Definition 1.4, for a function $f(x)$ **to be continuous except on a set of measure 0** means that $\lim_{x \to x_0} f(x) = f(x_0)$ except for a collection of exceptional points which have measure 0. A simple example of such a function defined on $[-1, 1]$ is:*

$$f_1(x) = \begin{cases} 0, & -1 \leq x \leq 0, \\ 1, & 0 < x \leq 1, \end{cases}$$

for which the exceptional set is $\{0\}$. Defined on $\mathbb{R}$, it has two exceptional points.

*An example of a function defined on $[0, 1]$ with a countable number of discontinuities is **Thomae's function**, named for **Carl Johannes Thomae** (1840–1921):*

$$f_2(x) = \begin{cases} 1, & x = 0, \\ 1/n, & x = m/n \text{ in lowest terms}, \\ 0, & x \text{ irrational}. \end{cases}$$

Defined on $[0, 1]$ this function is continuous except on the rationals because $\lim_{x \to x_0} f_2(x) = f_2(x_0) = 0$ for all irrational x_0. To see this, note that for any integer N, choose δ_N so that:

$$\delta_N < \min\{|x_0 - m/n| \, | \, n \leq N\}.$$

This minimum exists because only finitely many rationals need be considered, and this minimum is greater than 0 because x_0 is irrational. Then since $f_2(x) = 0$ for irrational x and $f_2(x) < 1/N$ for rationals in this interval by construction, it follows that $|f_2(x) - f_2(x_0)| = f_2(x) < 1/N$ for all x with $|x - x_0| < \delta$. That $f_2(x)$ is discontinuous on the rationals follows from the fact that given m/n and ϵ, there are infinitely many irrationals x with $|m/n - x| < \epsilon$ and so $|f_2(m/n) - f_2(x)| = 1/n$.

*For a function $f(x)$ **to equal a continuous function** $g(x)$ **except on a set of measure 0** means that $\lim_{x \to x_0} g(x) = g(x_0)$ for all x_0, and $f(x) = g(x)$ except on a set of measure 0. A simple example is*

$$f_3(x) = \begin{cases} 0, & x = 0, \\ 1, & -1 \leq x \leq 1, x \neq 0, \end{cases}$$

and $g_3(x) = 1$ *on* $[-1,1]$. *An example with countably many exceptional points from Example 3.14:*

$$f_4(x) = \begin{cases} 0, & x \text{ rational,} \\ 1, & x \text{ irrational,} \end{cases}$$

with $g_4(x) = 1$ *on* $[0,1]$.

To contrast the two notions of continuity above, note that:

a. There is no continuous $g_1(x)$ so that $f_1(x) = g_1(x)$ except on a set of measure 0;

b. For $f_2(x)$ this property is satisfied with $g_2(x) = 0$ on $[0,1]$.

Thus a function that satisfies item 1 may satisfy item 2 or not.
Going the other way:

c. The function $f_3(x)$ is continuous except on the singleton set $\{0\}$ of measure 0;

d. The function $f_4(x)$ is continuous nowhere.

Thus a function that satisfies item 2 can satisfy item 1 or not.
In summary, neither notion implies the other.

3.3.2 Characteristic or Indicator Functions

We next consider another collection of Lebesgue measurable functions. These functions are implicitly used in both the definition of Riemann integral in (1.2), and for Lebesgue integrals in (1.10). In both cases, terminology and notation are simplified by the intro-duction of what is called an **indicator function of a set** A, or **characteristic function of a set** A, defined as follows.

Definition 3.20 (Indicator/characteristic function of A) *Given a set A, the **indicator function of A**, or **characteristic function of A**, denoted $\chi_A(x)$ and sometimes $1_A(x)$, is defined by:*

$$\chi_A(x) = \begin{cases} 1, & x \in A, \\ 0, & x \notin A. \end{cases} \tag{3.2}$$

Exercise 3.21 (On unions/intersections) *Prove that in a general measure space $(X, \sigma(X), \mu)$, a set $A \subset X$ is $\sigma(X)$-measurable if and only if $\chi_A(x)$ defined in (3.2) is a $\sigma(X)$-measurable function.*

Also, for measurable sets, show that:

1. $\chi_{A \cap B}(x) = \chi_A(x)\chi_B(x)$,

2. $\chi_{A \cup B}(x) = \chi_A(x) + \chi_B(x) - \chi_{A \cap B}(x)$.

Prove that both results generalize to unions and intersections of n measurable sets by induction, noting that item 2 can be expressed as:

$$1 - \chi_{A \cup B}(x) = (1 - \chi_A(x))(1 - \chi_B(x)).$$

Finally, show that if $f : X \to \mathbb{R}$ is $\sigma(X)$-measurable and $D \in \sigma(X)$, then $\chi_D(x)f(x)$ is $\sigma(X)$-measurable. Note that this function is 0 outside D, and agrees with $f(x)$ on D.

Given an arbitrary collection of measurable sets $\{A_i\}_{i=1}^n$, it will be proved below that the function:

$$f(x) = \sum_{i=1}^n a_i \chi_{A_i}(x), \tag{3.3}$$

is measurable for any collection of real numbers $\{a_i\}_{i=1}^n$. In many applications, such as in the definitions of Riemann and Lebesgue integrals, it will be the case that $\bigcup_{i=1}^n A_i = [a, b]$, and it is often convenient to be able to assume that the collection of measurable sets $\{A_i\}_{i=1}^n$ is pairwise disjoint, meaning $A_i \cap A_j = \emptyset$ for $i \neq j$.

Exercise 3.22 (Disjoint sets) *Prove that it is always possible to express the function in (3.3) in terms of pairwise disjoint measurable sets and with distinct values of the coefficients. In other words, such $f(x)$ can be expressed as*

$$f(x) = \sum_{i=1}^N b_i \chi_{B_i}(x),$$

where $\{B_i\}_{i=1}^N$ are pairwise disjoint, and $\{b_i\}_{i=1}^N$ are distinct. Hint: Let $A = \bigcup_{i=1}^n A_i$ and define $C_i = A - A_i = A \cap \tilde{A}_i$. Then $A_i \cup C_i = A$ for all i and so $\bigcap_{i=1}^n [A_i \cup C_i] = A$. Use the prior exercise applied to $\chi_A(x)$, noting that $\chi_{A_i \cap C_i}(x) \equiv 0$, and show this implies A can be partitioned into 2^n disjoint measurable sets, though many of these sets may be empty. Define $\{B_i\}_{i=1}^N$ as the nonempty sets, so $N \leq 2^n$. Then since each $B_i \subset \bigcap_{j=1}^{n(i)} A_{i(j)}$, define $\{b_i\}_{i=1}^N$ by $b_i = \sum_{j=1}^{n(i)} a_{i(j)}$. Any B_i-sets with the same b_i can be unioned.

Definition 3.23 (Simple/step functions) *Given a collection of measurable sets $\{A_i\}$, the function in (3.3) is called a **simple function**. When the $\{A_i\}_{i=1}^n$ form a **partition** of the interval $[a, b]$ into subintervals $A_i = [x_{i-1}, x_i]$ where:*

$$a = x_0 < x_1 < \dots < x_{n-1} < x_n = b,$$

*the function in (3.3) is sometimes called a **step function**.*

Proposition 3.24 *Simple functions are measurable.*

Proof. *The demonstration that simple functions are measurable is notationally streamlined by assuming that the collection $\{a_i\}_{i=1}^n$ is distinct and indexed in increasing order: $a_i < a_{i+1}$, and that the collection $\{A_i\}_{i=1}^n$ is pairwise disjoint. This is always possible by Exercise 3.22. Then:*

$$f^{-1}((-\infty, y)) = \begin{cases} \emptyset, & y < a_1, \\ A_1, & a_1 \le y < a_2, \\ A_1 \bigcup A_2, & a_2 \le y < a_3, \\ \vdots & \vdots \\ \bigcup_{j \le k} A_j, & a_k \le y < a_{k+1}, \\ \mathbb{R}, & a_n \le y. \end{cases}$$

The measurability of $f(x)$ now follows from the assumed measurability of the A_i-sets. ∎

Step and simple functions are essential in the definitions of Riemann and Lebesgue integrals, respectively. For the Riemann integral, in (1.2) the function $f(x)$ is approximated on $[a, b]$ by upper and lower step functions:

$$\sum_{i=1}^{n} m_i' \chi_{A_i}(x) \le f(x) \le \sum_{i=1}^{n} M_i' \chi_{A_i}(x), \tag{3.4}$$

where m_i' and M_i' are defined in (1.9) as the greatest lower bound and least upper bound of $f(x)$ on $A_i = [x_{i-1}, x_i]$. The Riemann integrals of these step functions are defined as implied by the summations in (1.2). The Riemann integral of $f(x)$ is then defined when the upper and lower step function integrals approach the same limit as the mesh size of the partition of the domain, $\mu \equiv \max_{1 \le i \le n}\{x_i - x_{i-1}\}$, converges to 0. This definition of the Riemann integrable also requires that this limit is independent of the step functions used.

For the Lebesgue integral, simple functions underlie the bounds in (1.10). Here, the function is approximated on $[a, b]$ by upper and lower simple functions:

$$\sum_{i=1}^{n} m_i \chi_{A_i}(x) \le f(x) \le \sum_{i=1}^{n} M_i \chi_{A_i}(x), \tag{3.5}$$

where now m_i and M_i are defined in (1.9) as the greatest lower bound and least upper bounds of $f(x)$ on the level sets of the function defined in (1.8). The values of the Lebesgue integrals of these simple functions are defined as implied by the summations in (1.10), but where now $|A_i| \equiv m(A_i)$ is the Lebesgue measure of the set A_i. In Book III the Lebesgue integral of $f(x)$ will be defined in terms of such simple function integrals when the supremum of the lower integrals and the infimum of the upper integrals agree.

3.3.3 A Nonmeasurable Function

Our inventory of measurable functions now includes those identified in Summary 3.18, as well as all simple functions. Before continuing the development of this inventory, perhaps it is worth addressing a question. Are all functions measurable? With the work of the prior chapter, the answer is easily seen to be "no."

Example 3.25 ($\chi_{A_0}(x)$) *Let A_0 denote the set constructed in Proposition 2.31 and noted to be nonmeasurable in Remark 2.37. Specifically, $A_0 \notin \mathcal{M}_L$ and thus by definition $A_0 \notin \mathcal{B}(\mathbb{R})$. Define:*

$$\chi_{A_0}(x) = \begin{cases} 1, & x \in A_0, \\ 0, & x \notin A_0. \end{cases}$$

Then $\chi_{A_0}(x)$ is not Lebesgue measurable by Exercise 3.21, and explicitly since $\chi_{A_0}^{-1}([1, \infty)) = A_0$.

3.4 Properties of Measurable Functions

In this section we develop several properties of Lebesgue measurable functions which allow us to add to the collection of examples developed in the last sections. All of the results below can be generalized to apply to measurable functions defined on general measure spaces $(X, \sigma(X), \mu)$, but the terminology for the main results will refer to Lebesgue measurable functions for specificity. The general statements will follow with adapted proofs left as exercises.

The first result shows that for Lebesgue measurable functions, much can be said about the measurability of $f^{-1}(A)$ for sets A other than $(-\infty, y), [y, \infty), (y, \infty), (-\infty, y],$ and $\{y\}$. But note that in the statement of this result, Borel sets play two roles. First as is virtually always the case, $\mathcal{B}(\mathbb{R})$ is taken as the sigma algebra for the **range space** of a function. In addition, we consider the effect of using the Lebesgue or Borel sigma algebra in the **domain space**.

Proposition 3.26 (On Borel/Lebesgue measurable functions) *If $f : \mathbb{R} \to \mathbb{R}$ is a Lebesgue measurable function, then $f^{-1}(A)$ is Lebesgue measurable for every Borel set $A \in \mathcal{B}(\mathbb{R})$. That is:*

$$f^{-1}(\mathcal{B}(\mathbb{R})) \subset \mathcal{M}_L(\mathbb{R}).$$

If $f : \mathbb{R} \to \mathbb{R}$ is a Borel measurable function, then $f^{-1}(A)$ is Borel measurable for every Borel set $A \in \mathcal{B}(\mathbb{R})$. That is:

$$f^{-1}(\mathcal{B}(\mathbb{R})) \subset \mathcal{B}(\mathbb{R}).$$

In either case, $f^{-1}(\mathcal{B}(\mathbb{R}))$ is a sigma algebra.

Proof. *The proofs are effectively identical until the last steps. Define the collection:*

$$S \equiv \{f^{-1}(A) | A \in \mathcal{B}(\mathbb{R})\}.$$

To show that S is a sigma algebra, first note that $\emptyset, \mathbb{R} \in S$ since $\emptyset, \mathbb{R} \in \mathcal{B}(\mathbb{R}), f^{-1}(\emptyset) = \emptyset$, and $f^{-1}(\mathbb{R}) = \mathbb{R}$.

Now if $B = f^{-1}(A) \in S$ *for* $A \in \mathcal{B}(\mathbb{R})$, *the complement* $\widetilde{B} \in S$ *since:*

$$\widetilde{B} = \left[f^{-1}(A) \right]^c$$
$$= \{ x | f(x) \notin A \}$$
$$= f^{-1} \left(\widetilde{A} \right).$$

As $\mathcal{B}(\mathbb{R})$ *is closed under complementation,* $\widetilde{A} \in \mathcal{B}(\mathbb{R})$ *and thus* $\widetilde{B} \in S$.

Similarly, if $B = \bigcup_i B_i$ *is a countable union of sets with* $B_i = f(A_i) \in S$ *for all i, with all* $A_i \in \mathcal{B}(\mathbb{R})$, *then* $B \in S$ *since:*

$$B = \bigcup_i f^{-1}(A_i)$$
$$= \bigcup_i \{ x | f(x) \in A_i \}$$
$$= \{ x | f(x) \in \bigcup_i A_i \}$$
$$= f^{-1} \left(\bigcup_i A_i \right).$$

As $\mathcal{B}(\mathbb{R})$ *is closed under countable unions,* $\bigcup_i A_i \in \mathcal{B}(\mathbb{R})$ *and thus* $B \in S$.

Combining the above, S is a sigma algebra by Definition 2.5.

To complete the proof, note that for all $a, b \in [-\infty, \infty]$, *the open interval* $(a, b) \in \mathcal{B}(\mathbb{R})$. *Further, this collection of intervals generates this sigma algebra by Definition 2.13. That is, every set in* $\mathcal{B}(\mathbb{R})$ *is obtained by countably many operations of complementation, unions and intersections of such open intervals.*

If f is Lebesgue measurable, then $f^{-1}(a, b) \in \mathcal{M}_L$ *for any such interval. For example if a,* $b \in (-\infty, \infty)$ *then:*

$$f^{-1}(a, b) = f^{-1}(a, \infty) \bigcap f^{-1}(-\infty, b),$$

and this is Lebesgue measurable as an intersection of Lebesgue measurable sets by Definition 3.9. For $a = -\infty$ *or* $b = \infty$, *Lebesgue measurability follows by Definition 3.9.*

Then, by the above calculations, it follows that as $\mathcal{B}(\mathbb{R})$ *is generated by these intervals, so too is S generated by the collection* $\{ f^{-1}(a, b) \}$. *But since S is a sigma algebra and* $f^{-1}(a, b) \in \mathcal{M}_L$ *for all such* (a, b), *it follows that* $S \subset \mathcal{M}_L$ *since* $\mathcal{M}_L$ *is a sigma algebra.*

The same logic applies for Borel measurable f, since then $f^{-1}(a, b) \in \mathcal{B}(\mathbb{R})$ *for all such* (a, b). ∎

Corollary 3.27 (General property of measurable functions) *Let* $f(x)$ *be a* $\sigma(X)$-*measurable function defined on the measure space* $(X, \sigma(X), \mu)$. *Then, for every Borel set* $A \in \mathcal{B}(\mathbb{R})$, $f^{-1}(A)$ *is* $\sigma(X)$-*measurable. That is:*

$$f^{-1}(\mathcal{B}(\mathbb{R})) \subset \sigma(X).$$

Further, $f^{-1}(\mathcal{B}(\mathbb{R}))$ *is a sigma algebra.*

Proof. *That $f^{-1}(\mathcal{B}(\mathbb{R}))$ is a sigma algebra follows identically from the above proof, since the type of measurability played no role. The second half of the proof now uses the observation that $f^{-1}(a,b) \in \sigma(X)$ for all intervals (a,b).* ∎

Remark 3.28 (Nonmeasurable functions) *It is interesting to note that the above results also provide conclusions for general functions f defined on a measure space $(X, \sigma(X), \mu)$, where by "general" is meant, not necessarily $\sigma(X)$-measurable. The same is true for general functions defined on an arbitrary space X, which may not have a sigma algebra or measure defined on it.*

The first conclusion is that even in these cases, $f^{-1}(\mathcal{B}(\mathbb{R}))$ is again a sigma algebra on X. This follows from the above proof since neither the measurability of f nor the sigma algebra on X played a role. This result required only the sigma algebra structure of $\mathcal{B}(\mathbb{R})$.

*Secondly, in these general cases and the cases of the above proposition and corollary, $f^{-1}(\mathcal{B}(\mathbb{R}))$ is the **smallest sigma algebra on** X with respect to which f is measurable. Certainly in all cases f is $f^{-1}(\mathcal{B}(\mathbb{R}))$-measurable by definition, as $f^{-1}(A) \in f^{-1}(\mathcal{B}(\mathbb{R}))$ for all $A \in \mathcal{B}(\mathbb{R})$. If σ is a smaller sigma algebra on X, then f cannot be σ-measurable. Indeed, if $B \in f^{-1}(\mathcal{B}(\mathbb{R})) - \sigma$, then $B = f^{-1}(A)$ for some $A \in \mathcal{B}(\mathbb{R})$ and thus $f^{-1}(A) \notin \sigma$.*

Remark 3.29 (On $f^{-1}[\mathcal{M}_L]$) *This proposition states that for Lebesgue measurable functions:*

$$f^{-1}[\mathcal{B}(\mathbb{R})] \subset \mathcal{M}_L(\mathbb{R}). \tag{3.6}$$

It is natural to wonder if $f^{-1}(A)$ is Lebesgue measurable for all sets A in the larger sigma algebra of Lebesgue measurable sets $\mathcal{M}_L$, which contains $\mathcal{B}(\mathbb{R})$. In other words, is it the case that

$$f^{-1}[\mathcal{M}_L] \subset \mathcal{M}_L?$$

Because every Lebesgue measurable set A is the union of an $\mathcal{F}_\sigma$-set F', and a set Z of Lebesgue measure 0 by Proposition 2.42, a sufficient condition for this inclusion would be that $f^{-1}(Z) \subset \mathcal{M}_L$ for all sets Z of Lebesgue measure 0.

Now if Z is a set of Lebesgue measure 0, then for any $\epsilon > 0$, $Z \subset \bigcup_i B_i$, a union of open intervals with $\sum_i m(B_i) \leq \epsilon$. Taking $\epsilon = 1/n$, Z is contained in the intersection of all such unions. So with:

$$B \equiv \bigcap_n \{\bigcup_i B_i | Z \subset \bigcup_i B_i\},$$

it follows that $Z \subset B \in \mathcal{G}_\delta$ and thus:

$$f^{-1}(Z) \subset f^{-1}(B).$$

Hence $f^{-1}(Z)$ is contained in $f^{-1}(B)$, which is Lebesgue measurable since B is a Borel set. The Lebesgue measurability of $f^{-1}(Z)$ would then follow for example, from either a conclusion that $f^{-1}(B)$ has measure 0, applying Proposition 2.35, or that $f^{-1}(B) = f^{-1}(Z)$.

Although B has measure 0 by construction, it does not follow that $f^{-1}(B)$ has measure 0. One simple counterexample, $f(x) = 1$, demonstrates this point with $Z = \{1\}$, and $f^{-1}(Z) = \mathbb{R}$. It also does not follow that $f^{-1}(Z)$ is equal to $f^{-1}(B)$, because it need not be the case that $Z = B$. Recall

Proposition 2.46. Indeed, all that can be said in general from Proposition 2.42 is that $Z \subset B$ and that $m(B - Z) = 0$.

In summary, it cannot be concluded from the current analysis that $f^{-1}[\mathcal{M}_L] \subset \mathcal{M}_L$ for Lebesgue measurable f. In fact, this inclusion is not true, but since we do not need this result in forthcoming books, a demonstration of this will not be pursued.

3.4.1 Elementary Function Combinations

The next result demonstrates that Lebesgue measurability is preserved under all the usual mathematical combinations, with one notable exception of composition of functions. As will be seen from the proof, the conclusions of this proposition remain true when applied to Borel measurable functions, or more generally $\sigma(X)$-measurable functions. Initially we restrict attention to real-valued functions, then discuss generalizations to extended real-valued functions.

Proposition 3.30 (Elementary function combinations) *Let $f(x)$ and $g(x)$ be **real-valued** Lebesgue measurable functions defined on $\mathbb{R}$, and let $a, b \in \mathbb{R}$. Then the following are real-valued Lebesgue measurable functions:*

1. *$af(x) + b$,*
2. *$f(x) \pm g(x)$,*
3. *$f(x)g(x)$,*
4. *$f(x)/g(x)$ when defined on $\{x | g(x) \neq 0\}$.*

Proof. *To simplify notation, let the expression $\{x | f(x) < r\}$ be denoted by $\{f(x) < r\}$, and so forth. Taking these statements in turn:*

1. *For $a \neq 0$,*

$$\{af(x) + b < y\} = \{f(x) < (y - b)/a\}.$$

   This is Lebesgue measurable since $f(x)$ is a Lebesgue measurable function. If $a = 0$, the function $g(x) = b$ is continuous and hence Lebesgue measurable.

2. *Consider the sum since then by part 1, $-g(x)$ is measurable and this implies the result for $f(x) - g(x)$. Now for rational r, if $f(x) < r$ and $g(x) < y - r$ then $f(x) + g(x) < y$ and so taking a union over all rational r:*

$$\bigcup_r \left[\{f(x) < r\} \bigcap \{g(x) < y - r\} \right] \subset \{f(x) + g(x) < y\}.$$

   On the other hand, if $f(x) + g(x) < y$ then $f(x) < y - g(x)$, and by density of the rationals, there exists rational r so that $f(x) < r < y - g(x)$. This implies $f(x) < r$ and $g(x) < y - r$. Hence,

$$\{f(x) + g(x) < y\} = \bigcup_r \left[\{f(x) < r\} \bigcap \{g(x) < y - r\} \right].$$

This set is Lebesgue measurable as a countable union of intersections of Lebesgue measurable sets.

3. *For $f(x)g(x)$, first note that both $f^2(x)$ and $g^2(x)$ are Lebesgue measurable. For example,*

$$\{f^2(x) < y\} = \begin{cases} \{f(x) < \sqrt{y}\} \bigcap \{f(x) > -\sqrt{y}\}, & y \geq 0, \\ \emptyset, & y < 0, \end{cases}$$

so $f^2(x)$ is Lebesgue measurable. But then by this and part 2, so too is $[f(x) + g(x)]^2$. Then:

$$f(x)g(x) = 0.5 \left([f(x) + g(x)]^2 - f^2(x) - g^2(x) \right),$$

is Lebesgue measurable by part 2.

4. *First, $D \equiv \{g(x) \neq 0\}$ is a Lebesgue measurable domain, since:*

$$D = g^{-1}(-\infty, 0) \bigcup g^{-1}(0, \infty).$$

Now $1/g(x)$ is real-valued and well-defined on D. The Lebesgue measurability of $1/g(x)$ on D then follows since for $y > 0$:

$$\{1/g(x) < y\} = \{g(x) > 1/y\} \bigcup \{g(x) < 0\};$$

for $y = 0$:

$$\{1/g(x) < 0\} = \{g(x) < 0\};$$

while for $y < 0$:

$$\{1/g(x) < y\} = \{g(x) > 1/y\} \bigcap \{g(x) < 0\}.$$

Thus $1/g(x)$ is Lebesgue measurable as is $f(x)/g(x)$ by part 3.   ■

Remark 3.31 (Composition of functions) *There is one apparent omission from the above results, and that is a conclusion on the composition of measurable functions: $f(g(x))$. To be specific, assume $g(x)$ and $f(y)$ are Lebesgue measurable, and both real-valued:*

$$g(x) : \mathbb{R} \to \mathbb{R},$$
$$f(y) : \mathbb{R} \to \mathbb{R},$$

For $f(g)$ to be Lebesgue measurable requires for example that $[f(g)]^{-1}((-\infty, z))$ is Lebesgue measurable for all z. The inverse of this composition of functions is given by:

$$[f(g)]^{-1}((-\infty, z)) = g^{-1}[f^{-1}((-\infty, z))].$$

The Lebesgue measurability of f assures by Definition 3.9 that $f^{-1}((-\infty, z)) \in \mathcal{M}_L$, and so is Lebesgue measurable. But as was discussed in Remark 3.29, we cannot then conclude that g^{-1} applied to this Lebesgue measurable set is Lebesgue measurable.

Conclusion 3.32 *Composition of Lebesgue measurable functions need not be Lebesgue measurable.*

However, this analysis provides an insight to affirmative results which require some additional restrictions on f.

Proposition 3.33 (Measurable composite functions) *Let $g(x) : \mathbb{R} \to \mathbb{R}$, and $f(y) : \mathbb{R} \to \mathbb{R}$ be given real-valued functions. Then:*

1. *If f and g are Borel measurable, then $f(g)$ is Borel (and hence Lebesgue) measurable.*
2. *If f is Borel measurable and g is Lebesgue measurable, then $f(g)$ is Lebesgue measurable.*
3. *If f is continuous on $\mathbb{R}$ and g is Lebesgue measurable, then $f(g)$ is Lebesgue measurable.*

Proof. *As noted in Remark 3.31,*

$$[f(g)]^{-1}((-\infty, z)) = g^{-1}[f^{-1}((-\infty, z))].$$

In cases 1 and 2, $f^{-1}((-\infty, z))$ is Borel measurable, so 1 and 2 follow from Proposition 3.26.

In case 3, f is continuous and so $f^{-1}((-\infty, z))$ is an open set by Proposition 3.12 because $(-\infty, z)$ is open. Since open sets are Borel measurable, $g^{-1}[f^{-1}((-\infty, z))]$ is Lebesgue measurable if g is Lebesgue measurable. ∎

Remark 3.34 (Extended Real-Valued Functions) *As the results of Proposition 3.30 are stated in the more limited context of real-valued functions, we now consider their application to extended real-valued functions. If f and g are **extended real-valued** functions, we have to address the following definitional problems:*

1. *$af(x) + b$: when $a = 0$ and $f(x) = \pm\infty$;*
2. *$f(x) \pm g(x)$: when this expression becomes $\infty - \infty$ or $-\infty + \infty$;*
3. *$f(x)g(x)$: when this expression becomes $0 \cdot \infty$ or $0 \cdot (-\infty)$;*
4. *$f(x)/g(x)$: when this expression becomes or $\pm\infty/\pm\infty$ (noting that $\pm\infty/0$ is already eliminated by definition of domain to exclude $\{x|g(x) \neq 0\}$);*
5. *$f(g(x))$: if $g(x) = \pm\infty$.*

Of course if there are no values of x which produce the troubling expressions, the above measurability results apply without revision.

More generally, the collection of x-values which produce these kinds of problems are measurable sets, being intersections of measurable sets (Proposition 3.7). For example, given $f(x) + g(x)$, the set $f^{-1}(\infty) \bigcap g^{-1}(-\infty)$ is measurable; or for $f(x)g(x)$, $f^{-1}(-\infty) \bigcap g^{-1}(\infty)$ is measurable; etc. Thus we can restrict the domain of these expressions to the measurable complement of such sets to avoid these problems. Often these exceptional sets have measure 0 and are easily avoided.

*A note of caution: Rather than restrict the domain, it may be tempting to try to **define** the problem away by simply defining $0 \cdot \infty = 0 \cdot (-\infty) = 0$, for example. Under such a definition, the resulting functional combinations will again be Lebesgue measurable. But it is difficult to assure consistent mathematics after such definitional assignments. For example, without perhaps much thought, one could readily use in a derivation that:*

$$f(x)g(x)/g(x) = f(x).$$

This is of course true for finite $f(x)$ and $g(x) \neq 0$, but need not be true in the extended reals given such definitional assignments.

As a general rule, any such assignments are to be avoided.

Exercise 3.35 (Measure 0 definitional problem sets) *Show that if the set on which any of the above definitional problems occur has measure 0, then the associated function combinations will be Lebesgue measurable independent of how the function is defined on this set.*

The above discussions have perhaps not yet provided much insight to the power and flexibility of the notion of Lebesgue measurability, even vis-à-vis that of continuous functions. Indeed, the proposition on simple arithmetic combinations readily applies to continuous functions, and moreover, unlike the result on measurable functions, compositions of continuous functions are continuous without additional restrictions.

One hint of the power of the notion of Lebesgue measurability is the result above that this property is preserved even when the function is arbitrarily redefined on a set of Lebesgue measure 0. This result required the completeness of the underlying Lebesgue measure space $\mathcal{M}_L$, and thus was not true for a Borel measurable function. For a continuous function, such a redefinition on a set of measure 0 can produce a function that is continuous except on this special set, or produce a nowhere continuous function, recalling Remark 3.19.

The next section's results add to the conclusion of the robustness of the notion of measurability, in that this property is preserved with various limiting operations.

3.4.2 Function Sequences

We begin with a few definitions which again utilize the notions of infimum and supremum reflected in (1.3) and (1.4). In that application, infimum and supremum were defined in terms of the values of a given function $f(x)$, as x varied over a given interval or set.

Here we are given a sequence of functions $\{f_n(x)\}$, and the goal is to apply these notions pointwise, at each value of x, as n varies.

Definition 3.36 (Infimum/supremum) *Given a finite or countable sequence of functions $\{f_n(x)\}$, the **infimum** and **supremum** of the sequence are functions **defined pointwise** as follows. For each $x \in D \equiv \bigcap_n Dmn\{f_n\}$, where $Dmn\{f_n\}$ denotes the domain of the function f_n:*

$$\inf_n f_n(x) = \begin{cases} -\infty, & \{f_n(x)\} \text{ unbounded below,} \\ \max\{y | y \leq f_n(x) \text{ all } n\}, & \{f_n(x)\} \text{ bounded below.} \end{cases} \tag{3.7}$$

$$\sup{}_n f_n(x) = \begin{cases} \infty, & \{f_n(x)\} \text{ unbounded above,} \\ \min\{y | y \geq f_n(x) \text{ all } n\}, & \{f_n(x)\} \text{ bounded above.} \end{cases} \quad (3.8)$$

*When $\{f_n(x)\}$ is a **finite collection**, $\inf_n f_n(x)$ is often denoted:*

$$\inf{}_n f_n(x) \equiv \min\{f_1(x), ..., f_n(x)\},$$

and $\sup_n f_n(x)$ is denoted:

$$\sup{}_n f_n(x) \equiv \max\{f_1(x), ..., f_n(x)\}.$$

Because it is usually clear from the context, the subscript n is almost always dropped from the inf *and* sup *notation.*

Remark 3.37 (Inf/sup symmetry) *It is sometimes convenient to convert an infimum into a supremum, and conversely. To do so, note that:*

$$\inf f_n(x) = -\sup[-f_n(x)], \qquad \sup f_n(x) = -\inf[-f_n(x)]. \quad (3.9)$$

Because $\inf f_n(x)$ and $\sup f_n(x)$ are defined pointwise, it is natural to identify additional properties of these point sequences other than the tight lower and upper bounds which the infimum and supremum provide. For each x in the common domain D:

$$\inf f_n(x) \leq f_m(x) \leq \sup f_n(x), \text{ for all } m, \quad (3.10)$$

and no tighter bounds are possible.

In the case when both the infimum and supremum are finite at a given value of x, recall from the study of bounded numerical sequences that there is at least one **cluster point** or **accumulation point**, and perhaps many such points and even potentially an infinite number. For some sequences, there is also a **limit point**, which is unique when it exists. These special points are defined as follows:

Definition 3.38 (Accumulation and limit points) *A real number y is an **accumulation point** or **cluster point** of a real numerical sequence $\{y_n\}$, if given any $\epsilon > 0$ there is a y_m so that $|y_m - y| < \epsilon$.*

*Also, y is the **limit** or **limit point** of a sequence $\{y_n\}$, if given any $\epsilon > 0$ there is an N so that $|y_n - y| < \epsilon$ for $n \geq N$.*

*The extended real number ∞ is the **limit point** of a sequence $\{y_n\}$ if given any $M \in \mathbb{R}^+$ there is an N so that $y_n > M$ for $n \geq N$. The extended real number $-\infty$ is the **limit point** of a sequence $\{y_n\}$ if given any $M \in \mathbb{R}^+$ there is an N so that $y_n < -M$ for $n \geq N$.*

Notation 3.39 *These limits of a sequence are denoted $\lim_{n \to \infty} y_n = y$, $\lim_{n \to \infty} y_n = \infty$ or $\lim_{n \to \infty} y_n = -\infty$, respectively, and often without $n \to \infty$ when clear from the context.*

When the sequence is a function sequence rather than a point sequence, there is a related notion of limit function, in fact, two common notions.

Definition 3.40 (Convergence of a function sequence) *Given a function sequence $\{f_n(x)\}$ defined on a common domain $D \equiv \bigcap_n Dmn\{f_n\}$, the sequence **converges pointwise** on D to a function $f(x)$ if given any x and $\epsilon > 0$, there is an N so that $|f_n(x) - f(x)| < \epsilon$ for $n \geq N$. In other words, for each x, $f(x)$ is a limit point of the point sequence $\{f_n(x)\}$.*

*The sequence **converges uniformly** on D to a function $f(x)$ if given any $\epsilon > 0$, there is an N so that $|f_n(x) - f(x)| < \epsilon$ for all $x \in D$ and $n \geq N$. In other words, $f(x)$ is a limit point of the point sequence $\{f_n(x)\}$ for each x, but additionally, the "speed" of convergence, defined in terms of N and ϵ, is independent of x.*

Example 3.41 (Sequence behaviors)

1. *Define $\{y_n\}$ as any enumeration of the rational numbers in the interval $[0,1]$. Then for all m:*

$$0 = \inf y_n \leq y_m \leq \sup y_n = 1,$$

   and every real number in $[0,1]$ is an accumulation point of $\{y_n\}$. This follows because the rationals are dense in this interval. In other words, for any real number $r \in [0,1]$ and $\epsilon > 0$, the interval $(r - \epsilon, r + \epsilon) \bigcap [0,1]$ contains infinitely many of the sequence $\{y_n\}$, and hence there are infinitely many such points which satisfy $|y_m - y| < \epsilon$. On the other hand, $\{y_n\}$ has no limit points, which can be directly demonstrated with a proof by contradiction.

   As a bounded sequence, observe that among all accumulation points there is a maximum and minimum accumulation point, here 1 and 0 respectively, which in this case equal $\sup y_n$ and $\inf y_n$, respectively. In general, the infimum and supremum need not be accumulation points.

2. *Define a sequence $\{y_n\}$ as an arbitrary enumeration of all rational numbers in the interval $[0,2]$ of the form $\{1 \pm 1/m\}$ with m a positive integer. Then again the sequence is bounded with $\sup y_n = 2$ and $\inf y_n = 0$, yet these are not accumulation points. As a bounded sequence, there must be at least one accumulation point, and in this case there is exactly one, the number 1, which is also a limit point.*

3. *If the sequence is unbounded, such as defining $\{y_n\}$ as an enumeration of the rationals in $\mathbb{R}$, then $\sup y_n = \infty$ and $\inf y_n = -\infty$. Again every real number is an accumulation point, there are no limit points, and in this case there is no maximum or minimum accumulation point.*

4. *Defining $\{y_n\}$ as an arbitrary enumeration of the integers $\{\pm m\}$ shows that in general an unbounded sequence need not have an accumulation point.*

5. *Define the function sequence $\{f_n(x)\} = \{x^n\}$ on the interval $[0, \infty)$. Then on $[0,1]$, $\{f_n(x)\}$ converges pointwise to the function*

$$f(x) = \begin{cases} 0, & 0 \leq x < 1, \\ 1, & x = 1. \end{cases}$$

   This sequence converges uniformly to $f(x)$ on any $D \subset [0,a]$ for $a < 1$, but does not converge pointwise on any $D \subset (1, \infty)$.

Of special interest in the study of numerical sequences, and an interest that carries forward to the present study, is the identification of the minimum and maximum accumulation points if they exist. For bounded sequences, as exemplified above, there always exists at least one accumulation point, and so one can define the minimum and maximum such accumulation point. These often differ as in item 1, but may agree as in item 2 of Example 3.41.

These extreme accumulation points, when they exist, are called the **limit inferior** and **limit superior**, respectively. Universally these are referred to as the **lim inf** and **lim sup** of the sequence, with or without the space.

As is the case for numerical sequences, it is not initially apparent that the definition below provides these extreme accumulation points, but this and other properties will be stated below and assigned as exercises.

Definition 3.42 (Limits inferior/superior) *Given a sequence of functions $\{f_n(x)\}$, the **limit inferior** and **limit superior** of the sequence are functions that are **defined pointwise** as follows. For each $x \in D \equiv \bigcap_n Dmn\{f_n\}$:*

$$\liminf_{n \to \infty} f_n(x) = \sup_n \inf_{k \geq n} f_k(x), \tag{3.11}$$

$$\limsup_{n \to \infty} f_n(x) = \inf_n \sup_{k \geq n} f_k(x). \tag{3.12}$$

When clear from the context, the subscript $n \to \infty$ is usually dropped from the $\liminf$ and $\limsup$ notation.

Notation 3.43 *The limit superior of a function sequence is alternatively denoted $\overline{\lim} f_n(x)$, and the limit inferior denoted $\underline{\lim} f_n(x)$, but we will use the above more descriptive notation throughout these books.*

Remark 3.44 (Observations on limsup/liminf)

1. *For given x, define a sequence:*

$$\{g_n(x)\} \equiv \inf_{k \geq n} f_k(x),$$

and note that this sequence is increasing (more formally, nondecreasing):

$$g_{n+1}(x) \geq g_n(x).$$

If $f_n(x)$ is unbounded below as a function of n, then $g_n(x) = -\infty$ for all n and then $\liminf f_n(x) = -\infty$. On the other hand, if $\lim_{n \to \infty} f_n(x) = \infty$, then $\liminf f_n(x) = \infty$.

Similarly, defining:

$$\{h_n(x)\} \equiv \sup_{k \geq n} f_k(x),$$

this sequence is decreasing (or, nonincreasing):

$$h_{n+1}(x) \leq h_n(x).$$

So if $f_n(x)$ is unbounded above then $h_n(x) = \infty$ for all n, and then $\limsup f_n(x) = \infty$. On the other hand, if $\lim_{n\to\infty} f_n(x) = -\infty$, then $\limsup f_n(x) = -\infty$.

2. *In the same way that (3.9) can be used to convert infimums into supremums and conversely, the same calculations allow the limits inferior and superior to be interchanged:*

$$\liminf f_n(x) = -\limsup[-f_n(x)], \tag{3.13}$$
$$\limsup f_n(x) = -\liminf[-f_n(x)].$$

Because $\liminf f_n(x)$ and $\limsup f_n(x)$ are defined pointwise, and for each x are defined consistently with the corresponding notion for a real numerical sequence $\{y_n\}$, recall the following results from that theory. The proof is left as an exercise, or see **Reitano** (2010), for example.

Proposition 3.45 (On accumulation points) *Given a sequence of functions $\{f_n(x)\}$:*

1. *When finite for given x, $\liminf f_n(x)$ is an accumulation point, and the smallest such point of the function sequence. Thus given $\epsilon > 0$, there is an N so that for all $m \geq N$:*

$$\liminf f_n(x) - \epsilon < f_m(x).$$

2. *When finite for given x, $\limsup f_n(x)$ is an accumulation point, and the largest such point of the function sequence. Thus given $\epsilon > 0$, there is an N so that for all $m \geq N$:*

$$f_m(x) < \limsup f_n(x) + \epsilon.$$

3. *If both are finite for given x, then given $\epsilon > 0$ there is an N so that the interval:*

$$(\liminf f_n(x) - \epsilon, \ \limsup f_n(x) + \epsilon)$$

contains $f_m(x)$ for all $m \geq N$.

Proof. *Left as an exercise.* ∎

A corollary of item 3 of Proposition 3.45 is then:

Corollary 3.46 (Pointwise limits) *Given a sequence of functions $\{f_n(x)\}$, if for given x:*

$$-\infty < \liminf f_n(x) = \limsup f_n(x) < \infty, \tag{*}$$

then $\lim_{n\to\infty} f_n(x)$ exists and equals this common value. Conversely, if $\lim_{n\to\infty} f_n(x)$ exists, then by (*) the limits inferior and superior are equal.

The above proposition and corollary are applications of the results from the theory of numerical sequences since $\liminf f_n(x)$ and $\limsup f_n(x)$ are defined pointwise. From the above definitions and results we can add to (3.10) to state that:

$$\inf f_n(x) \leq \lim\inf f_n(x) \leq \lim\sup f_n(x) \leq \sup f_n(x). \tag{3.14}$$

We cannot tighten the bounds for $\{f_m(x)\}$ in (3.10) using the limits superior and inferior. But if finite and we expand the interval defined by these values by arbitrary $\epsilon > 0$, then this interval will contain all but finitely many $f_m(x)$. That is, for any $\epsilon > 0$ there exists N so that:

$$\lim\inf f_n(x) - \epsilon \leq f_m(x) \leq \lim\sup f_n(x) + \epsilon, \tag{3.15}$$

for all $m \geq N$.

Given the definitions above, it is natural to now investigate measurability. For example, defined as functions of x, is the limit inferior and/or limit superior Lebesgue measurable if $f_n(x)$ is Lebesgue measurable for all n? The next result answers this question in the affirmative, and also shows that $\inf f_n(x)$ and $\sup f_n(x)$ are Lebesgue measurable.

Because the proofs in this section only rely on the sigma algebra structure of the collection of Lebesgue measurable sets, this same result is true in the Borel measure space, $(\mathbb{R}, \mathcal{B}(\mathbb{R}), m)$, as well as in more general measure spaces $(X, \sigma(X), \mu)$. The common conclusion is that the measurability property of the function sequence is preserved in the infimum and limit inferior, as well as the supremum and limit superior.

Thus the type of measurability is suppressed in the statement and proof.

Proposition 3.47 (Measurability results on function sequences) *Given a sequence of measurable functions $\{f_n(x)\}$, the following functions are also measurable:*

1. *$\min_{n \leq N}\{f_n(x)\}$, for all N;*

2. *$\max_{n \leq N}\{f_n(x)\}$, for all N;*

3. *$\inf f_n(x)$;*

4. *$\sup f_n(x)$;*

5. *$\lim\inf f_n(x)$;*

6. *$\lim\sup f_n(x)$;*

7. *$\lim f_n(x)$, when this exists.*

Proof. *Part 1 follows from part 3, and 2 from 4, defining $f_n(x) = f_N(x)$ for $n \geq N$.*

If $h(x)$ is defined by $h(x) = \inf f_n(x)$, then recalling (3.10):

$$h(x) > y \Rightarrow f_n(x) > y \text{ for all } n.$$

Thus:

$$\{x | h(x) > y\} = \bigcap_n \{x | f_n(x) > y\},$$

and this set is measurable as the intersection of measurable sets.

Similarly, with $g(x) = \sup f_n(x)$:

$$\{x|g(x) < y\} = \bigcap_n \{x | f_n(x) < y\},$$

and this set is again measurable as the intersection of measurable sets.

Now let $h(x) = \liminf f_n(x)$, which by (3.11) means $h(x) = \sup_n \inf_{k \geq n} f_k(x)$. Then for each n, $F_n(x) \equiv \inf_{k \geq n} f_k(x)$ is measurable by 1, and hence $h(x) = \sup F_n(x)$ is measurable by 2. The same approach proves that $\limsup f_n(x)$ is measurable using (3.12).

Finally, the measurability of $\lim f_n(x)$ follows from Corollary 3.46. ■

While the above proposition is true generally on measure spaces $(X, \sigma(X), \mu)$, the following corollary requires the completeness of the sigma algebra. Hence the next result is true in every complete measure space, but is not applicable in the Borel measure space or other incomplete measure spaces.

Corollary 3.48 (On a.e. limits) *Given a sequence of Lebesgue measurable functions $\{f_n(x)\}$ such that $f(x) \equiv \lim f_n(x)$ exists almost everywhere, then $f(x)$ is Lebesgue measurable. The same statement is true for a measurable function sequence on any complete measure space.*

Proof. *The assumption on $\lim f_n(x)$ implies by Corollary 3.46 that:*

$$\liminf f_n(x) = \limsup f_n(x), \ a.e.,$$

where a.e. (almost everywhere) means except on a set of Lebesgue measure 0. Since the limits superior and inferior are Lebesgue measurable by the above proposition, and $f(x)$ equals these functions a.e., it too is Lebesgue measurable by Proposition 3.16. ■

3.5 Approximating Measurable Functions

To work with measurable functions, it is sometimes necessary and often convenient, to be able to approximate such functions with functions that have special properties. For example, given a Lebesgue measurable function defined on an interval $[a, b]$, can this function be approximated with step functions? These are introduced in Definition 3.23 with $A_i = [x_{i-1}, x_i]$, where:

$$a = x_0 < x_1 < ... < x_{n-1} < x_n = b.$$

Alternatively, can such a function be approximated with continuous functions?

The goal of the current investigation is to approximate measurable $f(x)$ in various ways, and arbitrarily well, but perhaps only outside an arbitrarily small exceptional set. To be useful as an approximation, we want to be able to control both the size of this exceptional set, and how well the step or continuous functions approximate $f(x)$ outside this exceptional set.

Because measurable functions are allowed to assume the values $\pm\infty$ and neither step nor continuous functions are so allowed, it is apparent that this exceptional set will always include $\{x|f(x) = \pm\infty\}$. Consequently, to achieve the results we seek it must always be assumed that this set has measure 0.

Proposition 3.49 (Step/continuous approximations) *Let $f(x)$ be an extended real-valued Lebesgue measurable function defined on the interval $[a, b]$, and assume that $\{x|f(x) = \pm\infty\}$ has Lebesgue measure 0.*

Then for any $\epsilon > 0$ there is a step function $g(x)$ and associated measurable set G, and a continuous function $h(x)$ and associated measurable set H, both defined on $[a, b]$, so that:

1. *$m(G) < \epsilon$ and $\left|f(x) - g(x)\right| < \epsilon$ on $[a, b] - G$;*
2. *$m(H) < \epsilon$ and $\left|f(x) - h(x)\right| < \epsilon$ on $[a, b] - H$.*

If $f(x)$ is bounded, $m \leq f(x) \leq M$, then $g(x)$ and $h(x)$ can be chosen with the same bounds.

Proof. *This proof has three main steps:*

1. The existence of g and G that satisfy statement 1 assure the existence of h and H that satisfy statement 2.

Assume that statement 1 is true with $\epsilon/2$. That is, there exists G with $m(G) < \epsilon/2$, and:

$$g(x) = \sum\nolimits_{i=1}^{n} a_i \chi_{A_i}(x),$$

where $A_i = [x_{i-1}, x_i]$ as above, so that $\left|f(x) - g(x)\right| < \epsilon/2$ on $[a, b] - G$. We construct continuous $h(x)$ as follows.

Define intervals $A_1' = [a, x_1 - \epsilon\lambda/4n]$, $A_n' = [x_{n-1} + \epsilon\lambda/4n, b]$, and otherwise:

$$A_i' = [x_{i-1} + \epsilon\lambda/4n, x_i - \epsilon\lambda/4n].$$

Here $\lambda < 1$ is chosen so that each A_i' is nonempty and $A_i' \subset A_i$. Define $h(x) = g(x)$ on $\bigcup A_i'$, and extend the definition of $h(x)$ linearly and continuously between "steps" on $[a, b] - \bigcup A_i'$. By the triangle inequality:

$$\left|f(x) - h(x)\right| \leq \left|f(x) - g(x)\right| + \left|g(x) - h(x)\right|.$$

Now $\left|f(x) - g(x)\right| < \epsilon/2$ outside G by assumption, while $\left|g(x) - h(x)\right| = 0$ outside $[a, b] - \bigcup A_i'$, which has measure less than $\epsilon/2$ by construction. Hence with $H \equiv G \bigcup \left[[a, b] - \bigcup A_i'\right]$, statement 2 is proved since by subadditivity,

$$m(H) \leq m(G) + m\left[[a, b] - \bigcup A_i'\right] < \epsilon.$$

This construction also shows also that if $m \leq g(x) \leq M$, then the same bounds will apply to $h(x)$.

2. Statement 1 is valid with $g(x) = k(x)$, a simple function.

To show that $f(x)$ can be approximated as asserted in statement 1, but with a simple function $k(x)$ of Definition 3.23, let $\epsilon > 0$ be given. Define $G_1(M) = \{x| \left|f(x)\right| > M\}$ and we claim that *for any given $\epsilon > 0$, there exists $M \equiv M(\epsilon)$ so that $m[G_1(M)] < \epsilon/2$. To prove this, note that $G_1(0) < b - a$, $G_1(N + 1) \subset G_1(N)$ and:*

$$\{x | f(x) = \pm\infty\} = \bigcap\nolimits_N G_1(N).$$

Thus if $m[G_1(N)] \geq \epsilon/2$ for all N, then $m\{x | f(x) = \pm\infty\} \geq \epsilon/2$ by (2.24), contradicting the assumption that this set has measure 0. Hence there is an M so that $m[G_1(M)] < \epsilon/2$. With $G_1 \equiv G_1(M)$ we now construct a simple function $k(x)$ so that:

$$\left| f(x) - k(x) \right| < \epsilon \text{ on } [a, b] - G_1.$$

For $i = 1, 2, ...,$ define:

$$B_i = \{x | -M + (i - 1)\epsilon/2 \leq f(x) < -M + i\epsilon/2\}.$$

Observe that on $[a, b] - G_1$, there are only finitely many nonempty B_i-sets, say J, which is approximately equal to $4M/\epsilon$. With $b_i \equiv -M + i\epsilon/2$, define the simple function:

$$k(x) = \sum\nolimits_{i=1}^J b_i \chi_{B_i}(x).$$

Then $k(x)$ approximates $f(x)$ within $\epsilon/2$ on $[a, b] - G_1$, and $m[G_1] < \epsilon/2$ as claimed.

   3. **Statement 1 is valid with a step function $g(x)$.**
   The final step is to approximate this simple $k(x)$ with a step function $h(x)$, outside of a small exceptional set G_2. Since each set B_i is measurable in step 2 it follows from Proposition 2.42 that there are open sets O_i with $B_i \subset O_i$ and $m(O_i - B_i) < \epsilon/4J$, where J is the number of terms in the step 2 simple function construction.
   Now any open set in $\mathbb{R}$ is the disjoint union of open intervals by Proposition 2.12. If $O_i = \bigcup_{j \leq n_i} I_{i,j}$ a finite union, then replace the ith term in the definition of $k(x)$ with:

$$b_i \chi_{B_i}(x) = \sum\nolimits_{j=1}^{n_i} b_i \chi_{I_{i,j}}(x),$$

a step function. If O_i is a countable union of disjoint open intervals, then since $O_i \subset [a, b]$ it follows by countable additivity that $\sum_{j=1}^\infty m(I_{i,j}) \leq b - a$. Thus it is possible to select n_i open intervals so that with O_i' denoting this finite union, $m(O_i - O_i') < \epsilon/4J$. The term $b_i \chi_{B_i}(x)$ in the definition of $k(x)$ is again replaced by the associated step function.
   Combining, define:

$$h(x) = \sum\nolimits_{i=1}^J \sum\nolimits_{j=1}^{n_i} b_i \chi_{I_{i,j}}(x),$$

and note that $h(x) = k(x)$ except on G_2 defined by:

$$G_2 \equiv \bigcup\nolimits_i [O_i - B_i] \bigcup \bigcup\nolimits_i [O_i - O_i'].$$

By construction G_2 has total measure less than $\epsilon/2$.

There is a subtle point in this construction. Note that while $\{B_i\}$ is a finite collection of disjoint sets by construction, and for each i the collection $\{I_{i,j}\}$ is also disjoint, it may happen that the full collection $\{I_{i,j}\}$ is not disjoint. But as this is a finite collection of open intervals, the above step function can be modified so that the intervals are disjoint without changing the measure of the union. For example, if I_1 and I_2 are any two such intervals with $I_1 \cap I_2 \neq \emptyset$, then define with disjoint open intervals:

$$\sum_{=1}^{2} b_i \chi_{I_i}(x) = \sum_{=1}^{3} b'_i \chi_{I'_i}(x).$$

Here $I'_3 = I_1 \cap I_2$, $b'_3 = b_1 + b_2$, and the other I'_i are defined as the largest open subintervals in $I_i - I_1 \cap I_2$ and with $b'_i = b_i$. This can be iterated and the process will end in finitely many steps.

Then $G \equiv G_1 \bigcup G_2$ has measure less than ϵ, and on $[a,b] - G$, $f(x)$ is approximated within $\epsilon/2$ by the step function $h(x)$. This step function can then equivalently be defined in terms of the associate closed intervals, without changing the measure of the exceptional set.

Finally note that if $f(x)$ is bounded, $m \leq f(x) \leq M$, then $G_1(\max(|M|, |m|)) = \emptyset$, and the above construction leads to simple, step and continuous functions which satisfy the same bounds. ∎

Remark 3.50 (On real-valued $f(x)$) When $f(x)$ is a real-valued Lebesgue measurable function on the interval $[a,b]$ with $\{x|f(x) = \pm\infty\} = \emptyset$, the above approximation theorem certainly applies but it can be considerably strengthened using deeper results than have been developed to this point. The result is called **Lusin's Theorem** and sometimes **Luzin's Theorem**, and named for **Nikolai Nikolaevich Lusin** (1883–1950). It is presented here for completeness and proved in Section 4.3.

Lusin's result states that not only can such $f(x)$ be approximated by a continuous function within ϵ outside a set of measure less than ϵ, but in fact $f(x)$ **is** a continuous function outside such a set.

Proposition 3.51 (Lusin's theorem) *Let $f(x)$ be a real-valued Lebesgue measurable function on the interval $[a,b]$. Then given $\epsilon > 0$, there exists an open set G with $m(G) \leq \epsilon$, such that $f(x)$ is continuous on $[a,b] - G$.*

Proof. *See Proposition 4.10.* ∎

The next approximation result states that every Lebesgue measurable function is almost everywhere equal to a Borel measurable function. So, if a given application is indifferent to a redefinition of a function on a set of measure 0, this result allows us to replace the Lebesgue measurable function with a Borel measurable function. The result is proved for nonnegative functions, but since every measurable function can be expressed as a difference of nonnegative measurable functions, this proof applies in general.

Definition 3.52 (Positive/negative parts of $f(x)$) *Given $f(x)$, the **positive part** of $f(x)$, denoted $f^+(x)$, is defined by:*

$$f^+(x) = \max\{f(x), 0\}. \tag{3.16}$$

The **negative part of** $f(x)$, *denoted* $f^-(x)$, *is defined by:*

$$f^-(x) = \max\{-f(x), 0\}. \tag{3.17}$$

Both the positive and negative part of a function are nonnegative functions, and are measurable if $f(x)$ is measurable by item 2 of Proposition 3.47. In addition, both the original function and its absolute value can be recovered from these functions:

$$f(x) = f^+(x) - f^-(x), \qquad \left|f(x)\right| = f^+(x) + f^-(x). \tag{3.18}$$

Proposition 3.53 (Borel approximations) *Let $f(x)$ be a Lebesgue measurable function on a Lebesgue measurable domain $D \subset \mathbb{R}$, where $m\{x|f(x) = \pm\infty\} = 0$. Then there exists a Borel measurable function $g(x)$ so that $f(x) = g(x)$, a.e. That is, $f(x) = g(x)$ outside a subset of D of Lebesgue measure 0.*

Proof. *It is sufficient to prove this result for nonnegative measurable functions, as this then applies to both $f^+(x)$ and $f^-(x)$, and thus to $f(x)$ by (3.18).*

To this end, we first claim that there exists an increasing sequence of simple functions $\{\varphi_n(x)\}$ so that $\varphi_n(x) \to f(x)$ for all x. For given n, define $N \equiv n2^n + 1$ Lebesgue measurable sets $\{A_j^{(n)}\}_{j=1}^N$ by:

$$A_j^{(n)} = \begin{cases} \{x \in D | (j-1)2^{-n} \le f(x) < j2^{-n}\}, & 1 \le j \le N-1, \\ \{x \in D | n \le f(x)\}, & j = N. \end{cases}$$

Now let:

$$\varphi_n(x) = \sum_{j=1}^N (j-1)2^{-n} \chi_{A_j^{(n)}}(x).$$

Then $\{\varphi_n(x)\}_{j=1}^\infty$ is an increasing sequence of nonnegative simple functions with $\varphi_n(x) \to f(x)$ for all $x \in D$. Details are left as an exercise, but let's see why this works by comparing $\{A_k^{(n+1)}\}$ to $\{A_j^{(n)}\}$. First, each set in $\{A_j^{(n)}\}$ for $j \le N_n - 1$ is split into two sets in $\{A_k^{(n+1)}\}$ for $1 \le k \le 2(N_n - 1)$, where $N_n \equiv n2^n + 1$ as above. In addition, the last set:

$$A_{N_n}^{(n)} \equiv \{n \le f(x)\} = \{n \le f(x) < n+1\} \bigcup A_{N_{n+1}}^{(n+1)}.$$

This first set is split into 2^{n+1} level sets in $\{A_k^{(n+1)}\}$ for $2(N_n - 1) + 1 \le k \le N_{n+1} - 1$. Now note that on each set, this simple function is defined as the minimum of $f(x)$ in the set, so splitting sets increases the simple function.

Defining $\psi_n(x) = \varphi_{n+1}(x) - \varphi_n(x)$ obtains that $\{\psi_n(x)\}_{j=1}^\infty$ is a sequence of nonnegative simple functions with $f(x) = \sum_n \psi_n(x)$ for all x. By the above, each $\psi_n(x)$ is a simple function with N_{n+1} terms in the summation, and thus $f(x)$ can be rewritten in terms of the underlying characteristic functions of Lebesgue measurable sets:

$$f(x) = \sum_{j=1}^\infty a_j \chi_{A_j}(x).$$

Here all $a_j \geq 0$ and A_j is Lebesgue measurable, since each $\psi_n(x)$ can be so written with the $\{A_k^{(n+1)}\}$-sets.

By Proposition 2.42, for each j there is a Borel measurable set $F_j \subset A_j$, with $m(A_j - F_j) = 0$. Specifically, $F_j \in \mathcal{F}_\sigma$, the collection of countable unions of closed sets. Define:

$$g(x) = \sum\nolimits_{j=1}^{\infty} a_j \chi_{F_j}(x),$$

then $g(x)$ is Borel measurable and $g(x) = f(x)$ except possibly on the set $\bigcup_j (A_j - F_j)$, which has measure 0. ∎

3.6 Distribution Functions

An important property of measurable functions which is utilized in Probability Theory pertains to the measure of sets such as:

$$f^{-1}((-\infty, y)\} \text{ or } f^{-1}((-\infty, y]),$$

where f^{-1} is defined in (3.1) as:

$$f^{-1}(A) \equiv \{x \in X | f(x) \in A\}.$$

The measure of such sets underlies the notion of the distribution function of a random variable.

The results of this section are valid for measurable functions defined on general measure spaces, and are no more difficult to prove. So we will again develop results in this general context. If desired, the reader can interpret these results in the context of the Lebesgue and Borel measure spaces, $(\mathbb{R}, \mathcal{M}_L, m)$ and $(\mathbb{R}, \mathcal{B}(\mathbb{R}), m)$, which in general requires no more than replacing the measure μ with Lebesgue measure m.

Definition 3.54 (Cumulative level sets) *Let $f(x)$ be an extended real-valued function on the measure space $(X, \sigma(X), \mu)$, so $f : X \to \overline{\mathbb{R}}$, and define the **cumulative level sets**:*

$$L_f(y) = f^{-1}((-\infty, y]), \qquad L_f(y^-) = f^{-1}((-\infty, y)). \tag{3.19}$$

Remark 3.55 *Note that $L_f(y) \subset L_f(z)$ if $y < z$, and similarly for $L_f(y^-)$. From this, it follows that:*

$$L_f(y) = \bigcap\nolimits_{r > y} L_f(r), \qquad L_f(y^-) = \bigcup\nolimits_{r < y} L_f(r), \tag{3.20}$$

where this union and intersection are defined with rational r to ensure a countable number of operations. This both ensures that the resultant sets are members of the given sigma algebra, and that the measure of such sets can be obtained by an application of the continuity of measures in Proposition 2.45.

The identities in (3.20) imply that the cumulative level set defined by $L_f(y)$ can be approximated well "from above" with sets $L_f(z)$ where $z > y$, but not approximated well "from below" with the sets $L_f(z)$ where $z < y$. From below we can only reach $L_f(y^-)$, missing the **level set** $\{x \in X | f(x) = y\}$. The significance of this observation will be seen below in a property of the **distribution function** associated with a measurable function $f(x)$, defined as follows.

Definition 3.56 (Distribution function) *Let f be an extended real-valued measurable function on the measure space $(X, \sigma(X), \mu)$. Define the **distribution function associated with** $f(x)$, denoted $F(y) \equiv F_f(y)$ for $y \in \mathbb{R}$, as the extended real-valued function:*

$$F(y) = \mu[L_f(y)]. \tag{3.21}$$

In other words, $F(y)$ is the μ-measure of the cumulative level set $L_f(y)$.

By definition $F(y) \geq 0$ for all y, while if $L_f(y) = \emptyset$ then $F(y) = 0$. In this case $F(y') = 0$ for $y' \leq y$ since $L_f(y') \subset L_f(y)$ for $y' < y$. More generally, this nesting property assures below that $F(y)$ is an **increasing function**, meaning that $F(y') \leq F(y)$ if $y' < y$. At the other extreme it is possible that $L_f(y) = X$ and thus $F(y) = \infty$. This is the case for Lebesgue measure space, that if $L_f(y) = \mathbb{R}$ then $F(y) = \infty$.

In the special case of a **finite measure space** X, meaning $\mu[X] < \infty$, one has $0 \leq F(y) < \infty$ for all y. For this reason, distribution functions are most useful and commonly encountered in the case of spaces with finite measure. This is the context of probability theory, where $\mu[X] = 1$.

Notation 3.57 (Monotonic function convention) *There is no single convention for the terminology used to describe **monotone** or **monotonic functions**, defined as functions which are increasing or decreasing everywhere. The terminology favored in these texts is as follows:*

1.(a) *$f(x)$ is **increasing**: $f(y) \leq f(z)$ if $y < z$;*
   (b) *$f(x)$ is **strictly increasing**: $f(y) < f(z)$ if $y < z$.*

2.(a) *$f(x)$ is **decreasing**: $f(y) \geq f(z)$ if $y < z$;*
   (b) *$f(x)$ is **strictly decreasing**: $f(y) > f(z)$ if $y < z$.*

Some authors use "increasing" to mean case 1(b), and "nondecreasing" for case 1(a). Similarly, decreasing is reserved for case 2(b), and nonincreasing is used for case 2(a).

The following proposition summarizes some of the essential properties of $F(y)$, which will be seen again in the context of probability theory of Book II and later. For the statement of this result, recall the definition of **one-sided limits**, and the associated notation.

Definition 3.58 (One-sided limits) *Given $f(x)$, the **right limit**, denoted $\lim_{y \to x+} f(y)$, and **left limit**, denoted $\lim_{y \to x-} f(y)$, are defined as follows:*

1. *Right limit*:

$$\lim_{y \to x+} f(y) \equiv \lim_{z \to 0,\, z>0} f(x+z). \tag{3.22}$$

In other words, $\lim_{y \to x+} f(y) = L$ if given $\epsilon > 0$ there is a $\delta > 0$ so that $|f(y) - L| < \epsilon$ for $x < y < x + \delta$.

2. *Left limit*:

$$\lim_{y \to x-} f(y) \equiv \lim_{z \to 0,\, z<0} f(x+z). \tag{3.23}$$

In other words, $\lim_{y \to x-} f(y) = L$ if given $\epsilon > 0$ there is a $\delta > 0$ so that $|f(y) - L| < \epsilon$ for $x - \delta < y < x$.

Notation 3.59 *In many books, $f(x+)$ or $f(x^+)$ is used to denote the right limit of $f(y)$ at x, and correspondingly, $f(x-)$ or $f(x^-)$ is used to denote the left limit of $f(y)$ at x. This notation is often convenient and will be used as needed.*

Proposition 3.60 (Properties of $F(y)$) *Given an extended real-valued measurable function $f(x)$ defined on the measure space $(X, \sigma(X), \mu)$, the distribution function associated with $f(x)$ has the following properties:*

1. *$F(y)$ is a nonnegative, increasing function on $\mathbb{R}$ which is Borel, and hence, Lebesgue measurable.*

2. *If $F(y_0) < \infty$ for some $y_0 < \infty$, then:*

   (a) *For all $y < y_0$,*

   $$\lim_{z \to y+} F(z) = F(y). \tag{3.24}$$

   (b) *For all $y < y_0$,*

   $$\lim_{z \to y-} F(z) = F(y) - \mu(\{x \mid f(x) = y\}). \tag{3.25}$$

   (c) *The limit of $F(y)$ exists as $y \to -\infty$, and:*

   $$\lim_{y \to -\infty} F(y) = \mu(\{x \mid f(x) = -\infty\}). \tag{3.26}$$

3. *$F(y)$ has at most countably many discontinuities.*

4. *If $\mu(X) < \infty$, then the limit of $F(y)$ exists as $y \to \infty$:*

$$\lim_{y \to \infty} F(y) = \mu(X) - \mu(\{x \mid f(x) = \infty\}). \tag{3.27}$$

Proof. 1. *First, $F(y)$ is nonnegative by Definition 2.23, since all measures μ are nonnegative set functions. Further, $F(y)$ is increasing by monotonicity of measures, and the observation that $L_f(y) \subset L_f(z)$ if $y < z$. In detail, if $y < z$, then by finite additivity:*

$$F(y) = \mu[L_f(y)]$$
$$\leq \mu[L_f(y)] + \mu[L_f(z) - L_f(y)]$$
$$= \mu[L_f(z)]$$
$$= F(z).$$

For Borel measurability, we first prove that $F^{-1}((-\infty, z]) \in \mathcal{B}(\mathbb{R})$ for all z. To this end:

$$F^{-1}((-\infty, z]) = \{y \mid \mu[L_f(y)] \leq z\}$$
$$= \{y \mid \mu[f^{-1}((-\infty, y])] \leq z\}.$$

Now if $\mu[f^{-1}((-\infty, y])] > z$ for all y then $F^{-1}((-\infty, z]) = \emptyset$, a Borel set. Otherwise, there exists $y_0 \in F^{-1}((-\infty, z])$, and so $F(y_0) \leq z$. Thus $(-\infty, y_0] \subset F^{-1}((-\infty, z])$ since $F(y) \leq F(y_0)$ for $y < y_0$. Now either $F^{-1}((-\infty, z]) = (-\infty, y_0]$ or there is $y_0' \in F^{-1}((-\infty, z])$ with $y_0' > y_0$, and so $(-\infty, y_0'] \subset F^{-1}((-\infty, z])$. Continuing in this way, it follows that $F^{-1}((-\infty, z])$ is a union of all such $(-\infty, y_0']$-sets. This union can be replaced by a union of all such sets with rational y_0' since F is increasing, and the result follows. That $F(y)$ is Borel measurable, meaning $F^{-1}(A) \in \mathcal{B}(\mathbb{R})$ for all $A \in \mathcal{B}(\mathbb{R})$, now follows as in the proof of Proposition 3.26.

2. For (3.24), recall $L_f(y) = \bigcap_{r>y} L_f(r)$ from (3.20), a countable intersection of rationals $r > y$, which by the nested property can be further restricted to $y < r < z_0$. Since $\mu[L_f(z_0)] < \infty$, continuity from above in Proposition 2.45 obtains that for $y < z_0$:

$$\mu[L_f(y)] = \lim_{r \to y+} \mu[L_f(r)].$$

This is equivalent to (3.24) since $F(z)$ is increasing.

Using the same proposition and argument applied to the countable union from (3.20), $X_f(y^-) = \bigcup_{r<y} X_f(r)$, it follows that:

$$\mu[L_f(y^-)] = \lim_{z \to y-} \mu[L_f(z)].$$

By finite additivity:

$$\mu[L_f(y)] = \mu[L_f(y^-)] + \mu[\{x \mid f(x) = y\}],$$

(3.25) is obtained from the observation that $\mu[\{x \mid f(x) = y\}] < \infty$ since $y < z_0$ and $F(z_0) < \infty$. The same argument applied to $L_f(-\infty) = \bigcap_{r<z_0} L_f(r)$ implies that:

$$\mu[L_f(-\infty)] = \lim_{z \to -\infty} \mu[L_f(z)],$$

or (3.26).

3. Assume $F(y_0) < \infty$ for some $y_0 < \infty$. Then there can be at most countably many discontinuities for $y < y_0$ since by parts 2(a) and 2.(b), a discontinuity at y implies that $F(y+) > F(y-)$. Every such jump contains a rational number, and as $F(y)$ is increasing, each jump contains a different rational number. If $F(y_0) < \infty$ for all y_0, then this completes the proof.

Otherwise, there exists $y_0' < \infty$ with $F(y_0') = \infty$. If $F(y_0') = \infty$ for all y_0', then there is nothing to prove, so assume there exists $y_0 < y_0' < \infty$ with $F(y_0) < \infty$. As $F(y)$ is increasing, redefine y_0' by:

$$y_0' = \inf\{y | F(y) = \infty\},$$

so $F(y_0) < \infty$ for $y_0 < y_0' < \infty$.

Now $(-\infty, y_0'] = (-\infty, y_0') \bigcup \{y_0'\}$, and with r denoting rationals:

$$(-\infty, y_0') = \bigcup_{y_0 < r < y_0'} (-\infty, r].$$

Thus by continuity from below:

$$F(y_0') = \infty = \mu[\{x | f(x) = y_0'\}] + \lim_{r \to y_0'} F(r).$$

If this limit is finite, then there are at most countably many discontinuities below r for all such r, plus one discontinuity due to $\mu[\{x | f(x) = y_0'\}]$. If this limit is infinite, again there are at most countably many discontinuities below any r with $F(r) < \infty$. As $F(r) < \infty$ for all such r we are done.

4. Finally, (3.20) obtains $L_f(\infty^-) = \bigcup_{r < \infty} L_f(r)$, and thus:

$$\mu[L_f(\infty^-)] = \lim_{z \to \infty} \mu[L_f(z)],$$

by continuity from below. Then (3.27) follows from:

$$\mu(\{x | f(x) < \infty\}) + \mu(\{x | f(x) = \infty\}) = \mu(X),$$

and the finiteness of $\mu(\{x | f(x) = \infty\})$. ∎

Remark 3.61 (Observations on distribution functions)

*1. If $f(x)$ is a **real-valued** function:*

$$\{x | f(x) = -\infty\} = \{x | f(x) = \infty\} = \emptyset,$$

then in items 2(c) and 3 we obtain the simplified results:

$$\lim_{z \to -\infty} F(z) = 0, \tag{3.28}$$

$$\lim_{z \to \infty} F(z) = \mu(X), \tag{3.29}$$

where the limit in (3.29) need not be finite. However, in item 2 we must still in general insist that $F(y_0) < \infty$ for some y_0 for the desired results (see item 5 below).

2. *The result in (3.24) states that $F(y)$ is **right continuous at y**.*

3. *The result in (3.25) states that $F(y)$ has **left limits**, and from the resulting equation we see that $F(y)$ will be **left continuous** at y if and only if $\mu(\{x|f(x) = y\}) = 0$. In general, however, there can be "jumps" or discontinuities of size $\mu(\{x|f(x) = y\})$ in the graph of $F(y)$. There can be at most **countably many such jumps** at y with $-\infty < y < \infty$ as noted in item 3 of the proposition.*

4. *Combining items 2 and 3, it follows that distribution functions are continuous except for at most countably many discontinuities, and hence they are **continuous outside a set of measure 0**, or **continuous a.e.** But note that Proposition 3.16 does not apply to conclude that such functions are Lebesgue measurable. For that proposition to apply would require that a distribution function equal a continuous function except on a set of measure 0, which is a different notion as discussed in Remark 3.19. Hence the need for a direct proof in part 1.*

   We will see in Book III in the section on "Riemann Implies Lebesgue on Rectangles," that every bounded function that is continuous almost everywhere on a closed interval is in fact Lebesgue measurable. This will be accomplished by demonstrating that such functions are Lebesgue integrable, and that this property ensures Lebesgue measurability.

5. *The restriction for the conclusions of item 2, that $F(y_0) < \infty$ for some y_0, is automatically satisfied in the cases where $\mu(X) < \infty$. This is always true on probability spaces where $\mu(X) = 1$.*

In summary, the graph of the distribution function of every measurable function is an increasing, Borel measurable, right continuous function with left limits, and has at most countably many discontinuities. Further, the distribution function will be continuous if and only if $\mu(\{x|f(x) = y\}) = 0$ for all y. More specifically, $F(y)$ is continuous at y_0 if and only if $\mu(f^{-1}(y_0)) = 0$.

Notation 3.62 (On càdlàg) *Functions which are **"continuous on the right and with left limits,"** are sometimes referred to as **càdlàg**, from the French "continu à droite, limite à gauche." Thus all distribution functions are càdlàg. This notion arises in probability theory, as well as in the theory of stochastic processes.*

But in general a càdlàg function need not be increasing, and it is only in this more general case that the statement "with left limits" is a restriction. For any increasing function such as the distribution functions above, right continuity assures the existence of left limits. This follows because if $x_n \to x^-$, then since $f(x_n) \le f(x_{n+m}) \le f(x)$ for all m, we have that $\{f(x_n)\}$ is an increasing and bounded sequence and hence by Definition 3.42:

$$\lim_{x_n \to x^-} f(x_n) = \limsup_{n \to \infty} f(x_n).$$

Example 3.63 (Distribution function behaviors) *In this example we illustrate that a distribution function depends both on the function $f(x)$, as well as on the measure μ defined on the measure space. For a general characterization of distribution functions on $\mathbb{R}$, see Chapter 1 of Book IV.*

1. *On* $X = [0,1]$ *with Lebesgue measure* m, *define the Lebesgue measurable function* $f : X \to \mathbb{R}$:

$$f(x) = \begin{cases} 1, & x \text{ rational,} \\ 0, & x \text{ irrational.} \end{cases}$$

This is the **Dirichlet function**, *named for its discoverer,* **J. P. G. Lejeune Dirichlet** *and encountered earlier. The distribution function associated with* f *is then*

$$F(y) = \begin{cases} 0, & y < 0, \\ 1, & y \geq 0, \end{cases}$$

for which the càdlàg property is apparent. The graph of $F(y)$ *has one jump associated with the unique value of* $y = 0$ *for which* $m(\{x | f(x) = y\}) > 0$.

2. *On* $X = [0,1]$ *with Lebesgue measure, enumerate the rationals* $\{r_j\}_{j=1}^{\infty} \subset [0,1]$ *arbitrarily and define the function* $f : [0,1] \to \mathbb{R}$ *by:*

$$f(x) = \begin{cases} 1/2^n, & x = r_n \text{ rational,} \\ 0, & x \text{ irrational.} \end{cases}$$

To simplify notation in the description of $F(y)$, *note that for* $0 < y \leq 1/2$ *that* $f(x) \leq y$ *if and only if either* x *is irrational, or* $x = r_n$ *rational and* $1/2^n \leq y$, *which is equivalent to* $n \geq -\log_2 y$. *On* $(0,1]$ *define:*

$$n_y = \min\{n \in \mathbb{Z} | n \geq -\log_2 y\}.$$

Note that n_y *is a decreasing step function, unbounded as* $y \to 0$, *with* $n_y = n$ *for* $y \in [1/2^n, 1/2^{n-1})$ *and* $n_1 = 0$.

With this notation, it can be seen that f *is Borel and thus Lebesgue measurable. Letting* $\mathbb{J}$ *denote the irrationals in* $[0,1]$:

$$f^{-1}((-\infty, y]) = \begin{cases} \varnothing, & y < 0, \\ \mathbb{J}, & y = 0, \\ \mathbb{J} \cup \{r_n | n \geq n_y\}, & 0 < y < 1/2, \\ [0,1], & 1/2 \leq y. \end{cases}$$

Each of these pre-image sets is Borel measurable, and the distribution function associated with $f(x)$ *is:*

$$F(y) = \begin{cases} 0, & y < 0, \\ 1, & y \geq 0, \end{cases}$$

the same as in the first example.

In this example, all of the richness of the function $f(x)$ *has been lost in the associated distribution function because under Lebesgue measure, everything interesting that this function does is being done on a set of Lebesgue measure* 0.

3. *Consider the second example above, but now using a* **counting measure on the rationals** $\mu_{\mathbb{Q}}$ *on* $\sigma(P([0,1]))$. *This idea was originally introduced in (2.3) of Example 2.24, but here we modify it somewhat. For the rationals* $\{r_j\}_{j=1}^{\infty} \subset [0,1]$ *enumerated as in the second example, define* $\mu_{\mathbb{Q}}(r_j) = 4^{-j}$. *For any set* $A \subset [0,1]$, *define* $\mu(A) = \sum_{r_j \in A} 4^{-j}$, *and* $\mu(\emptyset) = 0$. *Then* $\mu_{\mathbb{Q}}$ *is countably additive by construction, and is hence a measure on the power sigma algebra* $\sigma(P([0,1]))$.

Now f defined in example 2 is $\mu_{\mathbb{Q}}$-*measurable because every set is in the power sigma algebra, and so too is* $f^{-1}((-\infty,y])$ *for any y. The distribution function associated with f now reflects far more of the structure of this function, because the details of this structure are "recognized" by the measure on* $[0,1]$. *In detail, where* n_y *is defined in the second example:*

$$F(y) = \begin{cases} 0, & y \leq 0, \\ \sum_{n=n_y}^{\infty} 4^{-n}, & 0 < y < 1/2, \\ 1/3, & 1/2 \leq y. \end{cases}$$

This distribution function now reflects more of this function's structure. This illustrates the point that the distribution function associated with a given measurable function reflects both the values of the function as well as the measure used on the domain to measure the function's pre-image sets.

Remark 3.64 (Proof that F(y) in the third example is càdlàg) *From the definition in the third example it follows that* $F(y)$ *has a discontinuity at every rational* $y = 1/2^n$, *as these are the values of y for which* n_y *changes. However, it is not immediately apparent that this function is càdlàg. On* $(-\infty,0) \bigcup (1/2,\infty)$, $F(y)$ *is continuous so we focus on* $[0,1/2]$.

By the definition of n_y, *a new term is added to the summation which defines* $F(y)$ *at each* $y = 1/2^n$. *As is the case for* n_y, $F(y)$ *is constant on each interval* $[1/2^n, 1/2^{n-1})$ *for* $n \geq 1$, *proving right continuity and left limits on* $(0,1)$. *Finally,* $F(y)$ *is in fact right (and left) continuous at 0 since for any* $\epsilon > 0$, *we have* $F(y) < \epsilon$ *if:*

$$\sum_{n=n_y}^{\infty} 4^{-n} = 4^{1-n_y}/3 < \epsilon.$$

This can be solved for the lower bound for n_y *required:*

$$n_y > 1 - 0.5 \log_2[3\epsilon].$$

Since $n_y \geq -\log_2 y$ *by definition, this can then be converted to a bound for y:*

$$-\log_2 y > 1 - 0.5 \log_2[3\epsilon].$$

This obtains that $F(y) < \epsilon$ *when* $y < [3\epsilon]^{0.5}/2$. *Hence* $F(y)$ *is right continuous, and in fact continuous, at* $y = 0$.

4

Littlewood's Three Principles

In his 1944 book, *Lectures on the Theory of Functions* (Oxford), **J. E. Littlewood** (1885–1977) set out three principles of real analysis which have come to be known by this chapter's title. The statement of these principles is quoted on p. 71 of **Royden** (1971), and can be summarized as follows:

1. Every Lebesgue measurable set of finite measure is "nearly" the finite union of open intervals.

2. Every Lebesgue measurable function is "nearly" continuous.

3. Every convergent sequence of Lebesgue measurable functions is "nearly" uniformly convergent.

Littlewood's meaning of "nearly" differed from the notion of "almost" which would have implied "almost everywhere" or "except on a set of measure 0." By "nearly" he meant to imply that these statements were true "except on a set of measure $\epsilon > 0$," where ϵ could be arbitrarily specified.

In this section these principles are developed in the order $1 - 3 - 2$, since the convergence result in 3 will be used for the continuity conclusion in 2.

4.1 Measurable Sets

The first principle is related to the result in Proposition 2.42 which states in part that for any Lebesgue measurable set $A \in \mathcal{M}_L$ and any $\epsilon > 0$, there is an open set $G \in \mathcal{G}$ with $A \subset G$ and:

$$m(G - A) \le \epsilon.$$

Now every open set G is the countable union of disjoint open intervals by Proposition 2.12, but Littlewood's first principle states that if $m(A) < \infty$, all but finitely many of these intervals can be discarded.

Consider the example:

$$A = \bigcup_{n \ge 0} \left(n + 2^{-n-1}, n + 2^{-n}\right),$$

DOI: 10.1201/9781003257745-4

which is a measurable set with $m(A) < \infty$, and one for which Proposition 2.42 allows the choice $G = A$. But there is no finite union of open intervals G' for which $A \subset G'$. However, defining

$$G_N = \bigcup_{n=0}^{N} \left(n + 2^{-n-1}, n + 2^{-n} \right),$$

we see that G_N is a finite union of open intervals which covers all of A except a subset of measure that can be made as small as desired.

This example motivates that for Littlewood's formulation, using only finitely many open intervals as G, we must forego the conclusion that A will be contained in this finite union. So by *"nearly a finite union of open intervals"* we will mean that G is a finite union with:

$$m(G \triangle A) \leq \epsilon.$$

Definition 4.1 (Symmetric set difference) *Given sets A and G, the **symmetric set difference** $G \triangle A$ is defined by:*

$$G \triangle A \equiv (G - A) \bigcup (A - G). \tag{4.1}$$

Thus $G \triangle A$ equals the union of the G-points not in A, and the A-points not in G, and is "symmetric" in the sense that:

$$G \triangle A = A \triangle G.$$

Also note that if A and G are measurable, then so too is $G \triangle A$ as justified with a few sigma algebra manipulations, recalling part 5 of Remark 2.3.

Proposition 4.2 (Littlewood's First Principle) *Let $A \in \mathcal{M}_L(\mathbb{R})$ with $m(A) < \infty$. Then for any $\epsilon > 0$, there is a finite collection of disjoint open intervals $\{I_j\}_{j=1}^{N}$, so that with $I \equiv \bigcup_{j \leq N} I_j$:*

$$m(G \triangle I) \leq \epsilon.$$

Proof. *Applying Proposition 2.42 with $\epsilon/2$, there is an open set G and thus by Proposition 2.12 a countable collection of disjoint open intervals $\{I_j\}_{j=1}^{\infty}$ so that $A \subset \bigcup_j I_j$ and $m(\bigcup_j I_j - A) \leq \epsilon/2$. Since*

$$\bigcup_j I_j = A \bigcup (\bigcup_j I_j - A),$$

and by assumption $m(A) < \infty$ and $m(\bigcup_j I_j - A) \leq \epsilon/2$, finite additivity yields that $m(\bigcup_j I_j) < \infty$. Then countable additivity obtains that $\sum_i m(I_j)$ is a convergent series and there is an N so that $\sum_{j=N+1}^{\infty} m(I_j) < \epsilon/2$.

Now $A \subset \bigcup_j I_j$ implies $A - \bigcup_{j \leq N} I_j \subset \bigcup_{j > N} I_j$, and so:

$$m(A - \bigcup_{j \leq N} I_j) < \epsilon/2.$$

Then since $\bigcup_{j \leq N} I_j - A \subset \bigcup I_j - A$, we derive

$$m(\bigcup_{j \leq N} I_j - A) \leq \epsilon/2,$$

and the result follows. ∎

4.2 Sequences of Measurable Functions

Recall the definition of two notions of convergence of a function sequence in Definition 3.40. Given a function sequence $\{f_n(x)\}$, a function $f(x)$, and set E, we say that $f_n(x)$ **converges pointwise** to $f(x)$ on E if given $x_0 \in E$ and $\epsilon > 0$, there is an N so that $|f_n(x_0) - f(x_0)| < \epsilon$ for $n \geq N$. If given $\epsilon > 0$ there is an N so that $|f_n(x) - f(x)| < \epsilon$ for all $n \geq N$ and for all $x \in E$, we say that $f_n(x)$ **converges uniformly** to $f(x)$ on E.

Notation 4.3 *The notation $f_n(x) \rightarrow f(x)$ is commonly used to express convergence, and one uses a descriptive phrase to identify which type of convergence is considered.*

Littlewood's third principle, that a convergent sequence of Lebesgue measurable functions is "nearly" uniformly convergent, is striking because the conclusion of uniform convergence, even "nearly," is a very powerful conclusion. We begin with an exercise that investigates part 5 of Example 3.41 more closely.

Exercise 4.4 *Given a continuous function sequence $\{f_n(x)\}$, a function $f(x)$, and E compact. We know from part 5 of Example 3.41 that $f(x)$ need not be continuous on E if $f_n(x) \rightarrow f(x)$ pointwise on E.*

Prove that if $f_n(x) \rightarrow f(x)$ uniformly on E, then $f(x)$ is continuous on E. Hint:

$$|f(x) - f(y)| \leq |f(x) - f_n(x)| + |f_n(x) - f_n(y)| + |f_n(y) - f(y)|.$$

To complete the proof, you must show that each $f_n(x)$ is in fact uniformly continuous on E (Definition 1.4), and this will follow from compactness.

We begin with a weaker version of Littlewood's third principle which is needed for the final result. A discussion will follow as to why this falls short of Littlewood's result even though it perhaps appears adequate on first reading.

Proposition 4.5 ("Almost" Littlewood's Third Principle) *Let $\{f_n(x)\}$ be a sequence of Lebesgue measurable functions defined on a Lebesgue measurable set E with $m(E) < \infty$, and let $f(x)$ be a real-valued function so that $f_n(x) \rightarrow f(x)$ pointwise for $x \in E$.*

Then, given $\epsilon > 0$ and $\delta > 0$, there is a measurable set $A \subset E$ with $m(A) < \delta$, and an N, so that for all $x \in E - A$:

$$\left| f_n(x) - f(x) \right| < \epsilon,$$

for all $n \geq N$.

Proof. *Given $\epsilon > 0$, define:*

$$G_n = \{x| \left| f_n(x) - f(x) \right| \geq \epsilon\},$$

and:

$$E_N = \bigcup_{n=N}^{\infty} G_n$$

$$= \{x| \left| f_n(x) - f(x) \right| \geq \epsilon \text{ for some } n \geq N\}.$$

Then $\{E_N\}$ is a nested sequence, $E_{N+1} \subset E_N$. Also $f_n(x) \to f(x)$ for each $x \in E$ implies that for every $x \in E$ there is an E_N with $x \notin E_N$.

Hence $\bigcap_N E_N = \emptyset$ and since $m(E) < \infty$, it follows from Proposition 2.44 that $\lim_{N\to\infty} m[E_N] \to 0$. Thus, given $\delta > 0$ there is an N with $m[E_N] < \delta$. With $A \equiv E_N$, we have $m(A) < \delta$ and by definition, if $x \notin A$ then $\left| f_n(x) - f(x) \right| < \epsilon$ for all $n \geq N$. ∎

Corollary 4.6 (Convergence a.e.) *The conclusion of the above proposition remains valid if $f_n(x) \to f(x)$ for each $x \in E$ outside a set of Lebesgue measure 0. In other words, if $f_n(x) \to f(x)$ a.e.*

Proof. *Repeating the above proof, it can only be concluded that for every $x \in E$ outside an exceptional set of measure 0, there is an E_N with $x \notin E_N$. Hence $\bigcap_N E_N$ equals this set of measure 0. But then again $\lim_{N\to\infty} m[E_N] \to 0$, and the proof follows as above.* ∎

Remark 4.7 (On uniform convergence) *The above proposition does **not** imply that $f_n(x)$ converges uniformly to $f(x)$ on $E - A$, because the set A depends on the given ϵ and δ. This result is close to, but not equivalent to, Littlewood's third principle. To improve this result to the conclusion of "nearly uniformly convergent," we must show that $f_n(x) \to f(x)$ uniformly on $E - A$.*

That is, we need to find a fixed set A with $m(A) < \delta$, such that for any $\epsilon > 0$ there is an N with $\left| f_n(x) - f(x) \right| < \epsilon$ for all $x \in E - A$ and all $n \geq N$.

*This is known as **Egorov's Theorem**, named for **Dmitri Fyodorovich Egorov** (1869–1931) and sometimes phonetically translated to **Egoroff**. It is also known as the **Severini-Egorov theorem** in recognition of the somewhat earlier and independent proof by **Carlo Severini** (1872–1951).*

The next result is also **Littlewood's third principle**.

Proposition 4.8 (Severini-Egorov theorem) *Let $\{f_n(x)\}$ be a sequence of Lebesgue measurable functions defined on a Lebesgue measurable set E with $m(E) < \infty$, and let $f(x)$ be a real-valued Lebesgue measurable function so that $f_n(x) \to f(x)$ pointwise for $x \in E$.*

Then given $\delta > 0$ there is a measurable set $A \subset E$ with $m(A) < \delta$, so that $f_n(x) \to f(x)$ uniformly on $E - A$. That is, for $\epsilon > 0$ there is an N so that $\left| f_n(x) - f(x) \right| < \epsilon$ for all $x \in E - A$ and $n \geq N$.

Proof. *Given $\delta > 0$, define $\epsilon_m = 1/m$ and $\delta_m = \delta/2^m$. Apply Proposition 4.5 to find a set A_m with $m(A_m) < \delta_m$, and an integer N_m, so that $\left| f_n(x) - f(x) \right| < \epsilon_m$ for $n \geq N_m$ and for all $x \in E - A_m$. Now let $A = \bigcup_m A_m$. By countable subadditivity, $m(A) \leq \sum m(A_m) = \delta$. To show that $f_n(x) \to f(x)$ uniformly on $E - A$, let $\epsilon > 0$ be given. There is an m so that $\epsilon_m < \epsilon$ and hence also an N_m so that $\left| f_n(x) - f(x) \right| < \epsilon_m < \epsilon$ for $n \geq N_m$ and for all $x \in E - A_m$. But then this statement is also true for $x \in E - A$ since $A_m \subset A$, and the result follows.* ∎

Corollary 4.9 (Severini-Egorov theorem) *The result above remains valid if $f_n(x) \to f(x)$ almost everywhere for $x \in E$.*

Proof. *Left as an exercise.* ∎

4.3 Measurable Functions

Finally, we turn to the Littlewood's second principle, which significantly strengthens Proposition 3.49. There it was shown that outside an arbitrarily small set, measurable functions were "close" to continuous functions. But this is not the meaning of Littlewood's concept of *"nearly continuous."* The following result is called **Lusin's theorem**, named for **Nikolai Nikolaevich Lusin** (1883–1950), and sometimes phonetically translated to **Luzin**.

The next result is **Littlewood's second principle**.

Proposition 4.10 (Lusin's theorem) *Let $f(x)$ be a real-valued Lebesgue measurable function on the interval $[a, b]$. Then given $\epsilon > 0$, there exists an open subset G with $m(G) \leq \epsilon$, such that $f(x)$ is continuous on $[a, b] - G$.*

Proof. *We prove this result by a careful application of a simpler result, as was the case for the Severini-Egorov theorem. Given $\epsilon > 0$, apply Proposition 3.49 with $\epsilon_m = \epsilon/2^{m+1}$. Then there is a continuous function $h_m(x)$ defined on $[a, b]$ and associated measurable set $H_m \subset [a, b]$, so that $m(H_m) < \epsilon/2^{m+1}$ and $\left| f(x) - h_m(x) \right| < \epsilon/2^{m+1}$ on $[a, b] - H_m$.*

Using Proposition 2.42, let G_m be an open set with $H_m \subset G_m$ and $m(G_m - H_m) \leq \epsilon/2^{m+1}$. By finite additivity:

$$m(G_m) = m(H_m) + m(G_m - H_m) \leq \epsilon/2^m.$$

Further, that $H_m \subset G_m$ assures that $\left| f(x) - h_m(x) \right| < \epsilon/2^{m+1}$ on $[a, b] - G_m$.

Now let $G \equiv \bigcup G_m$, and note that $m(G) \leq \epsilon$ by subadditivity. Since $G_m \subset G$ for all m, this implies that $\left| f(x) - h_m(x) \right| < \epsilon/2^{m+1}$ on $[a, b] - G$ for all m. In other words, the sequence $\{h_m(x)\}$ converges uniformly to $f(x)$ on $[a, b] - G$.

As a countable union of open sets G is open, and so:

$$[a, b] - G \equiv [a, b] \bigcap \widetilde{G},$$

is closed as an intersection of closed sets, and bounded. Thus $[a, b] - G$ is compact by the Heine-Borel theorem of Proposition 2.27.

Finally, uniform convergence of continuous functions on the compact set $[a, b] - G$ implies that $f(x)$ is continuous on this set by Exercise 4.4. ∎

Note that Lusin's theorem asserts that the set of discontinuity points of a real-valued Lebesgue measurable function defined on a closed interval can be contained in a set with arbitrarily small measure. This result does not imply that such a function is equal to a continuous function almost everywhere, nor that it is continuous almost everywhere, because in general the continuous function constructed in the above proof depends on ϵ.

Indeed, recall the discussion in Remark 3.19 where it was shown that:

1. A Lebesgue measurable function need not agree with a continuous function almost everywhere. This was illustrated by the example:

$$f_1(x) = \begin{cases} 0, & -1 \leq x \leq 0, \\ 1, & 0 < x \leq 1, \end{cases}$$

though this function is continuous almost everywhere.

2. A Lebesgue measurable function need not be continuous almost everywhere, and in fact need not be continuous anywhere as is illustrated by:

$$f_4(x) = \begin{cases} 0, & x \text{ rational}, \\ 1, & x \text{ irrational}. \end{cases}$$

3. In contrast, every function that agrees with a continuous function almost everywhere is Lebesgue measurable. This follows by Proposition 3.11, that continuous functions are so measurable, and Proposition 3.16, that a function is Lebesgue measurable if equal to a Lebesgue measurable function almost everywhere.

4. In addition, it will be seen in Book III on Lebesgue integration theory, that if a function is bounded and continuous almost everywhere on a closed interval, it is Lebesgue measurable.

5

Borel Measures on $\mathbb{R}$

Although many earlier results generalize as noted to $(X, \sigma(X), \mu)$, this book has so far largely focused on the development of Lebesgue measure on $\mathbb{R}$, with particular emphasis on the Lebesgue measure space $(\mathbb{R}, \mathcal{M}_L(\mathbb{R}), m)$ and the Borel measure space $(\mathbb{R}, \mathcal{B}(\mathbb{R}), m)$. What characterized Lebesgue measure m was that for a given interval $\{a, b\}$, this notation implying that this interval is open, closed, or half-open, that Lebesgue measure obtained:

$$m(\{a, b\}) = b - a.$$

This property entered the development of m as a measure in the definition of m^*, the Lebesgue outer measure. In Definition 2.25, $m^*(A)$ is defined for given $A \subset \mathbb{R}$ by:

$$m^*(A) = \inf \left\{ \sum_n |I_n| \ \Big| \ A \subset \bigcup_n I_n, \right\},$$

where I_n is an open interval, and $|I_n|$ denotes its interval length. Thus $m^*(I)$ is explicitly defined as interval length for an open interval, and in Proposition 2.28 it is proved that $m^*(I)$ equals interval length for all intervals.

It is only natural to wonder, what if we began this development with a different definition of $|I_n|$ than interval length? Some questions arising from this generalization are:

1. What collection of basic sets $\{I_n\}$ could be used in addition to, or instead of, the open intervals?

2. What definitions of $|I_n|$ would make sense, other than interval length?

3. Given a reasonable definition of $|I_n|$, could the entire development underlying Lebesgue measure be repeated, producing a new measure μ, and a measure space, $(\mathbb{R}, \mathcal{M}_\mu(\mathbb{R}), \mu)$ with $\{I_n\} \subset \mathcal{M}_\mu(\mathbb{R})$?

4. If the first part of Question 3 has an affirmative answer, would this space be complete?

If the development implied by Question 3 has an affirmative answer, and if this collection $\{I_n\}$ of basic sets is defined in a way that it generates the intervals using sigma algebra manipulations, then $\{I_n\} \subset \mathcal{M}_\mu$ will assure that $\mathcal{B}(\mathbb{R}) \subset \mathcal{M}_\mu(\mathbb{R})$. So such a construction would also lead to a new Borel measure space, $(\mathbb{R}, \mathcal{B}(\mathbb{R}), \mu)$, and μ a new Borel measure.

In this chapter we investigate the relationship between Borel measures on $\mathbb{R}$ and the associated definitions of $|I_n|$ for an appropriate collection $\{I_n\}$. In essence, $|I_n|$ will be a **generalized interval length**.

DOI: 10.1201/9781003257745-5

Before continuing further we formalize the definition of a **Borel measure**, which will be seen to generalize the properties of m on $\mathcal{B}(\mathbb{R})$. This measure is named for **Émile Borel** (1871–1956).

Definition 5.1 (Borel measure on $\mathbb{R}$) *A Borel measure is a nonnegative **set function** μ defined on the Borel sigma algebra $\mathcal{B}(\mathbb{R})$, taking values in the nonnegative **extended real numbers** $\overline{\mathbb{R}}^+ \equiv \mathbb{R}^+ \bigcup \{\infty\}$, and which satisfies the following properties:*

1. $\mu(\emptyset) = 0$.
2. ***Countable Additivity:*** *If $\{A_j\} \subset \mathcal{B}(\mathbb{R})$ is a countable collection of pairwise disjoint sets then:*

$$\mu\left(\bigcup_j A_j\right) = \sum_j \mu(A_j).$$

3. *For any **compact** set $A \in \mathcal{B}(\mathbb{R})$, $\mu(A) < \infty$.*

*Then $(\mathbb{R}, \mathcal{B}(\mathbb{R}), \mu)$ is called a **Borel measure space**.*

Remark 5.2 (On Borel measures) *Recall from Definition 2.26 that by compact it is meant that every open cover of A has a finite subcover. By the Heine-Borel theorem, a set $A \subset \mathbb{R}$ is compact if and only if it is closed and bounded.*

As noted in Definition 2.21, Lebesgue measure is an example of a Borel measure. This follows because if A is compact, then $A \subset [a, b]$ by boundedness, and thus $m(A) \leq b - a$ by monotonicity of m.

A Borel measure by Definition 5.1 is not simply a measure on the Borel sigma algebra $\mathcal{B}(\mathbb{R})$ due to the added restriction in property 3. For example, if $\{r_j\}_{j=1}^{\infty}$ is an enumeration of the rationals, define $\mu(r_j) = 1$ and in general $\mu(A) = \sum_{r_j \in A} \mu(r_j)$. Then μ is a measure on the power sigma algebra $\sigma(P(\mathbb{R}))$ of Example 2.7, and thus also a measure on $\mathcal{B}(\mathbb{R})$. Certainly μ does not satisfy property 3 of Definition 5.1 since the only compact sets with finite measure are the sets that contain finitely many rationals. So μ is not a Borel measure.

The addition of the restriction in property 3 is not universally required by all authors in their definitions of Borel measure, but this restriction eliminates measures on $\mathcal{B}(\mathbb{R})$ with behaviors far outside the applications of interest in these books. In particular, all probability measures satisfy this requirement since then $\mu(\mathbb{R}) = 1$.

Comparing this definition to that for a Lebesgue measure it is immediately clear that there are two differences:

1. Replacing an unspecified sigma algebra $\sigma(\mathbb{R})$ with the Borel sigma algebra $\mathcal{B}(\mathbb{R})$;
2. Eliminating the requirement on the value of the measure of intervals, but replacing it with the requirement that bounded sets have finite Borel measure. While property 3 specifically addresses compact sets, it applies to all bounded sets. If $A \in \mathcal{B}(\mathbb{R})$ is bounded, then its closure $\bar{A}$ is compact by the Heine-Borel theorem and by monotonicity of measures, $\mu(A) \leq \mu(\bar{A}) < \infty$.

In the Lebesgue development, it turned out that the unspecified sigma algebra $\sigma(\mathbb{R})$ contained $\mathcal{B}(\mathbb{R})$ because it was shown to contain the intervals, but otherwise $\sigma(\mathbb{R})$ was allowed to be arbitrary. Though not proved, it was noted in that development that $\mathcal{B}(\mathbb{R})$ was in fact a proper subset of the complete sigma algebra $\mathcal{M}_L(\mathbb{R})$ that was constructed.

In the general Borel measure case, it is now required that any constructed sigma algebra contain $\mathcal{B}(\mathbb{R})$ and this will be assured by properly identifying the collection $\{I_n\}$. As in the Lebesgue case there will again be an associated complete sigma algebra $\mathcal{M}_\mu$, which will be strictly bigger than $\mathcal{B}(\mathbb{R})$.

Remark 5.3 (General Borel measures) *The notion of a Borel measure can be generalized to any topological space X by defining* $\mathcal{B}(X)$ *as in Definition 2.13, as the smallest sigma algebra that contains all the open sets of X. The notion of compactness is then well defined in terms of open covers, that an arbitrary "open cover" of the compact set has a finite subcover. In the special case of* $X = \mathbb{R}^n$ *and the usual topology generated by open balls* $B_r(x) = \{y|\,|x - y| < r\}$, *this characterization of compact is equivalent to being closed and bounded by the Heine-Borel theorem.*

We will show in this chapter that there is essentially a one-to-one correspondence between Borel measures and general interval lengths defined by increasing, right continuous functions. For example, it will follow that for every **distribution function** $F(y)$ of Section 3.6, there is an essentially unique and naturally associated Borel measure, μ_F. By "essentially unique" it is meant that $F(y) + c$ will lead to the same Borel measure for any constant c.

Going the other way and starting with a Borel measure μ, an essentially unique distribution function $F_\mu(y)$ can be defined which is increasing and right continuous. When μ is a finite Borel measure, $F_\mu(y)$ can be defined to satisfy the constraint that $F_\mu(-\infty) = 0$ and $F_\mu(\infty) < \infty$.

We address the second results first.

5.1 Functions Induced by Measures

Given a Borel measure μ on $\mathbb{R}$, we begin by defining a **function induced by** μ, also called the **distribution function associated with** μ. That such functions have the properties of the distribution functions of Section 3.6 is yet to be proved. See Proposition 5.7.

Definition 5.4 (Distribution function of Borel μ**)** *Let* μ *be a Borel measure on* $\mathbb{R}$. *A distribution function associated with* μ, *denoted* $F_\mu(y)$, *is defined by:*

$$F_\mu(y) = \begin{cases} -\mu((y, 0]), & y < 0, \\ 0, & y = 0, \\ \mu((0, y]), & y > 0. \end{cases} \tag{5.1}$$

Remark 5.5

1. *Note that* $F_\mu(y)$ *is well defined for a Borel measure because* $(0, y], (y, 0] \in \mathcal{B}(\mathbb{R})$ *for all y. Further, because Borel measures are finite on bounded sets,* $F_\mu(y)$ *is real-valued (i.e., in contrast to extended real-valued).*

2. *This is "a" distribution function and not "the" distribution function associated with μ because as was noted above, all of the properties of $F_\mu(y)$ of Proposition 5.7, and in particular (5.2) below, remain true with $F_\mu(y) + c$ for all real c.*

3. *If μ is a finite Borel measure, meaning $\mu(\mathbb{R}) < \infty$, then $F_\mu(y)$ is closely related to the distribution functions $F(y)$ introduced in Section 3.6. Define Borel measurable:*

$$f : (\mathbb{R}, \mathcal{B}(\mathbb{R}), \mu) \to (\mathbb{R}, \mathcal{B}(\mathbb{R}), m),$$

by $f(x) = x$. Then since $f^{-1}((-\infty, y]) = (-\infty, y]$, the associated distribution function of (3.21) $F(y) \equiv \mu(-\infty, y] < \infty$. It then follows that:

$$F_\mu(y) = F(y) - F(0).$$

Example 5.6 (Lebesgue measure) *If $\mu = m$, Lebesgue measure, then $F_m(y) = y$ the identity function.*

Proposition 5.7 (Properties of $F_\mu(y)$) *Let μ be a **Borel measure** and $F_\mu(y)$ defined in (5.1). Then:*

1. *$F_\mu(y)$ is increasing and Borel measurable, and hence Lebesgue measurable.*

2. *$F_\mu(y)$ is continuous on the right and has left limits.*

3. *$F_\mu(y)$ has at most countably many discontinuities.*

4. *$F_\mu(y)$ is continuous at b if and only if $\mu(\{b\}) = 0$.*

5. *Under μ, the measure of a right semi-closed interval $(a, b]$ is given by:*

$$\mu((a, b]) = F_\mu(b) - F_\mu(a). \tag{5.2}$$

*If μ is a **finite Borel measure**, meaning $\mu(\mathbb{R}) < \infty$, then a distribution function associated with μ can also be defined by:*

$$\bar{F}_\mu(y) = \mu((-\infty, y]), \tag{5.3}$$

and properties 1–5 remain true. Also in this case:

$$\lim_{y \to -\infty} \bar{F}_\mu(y) = 0; \quad \lim_{y \to \infty} \bar{F}_\mu(y) = \mu(\mathbb{R}).$$

Proof. *When μ is a finite Borel measure, then by part 3 of Remark 5.5 many of these conclusions follow from Proposition 3.60 since $F_\mu(y) = F(y) - F(0)$. But to accommodate the general case we provide direct proofs.*

1. That $F_\mu(y)$ is increasing follows from monotonicity of μ and the observation that if $0 < y' < y$ then $(0, y'] \subset (0, y]$ and so $F_\mu(y') \leq F_\mu(y)$. The other case of $y' < y < 0$ is identical, while $y' < 0 < y$ follows by definition since $F_\mu(y') \leq 0 \leq F_\mu(y)$. That an increasing function F is Borel measurable follows as in the proof of Proposition 3.60 that for any y, $F^{-1}((-\infty, y])$ equals $(-\infty, x)$ or $(-\infty, x]$ for some x.

2. *For right continuity, consider* $\lim_{z \to y+} F_\mu(z)$. *Because* $F_\mu(y)$ *is increasing from part 1, this limit can be restricted to a rational sequence of z-values, say* $\{z_n\}$. *If* $y > 0$, $\mu((0, z_1]) < \infty$ *by definition and by monotonicity of a Borel measure, since* $(0, z_1] \subset [0, z_1]$, *which is compact. Also,* $(0, y] = \bigcap_n (0, z_n]$, *so by continuity from above in (2.26):*

$$\mu((0, y]) = \lim_{z_n \to y+} \mu((0, z_n]),$$

proving right continuity. The case $y < 0$ *is handled analogously, while for* $y = 0$ *the result follows from continuity from above and* $\bigcap_n (0, z_n] = \emptyset$.

To demonstrate the existence of left limits for $y > 0$, *first note that* $(0, y) = \bigcup_n (0, z_n]$ *where* $\{z_n\}$ *is a rational sequence with* $z_n \to y-$, *meaning* $z_n < y$ *and* $z_n \to y$. *Now because*

$$\mu((0, z_n]) \le \mu((0, y)) \le \mu((0, y]),$$

it follows that $\{\mu((0, z_n])\}$ *is an increasing, bounded sequence. This sequence then has a limit point, which implies the existence of left limits. The cases* $y = 0$ *and* $y < 0$ *are handled analogously.*

3. *If* $F_\mu(y)$ *is discontinuous at given* y, *then this implies by items 1 and 2 that:*

$$\lim_{z_n \to y-} F_\mu(z_n) \equiv F_\mu(y^-) < F_\mu(y).$$

The interval $[F_\mu(y^-), F_\mu(y)]$ *then contains a rational number, and any two such intervals contain different rationals since* F *is increasing. Hence the number of such discontinuities is at most countable, proving item 3.*

4. *Since any* $\{b\} = \bigcap_n (b - 1/n, b]$, *continuity from above in (2.26) applies because* $\mu((b - 1, b]) < \infty$ *as above, and thus:*

$$\mu(\{b\}) = \lim_{n \to \infty} \mu((b - 1/n, b])$$
$$= \lim_{n \to \infty} \left[F_\mu(b) - F_\mu(b - 1/n) \right].$$

Thus $\mu(\{b\}) = 0$ *if and only if* $F_\mu(y)$ *is continuous from the left at* b, *and since* $F_\mu(y)$ *is always continuous from the right, the conclusion of item 4 follows.*

5. *For (5.2) assume* $b > a > 0$. *Then since* $(0, a] \bigcup (a, b] = (0, b]$ *the result follows by finite additivity. For* $a < b < 0$ *and* $a < 0 < b$ *the approach is analogous. The other cases of* $a = 0$ *or* $b = 0$ *follow by definition.*

Finally, the addition of any constant c *to the definition of* $F_\mu(y)$ *in (5.1) has no effect on the validity of properties* $1 - 5$. *If* μ *is a finite Borel measure and thus* $\mu(\mathbb{R}) < \infty$, *then also* $\mu((-\infty, 0]) < \infty$ *by monotonicity. Now let* $c = \mu((-\infty, 0])$, *and note that by finite additivity:*

$$F_\mu(y) + c = \begin{cases} -\mu((y, 0]) + \mu((-\infty, 0]), & y < 0, \\ +\mu((-\infty, 0]), & y = 0, \\ \mu((0, y]) + \mu((-\infty, 0]), & y > 0. \end{cases}$$
$$= \mu((-\infty, y]),$$

which is $\bar{F}_\mu(y)$ of (5.3).

Then $\lim_{y \to -\infty} F_\mu(y) = 0$ follows from the finiteness of μ and continuity from above since $\bigcap_n (-\infty, -n] = \emptyset$, while $\lim_{y \to \infty} \bar{F}_\mu(y) = \mu(\mathbb{R})$ by continuity from below since $\bigcup_n (-\infty, n] = \mathbb{R}$. ∎

5.2 Measures from Distribution Functions

The prior section's results state that every Borel measure gives rise to a Lebesgue measurable distribution function which is uniquely defined up to an additive constant. This function is increasing, continuous from the right and with left limits, and has at most countably many discontinuities. Moreover, the length of every interval, $\mu((a, b])$, is given by (5.2) by:

$$\mu((a, b]) = F_\mu(b) - F_\mu(a).$$

Consequently, if we seek to generate a Borel measure starting with a function F, it is apparent that the lengths of right semi-closed intervals $(a, b]$ should be defined by this formula. It is also apparent that F should possess the properties we derived for general F_μ.

The approach taken in this section will largely follow, yet also generalize, the approach taken for Lebesgue measure:

1. Identify an increasing, right continuous function $F(x)$, which is then continuous on $\mathbb{R}$ expcept for at most countably many discontinuities.

   For this purpose, Proposition 3.60 states that we can choose the distribution function associated with any real-valued measurable function $f(x)$ defined on any measure space $(X, \sigma(X), \mu)$. In other words, for any such $f(x)$ we could simply define $F(x)$ as in Definition 3.56.

   Indeed, examples from probability theory are all within this class, where $(X, \sigma(X), \mu)$ is a **probability space**, meaning $\mu(X) = 1$, and the measurable function $f(x)$ is now called a **random variable**. We are not required to define $F(x)$ in this way, but it is good to keep in mind that this provides infinitely many choices for $F(x)$ just within this class.

2. Given such $F(x)$, define the F-**length** of a right semi-closed interval consistently with (5.2):

$$|(a, b]|_F = F(b) - F(a). \tag{5.4}$$

   This F-length is finite for bounded intervals since $F(x)$ is increasing and right continuous, but can be infinite for unbounded intervals when $F(x)$ is an unbounded function.

3. Extend $|(a, b]|_F$ from the collection of right semi-closed intervals to a measure $\mu_{\mathcal{A}}$ on the algebra $\mathcal{A}$, generated by such intervals, recalling Example 2.4.

4. Define the $\mu_{\mathcal{A}}^*$-outer measure on the power sigma algebra $\sigma(P(\mathbb{R}))$ as in (2.4), only using $\mu_{\mathcal{A}}$ and sets $A \in \mathcal{A}$ instead of open intervals and interval length.

5. Prove that if we restrict the collection of sets as in (2.14), that $\mu_{\mathcal{A}}^*$-outer measure is a measure on a complete sigma algebra $\mathcal{M}_{\mu_F}$, which will be denoted μ_F, and that this sigma algebra contains the Borel sigma algebra, $\mathcal{B}(\mathbb{R})$.

6. Conclusion: μ_F is then a Borel measure.

In step 1 we required $F(x)$ to be increasing and right continuous, and thus continuous except at countably many points. The following proposition expands the possibilities to any increasing function, once suitably redefined on a set of measure 0.

Proposition 5.8 (Modify increasing $F(x)$ to be a distribution) *If $F(x)$ is an increasing real-valued function, then it has at most countably many discontinuities, $\{x_j\}_{j=1}^{\infty}$, and can be made right continuous with left limits by redefining $F(x)$ on this set of measure 0:*

$$F(x_j) \equiv \lim_{x \to x_j+} F(x).$$

Proof. *We first show that every such increasing function has left and right limits at every point, recalling Definition 3.58. For right limits at some y say, let $\epsilon_n \to 0$ and define $a_n = \inf_{I_n} F(x)$ where $I_n = (y, y + \epsilon_n)$. Since $F(x)$ is increasing and real-valued, $\{a_n\}$ is a decreasing sequence that is bounded from below by $F(y)$. Hence $a \equiv \lim a_n$ exists. By construction $a = \lim_{x \to y_j+} F(x)$. The same argument applies for left limits with details left as an exercise.*

Given the everywhere existence of left and right limits, a point y is a disconti-nuity point if:

$$\lim_{x \to y+} F(x) \neq \lim_{x \to y-} F(x).$$

Since $F(x)$ is increasing, this obtains that $F(y+) > F(y-)$, and thus there is a rational in the interval $[F(y-), F(y+)]$. As $F(x)$ is increasing, there is a different rational number in each such interval, and hence there can be at most countably many discontinuities.

At any such discontinuity, we can redefine $F(x_j)$ as above and right continuity follows. ∎

5.2.1 F-Length to a Measure on an Algebra

We begin with a proof that F-length as defined on right semi-closed intervals in (5.4) can be extended to a measure on an algebra. Specifically, this algebra will be $\mathcal{A}$ in (2.1) of Example 2.4, and defined as the collection of all finite unions of right semi-closed intervals. As noted there, such intervals can be assumed to be disjoint. This algebra is defined to also include unbounded sets of the form $(-\infty, b]$, (a, ∞) and $(-\infty, \infty)$.

As a first step, we must define what is meant by a measure on an algebra. Compare Definition 2.23.

Definition 5.9 (Measure on an algebra) *A **measure on an algebra**, sometimes called a **pre-measure on an algebra**, is a nonnegative, extended real valued set function μ defined on the sets of an algebra $\mathcal{A}$ such that:*

1. $\mu(\emptyset) = 0$,

2. ***Countable Additivity:*** *If $\{A_j\}$ is a countable collection of **pairwise disjoint** sets in $\mathcal{A}$ with $\bigcup_j A_j \in \mathcal{A}$, then:*

$$\mu\left(\bigcup_j A_j\right) = \sum_j \mu(A_j). \tag{5.5}$$

Remark 5.10 (Compare Definition 2.23) *This definition of measure is consistent with Definition 2.23 of measure on a sigma algebra except that here we cannot assume that $\bigcup_j A_j \in \mathcal{A}$ just because $\{A_j\} \subset \mathcal{A}$.*

*This is perhaps the motivation of some authors to call this a **pre-measure**, to highlight that this set function only satisfies countable additivity in certain cases. As a counterpoint, there is rarely any confusion since if $\bigcup_j A_j \notin \mathcal{A}$, then $\mu(\bigcup_j A_j)$ need not even be defined and so there is no expectation as to the equality in (5.5).*

*Note that if this collection is finite, the union is always a member of $\mathcal{A}$ because algebras are closed under finite unions. Hence, (5.5) demands **finite additivity** on $\mathcal{A}$ due to property 1, since a finite union can be represented as a countable union with empty sets.*

We will now show that F-length can be extended to a measure on the algebra $\mathcal{A}$, and do this in (5.6). An important and necessary step in the development is to show that this extension is well-defined on $\mathcal{A}$. Though this seems obvious, it requires an investigation in Proposition 5.11 since elements of this algebra can be represented in infinitely many ways.

Proposition 5.11 (Extend F-length to $\mathcal{A}$) *Let $F(x)$ be an increasing, right continuous function and define the F-length of a right semi-closed interval by (5.4). Then this definition extends to a well-defined, nonnegative, extended real-valued set function $\mu_{\mathcal{A}}$ on the algebra $\mathcal{A}$ of all finite unions of right semi-closed intervals.*

Proof. *We first extend the definition of F-length from an interval in (5.4) to a set of $\mathcal{A}$. Because any set of $\mathcal{A}$ has infinitely many such representations, we must then demonstrate that the value obtained by the proposed extension does not depend on which representation is used.*

To this end, let $A = \bigcup_{j=1}^n I_j \in \mathcal{A}$ for disjoint intervals $I_j = (a_j, b_j]$. That elements of $\mathcal{A}$ can always be so represented was noted in Example 2.4, although such representations are not unique. Define

$$\mu_{\mathcal{A}}(A) \equiv \sum_{j=1}^n \left[F(b_j) - F(a_j)\right], \tag{5.6}$$

and so $\mu_{\mathcal{A}}(A)$ is defined as a summation of the F-lengths of the disjoint intervals that compose A.

As disjoint intervals, all but at most two of the I_j are bounded, and there is the possibility of one interval of each type: (a_j, ∞) and $(-\infty, b_j]$. Consequently as noted in comment 2 above on *the definition of* $|(a,b]|_F$ *, when A contains an unbounded interval this summation will have one or both terms of the form:*

$$F(\infty) - F(a_j), \ F(b_j) - F(-\infty).$$

In this case, the measure of A will be infinite when F(x) is unbounded.

To demonstrate that this definition is independent of the collection of disjoint intervals used to define A, assume also that $A = \bigcup_{k=1}^{m} J_k$, where $J_k = (c_k, d_k]$ are also disjoint. For notational ease assume that both representations of A as unions have been parametrized in increasing order of the interval endpoints. Define $K_{jk} = I_j \cap J_k$ for $1 \leq j \leq n, 1 \leq k \leq m$, and note that $\{K_{jk}\} \equiv \{(e_{jk}, f_{jk}]\}$ must also be disjoint, though potentially many such intervals will be empty. Also, $\bigcup_j K_{jk} = J_k$, $\bigcup_k K_{jk} = I_j$ and $\bigcup_{jk} K_{jk} = A$.

Since all intervals are disjoint, this implies that:

$$F(b_j) - F(a_j) = \sum_{k=1}^{m} \left[F(f_{jk}) - F(e_{jk})\right],$$
$$F(d_k) - F(c_k) = \sum_{j=1}^{n} \left[F(f_{jk}) - F(e_{jk})\right].$$

Thus ignoring unbounded intervals for the moment:

$$\mu_{\mathcal{A}}(A) = \sum_{j=1}^{n} \left[F(b_j) - F(a_j)\right]$$
$$= \sum_{j=1}^{n} \sum_{k=1}^{m} \left[F(f_{jk}) - F(e_{jk})\right]$$
$$= \sum_{k=1}^{m} \sum_{j=1}^{n} \left[F(f_{jk}) - F(e_{jk})\right]$$
$$= \sum_{k=1}^{m} \left[F(d_k) - F(c_k)\right].$$

In these manipulations, if either collection which defines A contains an unbounded interval, then both must contain such an interval and hence the derivation above may involve either or both $F(-\infty)$ and $F(\infty)$ terms. When F(x) is bounded, these are simply real values and the calculations are valid as presented.

When F(x) is unbounded, one or both of these values may be infinite and the above derivation which reverses summations needs further investigation. However, with either partition of A, we would conclude that $\mu_{\mathcal{A}}(A) = \infty$, so μ_A is again well defined on $\mathcal{A}$.

That $\mu_{\mathcal{A}}$ is nonnegative follows from F(x) increasing, and that it will in general be extended real-valued follows from the observation that at least some unbounded intervals will have infinite measure when F(x) is an unbounded function. ∎

Corollary 5.12 (Monotonicity of $\mu_{\mathcal{A}}$ on $\mathcal{A}$) *The set function $\mu_{\mathcal{A}}$ defined on $\mathcal{A}$ is monotonic, meaning if $A', A \in \mathcal{A}$ and $A' \subset A$, then $\mu_{\mathcal{A}}(A') \leq \mu_{\mathcal{A}}(A)$.*

Proof. *If A' and A are expressed in terms of disjoint right semi-closed intervals, the above proposition assures that $\mu_{\mathcal{A}}(A')$ and $\mu_{\mathcal{A}}(A)$ are each well defined as a summation of the F-lengths of their respective disjoint subintervals, given any such subinterval decompositions.*

Adapting the notation of the prior proof, let $A' = \bigcup_{j=1}^{n} I_j = \bigcup_{j=1}^{n} (a_j, b_j]$, $A = \bigcup_{k=1}^{m} J_k = \bigcup_{k=1}^{m} (c_k, d_k]$, and again define $K_{jk} = I_j \cap J_k = (e_{jk}, f_{jk}]$. That $A' \subset A$ implies that $\bigcup_j K_{jk} \subset J_k$, *and thus $J_k - \bigcup_j K_{jk} = \bigcup_l K'_{jl}$ where $\{K'_{jl}\} = \{(e'_{jl}, f'_{jl})\}$ is a disjoint collection of m_j right semi-closed intervals. In addition, $J_k = \bigcup_j K_{jk} \bigcup \bigcup_l K'_{jl}$ is a union of disjoint intervals.*

Thus:

$$\mu_{\mathcal{A}}(A') = \sum_{j=1}^{n} \left[F(b_j) - F(a_j) \right]$$
$$= \sum_{j=1}^{n} \sum_{k=1}^{m} \left[F(f_{jk}) - F(e_{jk}) \right]$$
$$= \sum_{k=1}^{m} \sum_{j=1}^{n} \left[F(f_{jk}) - F(e_{jk}) \right]$$
$$\leq \sum_{k=1}^{m} \left(\sum_{k=1}^{m} \left[F(f_{jk}) - F(e_{jk}) \right] + \sum_{l=1}^{m_j} \left[F(f'_{jl}) - F(e'_{jl}) \right] \right)$$
$$= \sum_{k=1}^{m} \left[F(d_k) - F(c_k) \right]$$
$$= \mu_{\mathcal{A}}(A). \qquad \blacksquare$$

We next prove that $\mu_{\mathcal{A}}$ as defined in (5.6) is in fact a measure on this algebra, by Definition 5.9.

Proposition 5.13 ($\mu_{\mathcal{A}}$ is a measure on $\mathcal{A}$) *Let $\mu_{\mathcal{A}}$ be the set function defined in (5.6) on the algebra $\mathcal{A}$ of finite unions of right semi-closed intervals. Then $\mu_{\mathcal{A}}$ is a measure on $\mathcal{A}$.*

Proof. *That $\mu_{\mathcal{A}}$ is finitely additive on $\mathcal{A}$ follows Proposition 5.11. Let $\{A_j\}_{j=1}^{n}$ be a finite collection of disjoint sets from $\mathcal{A}$, each expressed as a finite disjoint union of right semi-closed subintervals as in Example 2.4. Then, none of these subintervals intersect, and hence $A = \bigcup_j A_j$ equals a union of finitely many disjoint right semi-closed intervals.*

If $\mu_{\mathcal{A}}(A_j) = \infty$ for any j, then $\mu_{\mathcal{A}}(A) = \infty$ by monotonicity, proving finite additivity in this case, so assume all $\mu_{\mathcal{A}}(A_j) < \infty$. By the proof above, $\mu_{\mathcal{A}}(A)$ is well-defined and given by (5.6) applied to these intervals. But then this finite sum can be rearranged into the summations that make up $\mu_{\mathcal{A}}(A_j)$ for all j, and finite additivity follows.

Next, assume $\{A_j\}_{j=1}^{\infty} \subset \mathcal{A}$ is a disjoint countable collection of sets and that $A = \bigcup_j A_j \in \mathcal{A}$. By Example 2.4, each of the A_j sets equals a finite union of disjoint right semi-closed intervals, and by (5.6) the measure of such A_j is the sum of the measures of the associated intervals. Similarly, $A \in \mathcal{A}$ is a finite union of disjoint right semi-closed intervals and its measure the sum of the measures of such intervals. Consequently, each interval in A must equal a finite or countable union of disjoint intervals which define the collection $\{A_j\}_{j=1}^{\infty}$.

Hence, countable additivity:

$$\mu_{\mathcal{A}} \left(\bigcup_j A_j \right) = \sum_j \mu_{\mathcal{A}}(A_j),$$

will follow if this identity is valid for disjoint intervals:

$$\mu_{\mathcal{A}}(I) = \sum_{j=1}^{\infty} \mu_{\mathcal{A}}(I_j).$$

Here $I = \bigcup_{j=1}^{\infty} I_j$ is a countable union of disjoint semi-closed intervals which is also a semi-closed interval.

Now if $\sum_{j=1}^{\infty} \mu_{\mathcal{A}}(I_j) = \infty$, *then* $\mu_{\mathcal{A}}(I) = \infty$ *by monotonicity and finite additivity:*

$$\mu_{\mathcal{A}}(I) \geq \mu_{\mathcal{A}}\left(\bigcup_{j=1}^{n} I_j\right) = \sum_{j=1}^{n} \mu_{\mathcal{A}}(I_j).$$

So assume $\sum_{j=1}^{\infty} \mu_{\mathcal{A}}(I_j) < \infty$. *Then denoting* $I = (a, b]$ *and* $I_j = (a_j, b_j]$, *note that some* $b_k = b$ *since if* $b_j < b$ *for all* j, *then* $\bigcup_{j=1}^{\infty} I_j \subset (a, b)$ *a contradiction. We label the associated interval* $I_1' = (a_1', b]$, *and choose* I_2' *to be the interval* $(a_2', a_1']$. *Again, such an interval exists since if all other* $b_j < a_1'$ *then the union would again be open and contained in* (a, a_1'), *contradicting that* I *is an interval.*

Continuing in this way, we construct a reordering of $\{I_j\}$ *to* $\{I_j'\} = \{(a_j', b_j']\}$ *where* $b_1' = b$, *and* $b_{j+1}' = a_j'$. *In addition,* $\bigcup_{j=1}^{n} I_j' = (a_n', b]$ *for every* n, *and hence as* $n \to \infty$ *it follows that* $a_n' \to a$. *As* $\sum_{j=1}^{\infty} \mu_{\mathcal{A}}(I_j) < \infty$, *this series of positive values can be reordered and thus* $\sum_{j=1}^{\infty} \mu_{\mathcal{A}}(I_j) = \sum_{j=1}^{\infty} \mu_{\mathcal{A}}(I_j')$. *Then by finite additivity:*

$$\sum_{j=1}^{n} \mu_{\mathcal{A}}\left(I_j'\right) = \mu_{\mathcal{A}}\left(\bigcup_{j=1}^{n} I_j'\right) = F(b) - F(a_n').$$

Letting $n \to \infty$ *obtains that* $\sum_{j=1}^{n} \mu_{\mathcal{A}}(I_j') \to \sum_{j=1}^{\infty} \mu_{\mathcal{A}}(I_j)$, *and* $F(a_n') \to F(a)$ *by right continuity of* F. *The result follows since* $F(b) - F(a) = \mu_{\mathcal{A}}(I)$. ∎

Corollary 5.14 (Countable subadditivity of $\mu_{\mathcal{A}}$**)** *The measure* $\mu_{\mathcal{A}}$ *is **countably subadditive** on* $\mathcal{A}$. *That is, if* $\{A_j\}_{j=1}^{\infty} \subset \mathcal{A}$ *is a countable collection of sets, and* $A \subset \bigcup_{j=1}^{\infty} A_j$ *with* $A \in \mathcal{A}$, *then:*

$$\mu_{\mathcal{A}}(A) \leq \sum_{j} \mu_{\mathcal{A}}(A_j). \tag{5.7}$$

Proof. *Define:*

$$A_j' = A \bigcap A_j \bigcap \tilde{A}_{j-1} \bigcap \cdots \bigcap \tilde{A}_1.$$

Then $\{A_j'\}_{j=1}^{\infty}$ *are disjoint,* $A_j' \in \mathcal{A}$ *and* $A_j' \subset A_j$ *for all* j, *and* $A = \bigcup_{j=1}^{\infty} A_j'$.

Hence by the prior proposition and monotonicity of $\mu_{\mathcal{A}}$:

$$\mu_{\mathcal{A}}(A) = \sum_{j=1}^{\infty} \mu_{\mathcal{A}}\left(A_j'\right) \leq \sum_{j=1}^{\infty} \mu_{\mathcal{A}}(A_j). \qquad \blacksquare$$

Remark 5.15 *Note that if* $\bigcup_{j=1}^{\infty} A_j \in \mathcal{A}$, *then* (5.7) *would have been immediate from monotonicity of the prior corollary. Here we did not make this assumption and thus a proof was required.*

Also note that this proof made no use of the definition of $\mu_{\mathcal{A}}$ *and thus proved a very general statement:* **Every measure on an algebra is countably subadditive.**

5.2.2 To a Borel Measure

The final step is to extend the definition of $\mu_{\mathcal{A}}$ from the algebra $\mathcal{A}$ to a sigma algebra which contains $\mathcal{A}$. It is an exercise to check that any sigma algebra that contains $\mathcal{A}$ also contains the open intervals and thus too the Borel sigma algebra, $\mathcal{B}(\mathbb{R})$. Hence, this measure extension will be to a Borel measure.

We begin with a definition of $\mu_{\mathcal{A}}^*$-outer measure and $\mu_{\mathcal{A}}^*$-measurable sets. The notation and underlying approach will be familiar from Section 2.4 and was originally introduced by **Constantin Carathéodory** (1873–1950). It is common to use the terminology "**Carathéodory measurable**" in the general, non-Lebesgue applications as below.

Definition 5.16 ($\mu_{\mathcal{A}}^*$ outer measure) *Let $F(x)$ be an increasing, right continuous function, $\mathcal{A}$ the algebra of sets of Example 2.4, and $\mu_{\mathcal{A}}$ the measure on $\mathcal{A}$ induced by $F(x)$. Then for any set $A \subset \mathbb{R}$, the **outer measure of** A, or the **outer measure of** A **induced by** $\mu_{\mathcal{A}}$, denoted $\mu_{\mathcal{A}}^*(A)$, is defined by:*

$$\mu_{\mathcal{A}}^*(A) = \inf \left\{ \sum_n \mu_{\mathcal{A}}(A_n) \mid A \subset \bigcup_n A_n \right\}, \tag{5.8}$$

where $A_n \in \mathcal{A}$ and $\mu_{\mathcal{A}}(A_n)$ is defined in (5.6).

Remark 5.17 (Alternative formulation) *The outer measure of A induced by $\mu_{\mathcal{A}}$ can also be defined directly in terms of right semi-closed intervals rather than require each $A_n \in \mathcal{A}$. This is because each nonempty element of the algebra $\mathcal{A}$ is a finite union of right semi-closed intervals $A_n' = (a_n, b_n]$. These intervals can be taken to be disjoint as noted in Example 2.4, and hence by finite additivity this definition is equivalent to:*

$$\mu_{\mathcal{A}}^*(A) = \inf \left\{ \sum_n \mu_{\mathcal{A}}(A_n') \mid A \subset \bigcup_n A_n' \right\}. \tag{5.9}$$

In this formulation, the analogy with (2.4) for Lebesgue outer measure is more apparent since for such intervals:

$$\mu_{\mathcal{A}}(A_n') = F(b_n) - F(a_n).$$

Since $\{A_n'\}$ can be taken to be disjoint, so too the collection $\{A_n\} \subset \mathcal{A}$ in (5.8) can be taken to be disjoint.

Definition 5.18 (Carathéodory measurability w.r.t. $\mu_{\mathcal{A}}^*$) *Let $F(x)$ be an increasing, right continuous function and $\mu_{\mathcal{A}}^*$ the outer measure induced by $\mu_{\mathcal{A}}$ as in (5.8). A set $A \subset \mathbb{R}$ is **Carathéodory measurable with respect to** $\mu_{\mathcal{A}}^*$, or simply $\mu_{\mathcal{A}}^*$-measurable, if for any set $E \subset \mathbb{R}$:*

$$\mu_{\mathcal{A}}^*(E) = \mu_{\mathcal{A}}^*(A \cap E) + \mu_{\mathcal{A}}^*(\widetilde{A} \cap E). \tag{5.10}$$

The collection of $\mu_{\mathcal{A}}^*$-measurable sets is denoted $\mathcal{M}_{\mu_F}(\mathbb{R})$.

Remark 5.19 *Below we will show that* $\mu^*_{\mathcal{A}}$*-outer measure is countably subadditive, and hence it will always be the case that* $\mu^*_{\mathcal{A}}(E) \leq \mu^*_{\mathcal{A}}(A \bigcap E) + \mu^*_{\mathcal{A}}(\tilde{A} \bigcap E)$*. It therefore follows that* $A \subset \mathbb{R}$ *is* $\mu^*_{\mathcal{A}}$*-measurable if for any set* $E \subset \mathbb{R}$*:*

$$\mu^*_{\mathcal{A}}(E) \geq \mu^*_{\mathcal{A}}(A \bigcap E) + \mu^*_{\mathcal{A}}(\tilde{A} \bigcap E). \tag{5.11}$$

Because of this, many of the proofs that a given set is so measurable focus on demonstrating this single inequality rather than demonstrating equality as in (5.10).

Our goal is to show that $\mathcal{M}_{\mu_F}(\mathbb{R})$ is a complete sigma algebra which contains the Borel sigma algebra $\mathcal{B}(\mathbb{R})$, and that $\mu^*_{\mathcal{A}}$ restricted to this sigma algebra yields a measure. Here the notion of complete is the same as that of Definition 2.48. The restriction of $\mu^*_{\mathcal{A}}$ will be denoted μ_F to highlight the role of $F(x)$, and the associated measure space denoted $(\mathbb{R}, \mathcal{M}_{\mu_F}(\mathbb{R}), \mu_F)$. Thus for $A \in \mathcal{M}_{\mu_F}(\mathbb{R})$:

$$\mu_F(A) \equiv \mu^*_{\mathcal{A}}(A).$$

We first document a few of the properties of $\mu^*_{\mathcal{A}}$-outer measure.

Proposition 5.20 (Properties of $\mu^*_{\mathcal{A}}$**-outer measure)** *Let* $F(x)$ *be an increasing, right continuous function, and* $\mu^*_{\mathcal{A}}$ *the outer measure induced by* $\mu_{\mathcal{A}}$ *as in (5.8). Then:*

1. $\mu^*_{\mathcal{A}}(\emptyset) = 0$.
2. **Countable Subadditivity:** *Given* $A, \{A_n\}_{n=1}^{\infty}$, *if* $A \subset \bigcup_{n=1}^{\infty} A_n$, *then*

$$\mu^*_{\mathcal{A}}(A) \leq \sum_{n=1}^{\infty} \mu^*_{\mathcal{A}}(A_n). \tag{5.12}$$

3. **Monotonicity:** *If* $A \subset B$,

$$\mu^*_{\mathcal{A}}(A) \leq \mu^*_{\mathcal{A}}(B). \tag{5.13}$$

4. *If* $A \in \mathcal{A}$, *the algebra of finite unions of right semi-closed intervals defined in (2.1), then:*

$$\mu^*_{\mathcal{A}}(A) = \mu_{\mathcal{A}}(A), \tag{5.14}$$

where $\mu_{\mathcal{A}}$ *is defined as in Proposition 5.11.*

Proof. *Taking these properties in turn:*

1. *Applying (5.8) to sets* $A \equiv (a, a + \epsilon] \in \mathcal{A}$, *then since* $\emptyset \subset A_n$ *and* $\mu^*_{\mathcal{A}}$ *is defined as an infimum:*

$$\mu^*_{\mathcal{A}}(\emptyset) \leq \mu_{\mathcal{A}}(A) \equiv F(a + \epsilon) - F(a).$$

Since every such a is a point of right continuity of $F(x)$*, the result follows since the infimum on the right is 0.*

2. *If $A \subset \bigcup_n A_n$ and $\mu^*_{\mathcal{A}}(A_n) = \infty$ for any n, then (5.12) is satisfied. So assume $\mu^*_{\mathcal{A}}(A_n) < \infty$ for all n. Given $\epsilon > 0$, for each n let $\{A_{n_m}\}_{m=1}^\infty \subset \mathcal{A}$ with $A_n \subset \bigcup_m A_{n_m}$ and:*

$$\sum_m \mu_{\mathcal{A}}(A_{n_m}) < \mu^*_{\mathcal{A}}(A_n) + \epsilon/2^n.$$

This is possible because $\mu^*_{\mathcal{A}}(A_n)$ is equal to the infimum of all such summations by definition. Then since $A \subset \bigcup_{n,m} A_{n_m}$ and $\mu^*_{\mathcal{A}}(A)$ is defined as an infimum:

$$\mu^*_{\mathcal{A}}(A) \leq \sum_{n,m} \mu_{\mathcal{A}}(A_{n_m}) < \sum_n \mu^*_{\mathcal{A}}(A_n) + \epsilon.$$

Thus (5.12) is proved since ϵ is arbitrary.

3. *Monotonicity follows from countable subadditivity and property 1, by choosing $A_1 = B$ and all other $A_n = \emptyset$.*

4. *By Corollary 5.14, $\mu_{\mathcal{A}}$ is countably subadditive on $\mathcal{A}$. Hence, if $A \in \mathcal{A}$ and $A \subset \bigcup_n A_n$ with $A_n \in \mathcal{A}$, then:*

$$\mu_{\mathcal{A}}(A) \leq \sum_n \mu_{\mathcal{A}}(A_n).$$

Since this is true for all such collections $\{A_n\}$:

$$\mu_{\mathcal{A}}(A) \leq \inf \sum \mu_{\mathcal{A}}(A_n) \equiv \mu^*_{\mathcal{A}}(A).$$

But for $A \in \mathcal{A}$ choose all $A_n = A$ in (5.8) and it follows that $\mu^*_{\mathcal{A}}(A) \leq \mu_{\mathcal{A}}(A)$, proving (5.14). ∎

Remark 5.21 (Compare with Lebesgue results) *In contrast with the development for Lebesgue outer measure in Chapter 2, we note a few important distinctions:*

1. *For Lebesgue outer measure, $m^*(I) = b - a$ for any interval $I = \{a, b\}$, where this notation implies that the interval can be open, closed or semi-open. In the $\mu^*_{\mathcal{A}}$ case, we have as a corollary of item 4 above that $\mu^*_{\mathcal{A}}((a, b]) = F(b) - F(a)$. Thus Lebesgue outer measure of $(a, b]$ equals $\mu^*_{\mathcal{A}}$-outer measure with $F(x) = x$. But in general, the $\mu^*_{\mathcal{A}}$-measures of the other typses of I-intervals are no longer equal to this value.*

   *In Corollary 5.31 it will be proved that the Borel development with $F(x) = x$ reproduces Lebesgue measure m on the sigma algebra of Borel sets $\mathcal{B}(\mathbb{R})$. This will also be seen in Section 6.2.5 in the general discussion of uniqueness of extensions. Then in Section 6.5 it will be seen that these measures also agree on the **complete sigma algebras**.*

2. *While in the Lebesgue case every **singleton set** $\{a\}$, and every countable union of singleton sets, have Lebesgue outer measure 0, this will not be the case in general for $\mu^*_{\mathcal{A}}$-outer measure. Indeed, using a similar approach as was used to show that $\mu^*_{\mathcal{A}}(\emptyset) = 0$, we see that $\mu^*_{\mathcal{A}}(\{a\}) = 0$ for any point a which is a point of continuity of $F(x)$, and the same is true for any countable union of such points.*

In detail, since $\{a\} \subset (a - \epsilon, a + \epsilon] \in \mathcal{A}$:

$$\mu^*_{\mathcal{A}}(\{a\}) \leq F(a + \epsilon) - F(a - \epsilon).$$

Consequently, since $F(x)$ *is monotonic and continuous from the right:*

$$\mu^*_{\mathcal{A}}(\{a\}) \leq F(a) - F(a^-),$$

where $F(a^-)$ *denotes the left limit at a.*

On the other hand, for any $A \in \mathcal{A}$ *with* $\{a\} \subset A$, *it follows that* $(a - \epsilon_1, a + \epsilon_2] \subset A$ *for some* $\epsilon_1 > 0$ *and* $\epsilon_2 \geq 0$, *and so* $\mu^*_{\mathcal{A}}(A) \geq F(a + \epsilon_2) - F(a - \epsilon_1)$ *by monotonicity. Taking an infimum over all such A, and recalling that F is increasing, obtains:*

$$\mu^*_{\mathcal{A}}(\{a\}) = \inf_{\{a\} \subset A} \mu^*_{\mathcal{A}}(A) \geq \inf_{\epsilon_j}[F(a + \epsilon_2) - F(a - \epsilon_1)] = F(a) - F(a^-).$$

In summary,

$$\mu^*_{\mathcal{A}}(\{a\}) = F(a) - F(a^-). \qquad (5.15)$$

As the increasing function F can have at most countable many discontinuities, any outer measure $\mu^*_{\mathcal{A}}$ *induced by F can have at most countably many singleton sets with nonzero outer measure. Because of this, it cannot be the case in general that an interval has the same* $\mu^*_{\mathcal{A}}$-*outer measure whether open, closed or semi-open.*

Exercise 5.22 ($\mu^*_{\mathcal{A}}$ **on intervals)** *Show that* $\mu^*_{\mathcal{A}}$-*outer measure satisfies the following properties for bounded intervals:*

- $\mu^*_{\mathcal{A}}([a, b]) = F(b) - F(a^-)$
- $\mu^*_{\mathcal{A}}((a, b)) = F(b^-) - F(a)$
- $\mu^*_{\mathcal{A}}([a, b)) = F(b^-) - F(a^-)$

We now turn to the final result in the development of a Borel measure.

Proposition 5.23 (The Borel measure μ_F **induced by** $F(x)$**)** *The collection of* $\mu^*_{\mathcal{A}}$-*measurable sets defined in (5.10), denoted* $\mathcal{M}_F(\mathbb{R})$, *has the following properties:*

1. $\mathcal{A} \subset \mathcal{M}_F(\mathbb{R})$, *where* $\mathcal{A}$ *denotes the algebra of Example 2.4.*
2. $\mathcal{M}_F(\mathbb{R})$ *is a sigma algebra, and contains every set* $A \subset \mathbb{R}$ *with* $\mu^*_{\mathcal{A}}(A) = 0$.
3. $\mathcal{M}_F(\mathbb{R})$ *contains the Borel sigma algebra,* $\mathcal{B}(\mathbb{R}) \subset \mathcal{M}_F(\mathbb{R})$.
4. *If* μ_F *denotes the restriction of* $\mu^*_{\mathcal{A}}$ *to* $\mathcal{M}_F(\mathbb{R})$, *then* μ_F *is a Borel measure and* $(\mathbb{R}, \mathcal{M}_F(\mathbb{R}), \mu_F)$ *is a complete measure space.*

Proof. *We prove each property in turn.*

1. To show that $A \in \mathcal{A}$ is measurable and hence an element of $\mathcal{M}_F(\mathbb{R})$, it is enough to prove (5.11) since Proposition 5.20 provides the subadditive result. Let $E \subset \mathbb{R}$ be arbitrary. Then by definition of infimum, for any $\epsilon > 0$ there is a collection: $\{A_n\} \subset \mathcal{A}$ so that $E \subset \bigcup_n A_n$ and

$$\sum_n \mu_{\mathcal{A}}(A_n) < \mu_{\mathcal{A}}^*(E) + \epsilon.$$

As $\mu_{\mathcal{A}}$ is a measure on $\mathcal{A}$ and thus finitely additive, it follows that for any n:

$$\mu_{\mathcal{A}}(A_n) = \mu_{\mathcal{A}}(A_n \bigcap A) + \mu_{\mathcal{A}}(A_n \bigcap \widetilde{A})$$
$$= \mu_{\mathcal{A}}^*(A_n \bigcap A) + \mu_{\mathcal{A}}^*(A_n \bigcap \widetilde{A}),$$

applying (5.14).
  Now $E \subset \bigcup_n A_n$ and so:

$$E \bigcap A \subset \bigcup_n (A_n \bigcap A),$$

and

$$E \bigcap \widetilde{A} \subset \bigcup_n (A_n \bigcap \widetilde{A}).$$

By countable subadditivity and monotonicity of $\mu_{\mathcal{A}}^*$ proved in Proposition 5.20:

$$\mu_{\mathcal{A}}^*(E) + \epsilon > \sum_n \mu_{\mathcal{A}}^*(A_n \bigcap A) + \sum_n \mu_{\mathcal{A}}^*(A_n \bigcap \widetilde{A})$$
$$\geq \mu_{\mathcal{A}}^*\left(\bigcup_n (A_n \bigcap A)\right) + \mu_{\mathcal{A}}^*\left(\bigcup_n (A_n \bigcap \widetilde{A})\right)$$
$$\geq \mu_{\mathcal{A}}^*(E \bigcap A) + \mu_{\mathcal{A}}^*(E \bigcap \widetilde{A}).$$

This obtains (5.11) since $\epsilon > 0$ is arbitrary.
  2. First, $\mathcal{M}_F(\mathbb{R})$ is closed under complementation by the symmetry in the definition, that $A \in \mathcal{M}_F(\mathbb{R})$ if and only if $\widetilde{A} \in \mathcal{M}_F(\mathbb{R})$. To demonstrate that $\mathcal{M}_F(\mathbb{R})$ is closed under finite unions, it is sufficient by an induction argument to show that if $A_1, A_2 \in \mathcal{M}_F(\mathbb{R})$, then $A_1 \bigcup A_2 \in \mathcal{M}_F(\mathbb{R})$. By (5.11) this requires proof that for any $E \subset \mathbb{R}$:

$$\mu_{\mathcal{A}}^*(E) \geq \mu_{\mathcal{A}}^*(E \bigcap (A_1 \bigcup A_2)) + \mu_{\mathcal{A}}^*(E \bigcap \widetilde{A}_1 \bigcap \widetilde{A}_2),$$

noting that $\widetilde{(A_1 \bigcup A_2)} = \widetilde{A}_1 \bigcap \widetilde{A}_2$ by De Morgan's laws. Now since A_1 is measurable:

$$\mu_{\mathcal{A}}^*(E) = \mu_{\mathcal{A}}^*(E \bigcap A_1) + \mu_{\mathcal{A}}^*(E \bigcap \widetilde{A}_1),$$

and applying (5.10) with $E' \equiv E \bigcap \widetilde{A}_1$ to measurable A_2:

$$\mu_{\mathcal{A}}^*(E \bigcap \widetilde{A}_1) = \mu_{\mathcal{A}}^*(E \bigcap \widetilde{A}_1 \bigcap A_2) + \mu_{\mathcal{A}}^*(E \bigcap \widetilde{A}_1 \bigcap \widetilde{A}_2).$$

Combining:

$$\mu_{\boldsymbol{\mathcal{A}}}^{*}(E) = \mu_{\boldsymbol{\mathcal{A}}}^{*}(E \bigcap A_1) + \mu_{\boldsymbol{\mathcal{A}}}^{*}(E \bigcap \tilde{A}_1 \bigcap A_2) + \mu_{\boldsymbol{\mathcal{A}}}^{*}(E \bigcap \tilde{A}_1 \bigcap \tilde{A}_2).$$

But since $E \bigcap (A_1 \bigcup A_2) = (E \bigcap A_1) \bigcup (E \bigcap \tilde{A}_1 \bigcap A_2)$, *finite subadditivity obtains:*

$$\mu_{\boldsymbol{\mathcal{A}}}^{*}(E \bigcap A_1) + \mu_{\boldsymbol{\mathcal{A}}}^{*}(E \bigcap \tilde{A}_1 \bigcap A_2) \geq \mu_{\boldsymbol{\mathcal{A}}}^{*}(E \bigcap (A_1 \bigcup A_2)).$$

Combining results proves $A_1 \bigcup A_2 \in \boldsymbol{\mathcal{M}}_F(\mathbb{R})$. *This then extends to all finite unions by induction.*

To demonstrate that $\boldsymbol{\mathcal{M}}_F(\mathbb{R})$ *is closed under countable unions, let* $\{A_n\} \subset \boldsymbol{\mathcal{M}}_F(\mathbb{R})$ *and we prove that* $A \equiv \bigcup_n A_n \in \boldsymbol{\mathcal{M}}_F(\mathbb{R})$. *By Proposition 2.20, the collection* $\{A_n\}$ *can be replaced by disjoint sets* $\{A_n'\}$ *with the same partial and total unions. By the construction there,* $\{A_n'\} \subset \boldsymbol{\mathcal{M}}_F(\mathbb{R})$ *since* $\boldsymbol{\mathcal{M}}_F(\mathbb{R})$ *is closed under complementation and finite unions. We now relabel these disjoint sets for simplicity as* $\{A_n\}$.

Letting $B_n = \bigcup_{j \leq n} A_j$, *then by finite additivity* $B_n \in \boldsymbol{\mathcal{M}}_F(\mathbb{R})$. *Thus for* $E \subset \mathbb{R}$, *since* $B_n \subset A$ *implies* $\tilde{A} \subset \tilde{B}_n$, *we have by monotonicity of* $\mu_{\boldsymbol{\mathcal{A}}}^{*}$:

$$\mu_{\boldsymbol{\mathcal{A}}}^{*}(E) = \mu_{\boldsymbol{\mathcal{A}}}^{*}(E \bigcap B_n) + \mu_{\boldsymbol{\mathcal{A}}}^{*}(E \bigcap \tilde{B}_n)$$

$$\geq \mu_{\boldsymbol{\mathcal{A}}}^{*}(E \bigcap B_n) + \mu_{\boldsymbol{\mathcal{A}}}^{*}(E \bigcap \tilde{A}).$$

Now $B_n \bigcap A_n = A_n$ *and* $B_n \bigcap \tilde{A}_n = B_{n-1}$, *so by the measurability of* A_n *and* $E' \equiv E \bigcap B_n$:

$$\mu_{\boldsymbol{\mathcal{A}}}^{*}(E \bigcap B_n) = \mu_{\boldsymbol{\mathcal{A}}}^{*}(E \bigcap B_n \bigcap A_n) + \mu_{\boldsymbol{\mathcal{A}}}^{*}(E \bigcap B_n \bigcap \tilde{A}_n)$$

$$= \mu_{\boldsymbol{\mathcal{A}}}^{*}(E \bigcap A_n) + \mu_{\boldsymbol{\mathcal{A}}}^{*}(E \bigcap B_{n-1}).$$

By induction,

$$\mu_{\boldsymbol{\mathcal{A}}}^{*}(E) \geq \sum_{j=1}^{n} \mu_{\boldsymbol{\mathcal{A}}}^{*}(E \bigcap A_j) + \mu_{\boldsymbol{\mathcal{A}}}^{*}(E \bigcap \tilde{A}),$$

and hence since $E \bigcap A = \bigcup (E \bigcap A_j)$, *subadditivity yields:*

$$\mu_{\boldsymbol{\mathcal{A}}}^{*}(E) \geq \sum_{j=1}^{\infty} \mu_{\boldsymbol{\mathcal{A}}}^{*}(E \bigcap A_j) + \mu_{\boldsymbol{\mathcal{A}}}^{*}(E \bigcap \tilde{A})$$

$$\geq \mu_{\boldsymbol{\mathcal{A}}}^{*}(E \bigcap A) + \mu_{\boldsymbol{\mathcal{A}}}^{*}(E \bigcap \tilde{A}).$$

Hence A is measurable and $\boldsymbol{\mathcal{M}}_F(\mathbb{R})$ *is a sigma algebra.*

Finally, that $\mu_{\boldsymbol{\mathcal{A}}}^{*}(A) = 0$ *implies* $A \in \boldsymbol{\mathcal{M}}_F(\mathbb{R})$ *is identical to the proof in the Lebesgue measure case of Proposition 2.35. If* $\mu_{\boldsymbol{\mathcal{A}}}^{*}(A) = 0$ *then* $\mu_{\boldsymbol{\mathcal{A}}}^{*}(E \bigcap A) = 0$ *for any* $E \subset \mathbb{R}$ *since* $E \bigcap A \subset A$. *Then also* $\mu_{\boldsymbol{\mathcal{A}}}^{*}(E \bigcap \tilde{A}) \leq \mu_{\boldsymbol{\mathcal{A}}}^{*}(E)$ *since* $E \bigcap \tilde{A} \subset E$. *Combining yields* (5.11) *and hence* $A \in \boldsymbol{\mathcal{M}}_F(\mathbb{R})$.

3. That $\mathcal{M}_F(\mathbb{R})$ contains the Borel sigma algebra, $\mathcal{B}(\mathbb{R}) \subset M_F(\mathbb{R})$, is assigned as an exercise. Hint: Prove that any sigma algebra that contains $\mathcal{A}$ also contains $\mathcal{B}(\mathbb{R})$.

4. To show that μ_F so defined is a measure on $\mathcal{M}_F(\mathbb{R})$, we first prove finite additivity, which follows by induction if it is true for two disjoint sets. Let $A_1, A_2 \in \mathcal{M}_F(\mathbb{R})$ be disjoint. Then by definition of μ_F and then the measurability of A_1:

$$\mu_F(A_1 \bigcup A_2) \equiv \mu_{\mathcal{A}}^*(A_1 \bigcup A_2)$$

$$= \mu_{\mathcal{A}}^*(A_1 \bigcup A_2 \bigcap A_1) + \mu_{\mathcal{A}}^*(A_1 \bigcup A_2 \bigcap \tilde{A}_1)$$

$$= \mu_{\mathcal{A}}^*(A_1) + \mu_{\mathcal{A}}^*(A_2)$$

$$\equiv \mu_F(A_1) + \mu_F(A_2).$$

For countable additivity, let $A = \bigcup_n A_n \in \mathcal{M}_F(\mathbb{R})$ be a disjoint union of measurable sets. Then by monotonicity and finite additivity:

$$\mu_F(A) \geq \mu_F\left(\bigcup_{j \leq n} A_j\right) = \sum_{j \leq n} \mu_F(A_j).$$

This is true for any n and so:

$$\mu_F(A) \geq \sum_{j=1}^{\infty} \mu_F(A_j).$$

But the reverse inequality follows from countable subadditivity of $\mu_{\mathcal{A}}^*$, and the result follows.

Of course $\mu_F(\emptyset) = \mu_{\mathcal{A}}^*(\emptyset) = 0$ by Proposition 5.20, so μ_F is a measure on $\mathcal{M}_F(\mathbb{R})$.

Finally, μ_F is a Borel measure since if $A \in \mathcal{B}(\mathbb{R})$ is compact, then $A \subset (a, b]$ for a bounded interval by the Heine-Borel theorem. Thus, by monotonicity:

$$\mu_F(A) \leq \mu_{\mathcal{A}}^*((a, b]) = F(b) - F(a) < \infty. \qquad \blacksquare$$

Remark 5.24 (Borel measures and integration) *The measure space of Proposition 5.23, $(\mathbb{R}, \mathcal{M}_F(\mathbb{R}), \mu_F)$, is often referred to as a **Lebesgue-Stieltjes measure space**, and μ_F **the Lebesgue-Stieltjes measure induced by** $F(x)$, named for **Henri Lebesgue** (1875–1941) and **Thomas Stieltjes** (1856–1894). Integrals in such a measure space are then called **Lebesgue-Stieltjes integrals** and will be introduced in Book III and developed in Book V. **Lebesgue integration** is addressed in Book III.*

*Among many other contributions, Stieltjes is known for introducing a generalization of the **Riemann integral** introduced by **Bernhard Riemann** (1826–1866). This generalized integral is called the **Stieltjes integral** or **Riemann-Stieltjes integral**. In essence, Stieltjes replaced standard interval length of $b - a$ in the Riemann integral with $F(b) - F(a)$ for suitably defined functions $F(x)$. Riemann-Stieltjes integrals are addressed in Book III.*

*While Stieltjes did not propose the analogous generalization of the Lebesgue integral, it often bears his name because these integrals reflect the analogous generalization of interval measure as developed above. In fact much of the integration theory underlying Lebesgue-Stieltjes integrals is attributed to **Johann Radon** (1887–1956).*

As the above proposition demonstrates that $\mathcal{B}(\mathbb{R}) \subset \mathcal{M}_F(\mathbb{R})$, every Lebesgue-Stieltjes measure is a **Borel measure**, and it is common to call $(\mathbb{R}, \mathcal{B}(\mathbb{R}), \mu_F)$ a **Borel measure space**, named for Émile Borel.

Notation 5.25 Note that we can think of μ_F as either:

1. The **restriction** of $\mu_{\mathcal{A}}^*$ from the power sigma algebra $\sigma(P(\mathbb{R}))$ to the sigma subalgebra $\mathcal{M}_F(\mathbb{R})$, or,

2. The **extension** of $\mu_{\mathcal{A}}$ from the algebra $\mathcal{A}$ to the sigma algebras $\mathcal{B}(\mathbb{R})$ and $\mathcal{M}_F(\mathbb{R})$.

5.3 Consistency of Borel Constructions

In this section we discuss an important question regarding the above constructions which induced functions from measures, and measures from certain functions.

The question of consistency arises when one extends either of these constructions an additional step, so from a function we obtain a function, and from a measure we obtain a measure. One of these constructions can be fully addressed now, while the other will need a later result for a complete resolution. See also Section 8.3.1 for a generalization of this discussion.

To begin, what if the Borel measure μ illustrated in Section 5.1 was a measure μ_G, constructed in Section 5.2, from an increasing, right continuous function G? It makes sense to ask: Is it then the case that the increasing, right continuous function F_μ defined from μ:

$$G \to \mu_G \equiv \mu \to F_\mu,$$

would satisfy:

$$F_\mu = G?$$

The answer is "no" in general, although this conclusion is virtually correct and it will be easy to determine the minor modification that the reader may already anticipate.

Reversing the order of the constructions we could have started in Section 5.2 with G, where this increasing, right continuous function was in fact the distribution function F_μ of another Borel measure μ:

$$\mu \to F_\mu \equiv G \to \mu_G,$$

and then wonder, is it true that on $\mathcal{B}(\mathbb{R})$:

$$\mu_G = \mu?$$

Here, the answer is in the affirmative, but to formally prove this will require a result from Section 6.2.5 on Uniqueness of Extensions.

1. To investigate the first question requires only definitions. Given G, it follows from (5.4) that μ_G is defined on the collection of right semi-closed intervals, $\mathcal{A}'$:

$$\mathcal{A}' \equiv \{(a,b]\}, \tag{5.16}$$

by:

$$\mu_G[(a,b]] = G(b) - G(a).$$

Using μ_G as μ in (5.1) produces for all y:

$$F_\mu(y) = G(y) - G(0).$$

Thus it need not be the case that $G = F_\mu$ unless $G(0) = 0$. In the general case:

$$F_\mu = G + c.$$

This is the most affirmative result possible, since a constant addition to an increasing, right continuous function does not change the induced Borel measure. This is confirmed in (5.4), and this definition of interval length drives the entire Borel measure development.

2. For the second question, we begin with μ and then F_μ is defined in (5.1):

$$F_\mu(y) = \begin{cases} -\mu((y,0]), & y < 0, \\ 0, & y = 0, \\ \mu((0,y]), & y > 0. \end{cases}$$

With $G = F_\mu$, the set function μ_G is then defined on $\mathcal{A}'$ in (5.16) by $\mu_G[(a,b]] = F_\mu(b) - F_\mu(a)$. If $(a,b] \subset (0,\infty)$, then

$$\mu_G[(a,b]] = \mu((0,b]) - \mu((0,a]) = \mu[(a,b]].$$

The same result is produced in the other cases. Thus on $\mathcal{A}'$:

$$\mu_G = \mu,$$

and by finite additivity, $\mu_G = \mu$ on $\mathcal{A}$, the algebra generated by $\mathcal{A}'$.

The final step is subtle. By the construction in the prior section, μ_G is defined on $\mathcal{B}(\mathbb{R})$ as follows:

(a) The associated outer measure μ_G^* is defined on $\sigma(P(\mathbb{R}))$ using covering sets from $\mathcal{A}$, and μ_G as defined on these sets.

(b) The sigma algebra of Carathéodory measurable sets, $\mathcal{M}_G(\mathbb{R})$, is identified and this sigma algebra shown to contain $\mathcal{B}(\mathbb{R})$.

(c) On either of these sigma algebras, $\mu_G = \mu_G^*$.

If it was known that μ was also the result of such a process, using the associated outer measure μ^* and this algebra $\mathcal{A}$, then the equality $\mu_G = \mu$ on $\mathcal{A}$ would readily extend to $\mu_G^* = \mu^*$ and equality on $\mathcal{B}(\mathbb{R})$. But this is not known.

Just because μ is a measure on $\mathcal{A}$ that is extendable to a measure on $\mathcal{B}(\mathbb{R})$ does not imply that this extension was created by the Carathéodory process. Perhaps there are numerous ways to extend measures from algebras to sigma algebras that give different final results.

Perhaps.

But this is not the case. As will be seen in Section 6.2.5 on Uniqueness of Extensions, there is only one extension from a measure on $\mathcal{A}$ to a measure on the **smallest sigma algebra** that contains $\mathcal{A}$, which is $\mathcal{B}(\mathbb{R})$. This result requires that these measures be σ-finite, a property explained below, but σ-finiteness is true of all Borel measures by property 3 in Definition 5.1. Hence, with this powerful result below, it will be concluded that $\mu_G = \mu$ on $\mathcal{B}(\mathbb{R})$. See Example 6.15.

5.4 Approximating Borel Measurable Sets

In this section we extend the approximation results from Lebesgue measure to Borel measures. Because the algebra $\mathcal{A}$ as well as the collections $\mathcal{A}_\sigma$, $\mathcal{A}_\delta$, $\mathcal{A}_{\sigma\delta}$ and $\mathcal{A}_{\delta\sigma}$ are all families of Borel sets, the next proposition identifies the manner in which measurable sets, $B \in \mathcal{M}_F(\mathbb{R})$, are "close" to Borel sets.

The statement here appears quite different from that of Proposition 2.42 on Lebesgue measurable sets, and this is discussed below in Remark 5.27.

Proposition 5.26 (Approximations with Borel subsets/supersets) *Let $\mathcal{A}$ denote the algebra of finite disjoint unions of right semi-closed intervals. If $B \in \mathcal{M}_F(\mathbb{R})$, then given $\epsilon > 0$:*

1. *There is a set $A \in \mathcal{A}_\sigma$, the collection of countable unions of sets in the algebra $\mathcal{A}$, so that $B \subset A$ and:*

$$\mu_F(A) \leq \mu_F(B) + \epsilon, \qquad \mu_F(A - B) \leq \epsilon. \tag{5.17}$$

   Further, A can be taken as a disjoint union of sets in $\mathcal{A}'$, the collection of right semi-closed intervals that generates $\mathcal{A}$.

2. *There is a set $C \in \mathcal{A}_\delta$, the collection of countable intersections of sets in the algebra $\mathcal{A}$, so that $C \subset B$ and:*

$$\mu_F(B) \leq \mu_F(C) + \epsilon, \qquad \mu_F(B - C) \leq \epsilon. \tag{5.18}$$

3. *There is a set $A' \in \mathcal{A}_{\sigma\delta}$, the collection of countable intersections of sets in $\mathcal{A}_\sigma$, and $C' \in \mathcal{A}_{\delta\sigma}$, the collection of countable unions of sets in $\mathcal{A}_\delta$, so that $C' \subset B \subset A'$ and:*

$$\mu_F(A' - B) = \mu_F(B - C') = 0. \tag{5.19}$$

Proof. *We address these statements in turn.*

1. For $B \in \mathcal{M}_F(\mathbb{R})$, Proposition 5.23 obtains that $\mu_F(B) \equiv \mu^*_{\mathcal{A}}(B)$, and so:

$$\mu_F(B) = \inf\{\textstyle\sum_j \mu_F(A_j) | B \subset \bigcup_j A_j\},$$

with $\{A_j\} \subset \mathcal{A}$. This reformulation of the definition of $\mu^*_{\mathcal{A}}(B)$ reflects that $\mu_{\mathcal{A}}(A_j) = \mu^*_{\mathcal{A}}(A_j) = \mu_F(A_j)$ for such A_j by Propositions 5.20 and 5.23. By Remark 5.17 these sets can be assumed to be disjoint, and thus with $A \equiv \bigcup_j A_j$, it follows that $A \in \mathcal{A}_\sigma \subset \mathcal{M}_F(\mathbb{R})$ and $\mu_F(A) = \sum_j \mu_F(A_j)$ for any such collection $\{A_j\}$.

If $\mu_F(B) < \infty$, then by definition of infimum it follows that for any $\epsilon > 0$, there is a finite or countable disjoint collection $\{A_j\} \subset \mathcal{A}$ with $\mu_F(A) \leq \mu_F(B) + \epsilon$, where $A \equiv \bigcup_j A_j \in \mathcal{A}_\sigma$. This proves the first inequality in (5.17). In this case, finite additivity yields:

$$\mu_F(A) = \mu_F(B) + \mu_F(A - B),$$

and so $\mu_F(A - B) \leq \epsilon$ and the second inequality follows.

If $\mu_F(B) = \infty$, the first inequality of (5.17) is true by monotonicity of μ_F, using any collection $\{A_j\}$ with $B \subset A \equiv \bigcup_j A_j$, independent of ϵ. For the second inequality, define for $n = 1, 2, 3..$,

$$I_n \equiv [-n, -n+1) \bigcup [n-1, n),$$

$$B_n \equiv B \bigcap I_n.$$

Then $B_n \in \mathcal{M}_F(\mathbb{R})$, $\mu_F(B_n) < \mu_F(I_n) < \infty$, and so by the result just proved there exists $A_n \in \mathcal{A}_\sigma$ with $B_n \subset A_n$, and $\mu_F(A_n - B_n) \leq \epsilon/2^n$.

Letting $A \equiv \bigcup_n A_n \in \mathcal{A}_\sigma$ obtains $B = \bigcup_n B_n \subset A$, and using De Morgan's laws:

$$A - B \equiv \left(\bigcup_n A_n\right) \bigcap \left(\widetilde{\bigcup_n B_n}\right)$$

$$= \bigcup_n \left(A_n \bigcap \left(\bigcap_n \widetilde{B_n}\right)\right)$$

$$\subset \bigcup_n \left(A_n \bigcap \widetilde{B_n}\right).$$

From the monotonicity and subadditivity of μ_F:

$$\mu_F(A - B) \leq \mu_F \left[\bigcup_n \left(A_n \bigcap \widetilde{B_n}\right)\right]$$

$$\leq \sum \mu_F(A_n - B_n) \leq \epsilon,$$

which is the second inequality in (5.17) when $\mu_F(B) = \infty$.

What is left to prove is that any $A \in \mathcal{A}_\sigma$ can be taken as a disjoint union of sets in $\mathcal{A}'$, the collection of right semi-closed intervals that generates $\mathcal{A}$. By definition, each element of $\mathcal{A}$ is a finite union of elements of $\mathcal{A}'$, which as noted in Example 2.4 can be assumed to be a *finite disjoint union. So if* $\{A_j\} \subset \mathcal{A}$ *is a finite or countable collection, define* $\{A'_j\} \subset \mathcal{A}$ *by* $A'_j = A_j \bigcap \bigcap_{k<j} \widetilde{A}_k$. *Then each* A'_j *is at most finite union of sets in* $\mathcal{A}'$, *which can then be made disjoint as in Example 2.4.*

2. *Applying the second conclusion in part 1 to* $\widetilde{B} \in \mathcal{M}_F(\mathbb{R})$ *assures that there exists* $A \in \mathcal{A}_\sigma$ *with* $\widetilde{B} \subset A$ *and* $\mu_F(A - \widetilde{B}) \le \epsilon$. *Since A equals a countable union of elements of* $\mathcal{A}$, *defining* $C = \widetilde{A}$ *obtains that C is a countable intersection of complements of elements of* $\mathcal{A}$. *Since an algebra is closed under complementation, C is also a countable intersection of elements of* $\mathcal{A}$ *and thus* $C \in \mathcal{A}_\delta$. *Now* $C \subset B$ *by construction, and:*

$$B - C = B \bigcap \widetilde{C} = A - \widetilde{B}.$$

It then follows that $\mu_F(B - C) \le \epsilon$, *which is the second inequality in (5.18).*

The first inequality then follows from finite additivity since $B = C \bigcup (B - C)$.

3. *For (5.19), define* $A_n \in \mathcal{A}_\sigma$ *and* $C_n \in \mathcal{A}_\delta$ *so that the second inequalities in (5.17) and (5.18) are true with* $\epsilon = 1/n$. *In other words,* $C_n \subset B \subset A_n$ *and:*

$$\mu_F(A_n - B) \le 1/n, \quad \mu_F(B - C_n) \le 1/n. \tag{1}$$

Letting $A' \equiv \bigcap_n A_n$ *and* $C' \equiv \bigcup_n C_n$, *then* $A' \in \mathcal{A}_{\sigma\delta}$, $C' \in \mathcal{A}_{\delta\sigma}$, *and* $C' \subset B \subset A'$.

To prove (5.19), define $A'_n \equiv \bigcap_{j \le n} A_j$. *Then* $A' \subset A'_n \subset A_n$ *implies:*

$$A' - B \subset A'_n - B \subset A_n - B,$$

and hence $\mu_F(A' - B) = 0$ *by monotonicity and (1).*

The same argument applies to C', *defining* $C'_n \equiv \bigcup_{j \le n} C_j$. *Then* $C_n \subset C'_n \subset C'$ *implies:*

$$B - C' \subset B - C'_n \subset A - C_n. \qquad \blacksquare$$

Remark 5.27 (On Lebesgue Proposition 2.42) *The statement of results above initially appears at odds with the result for Lebesgue measure in Proposition 2.42. For Lebesgue measure, the collection of sets used in the outer measure definition was* $\mathcal{G}$, *the collection of open intervals, and it was seen that Lebesgue measurable sets could be approximated within* ϵ *by supersets in* $\mathcal{G}$, *as seen in (2.19), and approximated to measure 0 by supersets in* $\mathcal{G}_\delta$, *as seen in (2.20).*

Similarly, Lebesgue measurable sets could be approximated within ϵ *by closed subsets in* $\mathcal{F}$ *and approximated to measure 0 by subsets in* $\mathcal{F}_\sigma$. *The apparent difference in results stems from the fact that in the Lebesgue case,* $\mathcal{G}$ *is closed under countable unions so* $\mathcal{G}_\sigma = \mathcal{G}$, *and* $\mathcal{F}$ *is closed under countable intersections so that* $\mathcal{F} = \mathcal{F}_\delta$.

But note that neither $\mathcal{G}$ *nor* $\mathcal{F}$ *is an algebra since neither is closed under complementation.*

In the current setting and that of a general algebra $\mathcal{A}$, *this collection will not be closed under countable unions, only finite unions, and hence approximations with supersets requires* $\mathcal{A}_\sigma$. *Similarly, the collections of complements of* $\mathcal{A}$ *will not be closed under countable intersections, and so approximations with subsets requires* $\mathcal{A}_\delta$.

Improving approximations to within measure 0 now requires supersets in $\mathcal{A}_{\sigma\delta}$, *which is analogous with the role of* $\mathcal{F}_\delta$, *and subsets in* $\mathcal{A}_{\delta\sigma}$, *which is analogous with the role of* $\mathcal{G}_\sigma$.

5.5 Properties of Borel Measures

5.5.1 Continuity

As stated in Proposition 2.45, all measures and thus all Borel measures are **continuous from above** on finite sets, and **continuous from below** in general. But we restate this result here for completeness with only a notational change, and note that the proof requires only the same change of notation.

Proposition 5.28 (Continuity of Borel Measures) *Let* $\{A_i\} \subset \mathcal{M}_F(\mathbb{R})$. *Then:*

1. *Continuity from Below:* *If* $A_i \subset A_{i+1}$ *for all i, then:*

$$\mu_F \left(\bigcup\nolimits_{i=1}^{\infty} A_i \right) = \lim_{i \to \infty} \mu_F(A_i), \tag{5.20}$$

   where this limit may be finite or infinite.

2. *Continuity from Above:* *If* $A_{i+1} \subset A_i$ *for all i, and* $\mu_F(A_1) < \infty$, *then:*

$$\mu_F \left(\bigcap\nolimits_{i=1}^{\infty} A_i \right) = \lim_{i \to \infty} \mu_F(A_i). \tag{5.21}$$

Proof. *Identical to the proof of Proposition 2.44 with a change of notation.* ∎

5.5.2 Regularity

The next result relates to the **regularity** of Borel measures, defined here identically to the Lebesgue measure development in Section 2.7.1. That is:

1. **Outer regularity**: The measure of a measurable set equals the **infimum** of the measures of **open supersets**, and

2. **Inner regularity**: The measure of a measurable set equals the **supremum** of the measures of **compact subsets**.

A measure that is both outer and inner regular is called **regular**.

Unlike continuity, regularity is a property that cannot be true of all measures since at the minimum it requires the measure space X to have a topology (Definition 2.15) so that the notions of open and compact have meaning. As $\mathbb{R}$ does have a topology, one may wonder if all measures on $\mathbb{R}$ are regular, and thus the Borel result below is only a special case of this general result. The answer is in the negative.

For example, define μ on $\sigma(P(\mathbb{R}))$ by $\mu(\emptyset) = \mu(\{0\}) = 0$, and $\mu(A) = \infty$ otherwise. Then μ is inner regular but not outer regular:

$$0 = \mu(\{0\}) \neq \inf \mu((-\epsilon_1, \epsilon_2)) = \infty.$$

As another example, consider the measure defined in Remark 5.2. This measure is inner regular, but not outer regular.

Borel measure on $\mathbb{R}$ (as defined in Definition 5.1) **is** regular, and we will see in Section 8.3.3 that this result generalizes to general Borel measures on $\mathbb{R}^n$. The proof here is lengthier than for Lebesgue measure in Proposition 2.43 for two reasons.

1. Proposition 2.43 relied on the approximation results of Proposition 2.42 which used open supersets and closed subsets as approximating sets. In other words, approximating supersets are in $\mathcal{G}$ and subsets are in $\mathcal{F}$ of Notation 2.16. For Borel measures, the associated approximation results of Proposition 5.26 are defined in terms of supersets in $\mathcal{A}_\sigma$, the collection of countable unions of sets in the algebra $\mathcal{A}$, and subsets in $\mathcal{A}_\delta$, the collection of countable intersections of sets in $\mathcal{A}$. The proof of Proposition 2.42 actually used $\mathcal{G}_\sigma$ and $\mathcal{F}_\delta$ sets, but since $\mathcal{G}_\sigma = \mathcal{G}$ and $\mathcal{F} = \mathcal{F}_\delta$ as noted in Remark 5.27, we avoided this extra complication.

2. Recall that $\mathcal{G}$ is the collection of open sets generated by disjoint unions of open intervals $\{(a,b)\}$ by Proposition 2.12. By Definition 2.4, $\mathcal{A}$ is the algebra generated by the intervals $\{(a,b]\}$. Therein lies the second challenge. Unlike Lebesgue measure, Borel measures typically give different measures to open, closed and semi-open intervals by Exercise 5.22. Thus for supersets in the proof below, unions of $(a,b]$-sets in $\mathcal{A}_\sigma$ cannot simply be replaced by unions of (a,b)-sets in $\mathcal{G}$ without accounting for the change in measures. The same is true for subsets with $\mathcal{A}_\delta$-sets.

Proposition 5.29 (Regularity of Borel Measures) *Borel measure μ_F is **regular** on $\mathcal{M}_{\mu_F}(\mathbb{R})$. Specifically, if $A \in \mathcal{M}_F(\mathbb{R})$:*

1. μ_F *is outer regular:*

$$\mu_F(A) = \inf_{A \subset G} \mu_F(G), \ G \ open, \qquad (5.22)$$

2. μ_F *is inner regular:*

$$\mu_F(A) = \sup_{F \subset A} \mu_F(F), \ F \ compact. \qquad (5.23)$$

Proof. 1. Outer Regularity: *If $\mu_F(A) = \infty$, then (5.22) follows by monotonicity of μ_F, since then $\mu_F(G) = \infty$ for any G with $A \subset G$.*

So assume $A \in \mathcal{M}_F(\mathbb{R})$ and $\mu_F(A) < \infty$. If $A = (a,b]$ is a bounded semi-closed interval, let $G_m = (a, b + 1/m)$. Then G_m is open, $A \subset G_m$, and for any $\epsilon > 0$ there is an $m(\epsilon)$ so that

$$\mu_F(A) \leq \mu_F(G_{m(\epsilon)}) \leq \mu_F(A) + \epsilon.$$

The first inequality follows from monotonicity, and the second from the continuity from above property of the measure μ_F since $\bigcap_m G_m = A$.

This second inequality is also proved by right continuity of the increasing distribution function $F(x)$, since by Exercise 5.22:

$$\mu_F(G_{m(\epsilon)}) = F([b + 1/m]^-) - F(a)$$
$$\leq F(b + 1/m) - F(a)$$
$$= \mu_F(A) + [F(b + 1/m) - F(b)].$$

Hence, (5.22) is proved for bounded $A = (a, b]$. The same conclusion follows if $A = (a, \infty)$ or $A = (-\infty, b]$ with $\mu_F(A) < \infty$, using $G_m = (a, \infty)$ or $G_m = (-\infty, b + 1/m)$, respectively.

For general finite measurable A, Proposition 5.26 obtains that for any $\epsilon > 0$ there is a set $B \in \mathcal{A}_\sigma$, the collection of countable unions of sets in the algebra $\mathcal{A}$, so that $A \subset B$ and:

$$\mu_F(A) \leq \mu_F(B) \leq \mu_F(A) + \epsilon.$$

As obtained in item 3 of Example 2.4, any such $B \in \mathcal{A}_\sigma$ is a countable union of disjoint intervals:

$$B = \bigcup_n (a_n, b_n].$$

By the previous paragraph, given $\epsilon > 0$ there exists an open G_n with $(a_n, b_n] \subset G_n$ and:

$$\mu_F((a_n, b_n]) \leq \mu_F(G_n) \leq \mu_F((a_n, b_n]) + \epsilon/2^n.$$

With $G_\epsilon \equiv \bigcup_n G_n$ it follows that $A \subset B \subset G_\epsilon$ and thus $\mu_F(A) \leq \mu_F(G_\epsilon)$.
Also, by countable subadditivity:

$$\mu_F(G_\epsilon) \leq \sum_n \mu_F(G_n),$$

and recalling that $\{(a_n, b_n]\}$ are disjoint, countable additivity obtains:

$$\sum_n \mu_F(G_n) \leq \sum_n \mu_F((a_n, b_n]) + \epsilon$$
$$= \mu_F(B) + \epsilon$$
$$\leq \mu_F(A) + 2\epsilon.$$

Combining results proves that for all $\epsilon > 0$, there exists open G_ϵ with $A \subset G_\epsilon$ and:

$$\mu_F(A) \leq \mu_F(G_\epsilon) \leq \mu_F(A) + 2\epsilon. \tag{1}$$

This proves (5.22) for $\mu_F(A) < \infty$ with an infimum over such G_ϵ. But since $\mu_F(A) \leq \mu_F(G)$ for any G with $A \subset G$, (5.22) follows with an infimum over all open G.
Since $G_\epsilon = A \bigcup (G_\epsilon - A)$, this last inequality also shows that:

$$\mu_F(G_\epsilon - A) \leq 2\epsilon, \tag{2}$$

and this is needed next.

2. *Inner Regularity*: *To prove (5.23) apply (1) to* $\tilde{A}$ *obtaining an open* G_ϵ *so that* $\tilde{A} \subset G_\epsilon$ *and*:

$$\mu_F(\tilde{A}) \le \mu_F(G_\epsilon) \le \mu_F(\tilde{A}) + 2\epsilon.$$

Define closed $F_\epsilon = \widetilde{G_\epsilon}$. *Then* $F_\epsilon \subset A$ *with*:

$$A - F_\epsilon = A \bigcap G_\epsilon = G_\epsilon - \tilde{A},$$

and (2) yields:

$$\mu_F(A - F_\epsilon) \le 2\epsilon.$$

Thus since $A = F_\epsilon \bigcup (A - F_\epsilon)$ *and* $F_\epsilon \subset A$, *this obtains*:

$$\mu_F(A) - 2\epsilon \le \mu_F(F_\epsilon) \le \mu_F(A), \tag{3}$$

proving (5.23) with such closed F_ϵ. *This extends to all closed subsets of* A *by monotonicity.*

Now define $F_\epsilon^{(n)} = F_\epsilon \bigcap I_n$, *where* $I_n \equiv [-n, -(n-1)) \bigcup [n-1, n)$. *Then* $F_\epsilon^{(n)} \subset F_\epsilon$ *and* $F_\epsilon^{(n)}$ *is bounded for all* n. *Since* $F_\epsilon = \bigcup_n F_\epsilon^{(n)}$ *and* $F_\epsilon^{(n)}$ *are disjoint, countable additivity obtains*:

$$\mu_F(F_\epsilon) = \sum_{n=1}^{\infty} \mu_F\left(F_\epsilon^{(n)}\right). \tag{4}$$

If $\mu_F(A) < \infty$, *then* $\mu_F(F_\epsilon) < \infty$ *by (3), so there is an* N *with*:

$$\sum_{n=N+1}^{\infty} \mu_F\left(F_\epsilon^{(n)}\right) < \epsilon.$$

Again by countable additivity and monotonicity:

$$\begin{aligned} \epsilon &> \sum_{n=N+1}^{\infty} \mu_F\left(F_\epsilon^{(n)}\right) \\ &= \mu_F\left(F_\epsilon \bigcap \left\{(-\infty, -N) \bigcup [N, \infty)\right\}\right) \\ &\ge \mu_F\left(F_\epsilon \bigcap \left[(-\infty, -N) \bigcup (N, \infty)\right]\right). \end{aligned} \tag{5}$$

Defining $\bar{F}_\epsilon^N = F_\epsilon \bigcap [-N, N]$, *then* $\bar{F}_\epsilon^N \subset F_\epsilon$ *is closed and bounded and thus compact, and* $\mu_F(F_\epsilon) - \mu_F(\bar{F}_\epsilon^N) < \epsilon$ *by (5). Combining with (3) obtains*:

$$\mu_F(A) - 3\epsilon \le \mu_F(\bar{F}_\epsilon^N) \le \mu_F(A),$$

and (5.23) follows with such compact $\bar{F}_\epsilon^N$, *and then for all compact* $F \subset A$ *by definition.*

If $\mu_F(A) = \infty$ *then* $\mu_F(F_\epsilon) = \infty$ *by (3), and thus* $\sum_{n=1}^{N} \mu_F\left(F_\epsilon^{(n)}\right)$ *is unbounded in* N. *But if* $\bar{F}_\epsilon^N = F_\epsilon \bigcap [-N, N]$ *as above*:

$$\bigcup_{n=1}^{N} F_{\epsilon}^{(n)} = F_{\epsilon} \bigcap [-N, N)$$

$$\subset \bar{F}_{\epsilon}^{N}.$$

Thus by monotonicity, $\mu_F(\bar{F}_{\epsilon}^N)$ is unbounded in N with compact $\bar{F}_{\epsilon}^N$ and (5.23) is proved in this case. ∎

5.6 Differentiable $F(x)$

The assumption that F was increasing and right continuous was enough to assure that the definition of F-length of an interval $(a, b]$ could be extended to a measure on a sigma algebra that included $\mathcal{B}(\mathbb{R})$, the Borel sigma algebra. Although by no means obvious, we will show in Book III that every increasing function F is in fact **differentiable almost everywhere**. We already know from Proposition 5.8 that such functions are continuous almost everywhere, with at most countably many discontinuities, and can be modified on this set to be right continuous, and with left limits.

The following proposition states that if A is a Borel set on which the function F is differentiable and with bounded derivative, then $\mu_F(A)$ can be bounded by a multiple of the Lebesgue measure, $m(A)$.

Proposition 5.30 (Bounding $\mu_F(A)$ with $m(A)$ for differentiable F) *Let F be an increasing, right continuous function and μ_F the Borel measure induced by F. For any Borel set $A \in \mathcal{B}(\mathbb{R})$ on which F is differentiable with bounded derivative:*

$$k_A \leq F'(x) \leq K_A,$$

then:

$$k_A m(A) \leq \mu_F(A) \leq K_A m(A), \tag{5.24}$$

where $m(A)$ is the Lebesgue measure of A.

Proof. *Any Borel set A can be expressed as a disjoint union of bounded Borel sets as noted in the proof of Proposition 5.29. In detail, let $I_n \equiv [-n, -n+1) \bigcup [n-1, n)$ and define $A_n \equiv A \bigcap I_n$. Then $\{A_n\}$ is a disjoint collection of bounded Borel sets with $A = \bigcup_n A_n$. Thus it is enough to prove (5.24) under the assumption that A is bounded, and apply countable additivity for the final result.*

1. To prove that:

$$\mu_F(A) \leq K_A m(A),$$

fix $\epsilon > 0$. If F is differentiable at x there is a δ_x so that if $|b - a| < \delta_x$:

$$\left| \frac{F(y) - F(x)}{y - x} - F'(x) \right| < \epsilon.$$

Thus if $|b - a| < \delta_x$ *and* $a < x < b$:

$$\left| \frac{F(b) - F(a)}{b - a} - F'(x) \right|$$

$$\leq \frac{b - x}{b - a} \left| \frac{F(b) - F(x)}{b - x} - F'(x) \right| + \frac{x - a}{b - a} \left| \frac{F(x) - F(a)}{x - a} - F'(x) \right|$$

$$< \epsilon.$$

So for $x \in A$ *and* $a < x < b$ *with* $m((a, b]) < \delta_x$,

$$\mu_F((a, b]) \leq (K_A + \epsilon) \, m((a, b]), \tag{1}$$

because $\mu_F((a, b]) = F(b) - F(a)$, $m((a, b]) = b - a$, *and* $F'(x) \leq K_A$.

Next let $D = \{b_j\}$ *be a countable dense set and denote by* $I_{j,k}^{(n)} = (b_j, b_k]$ *any interval for which* $0 < m\left[(b_j, b_k]\right] < \min\{1/n, \delta_x\}$ *for some* $x \in (b_j, b_k)$ *with* $x \in A$, *and thus for which* (1) *is satisfied. Define* $A_n = \bigcup_{j,k} \left(A \cap I_{j,k}^{(n)} \right)$ *over all such intervals, and so by definition* $A_n \subset A$. *But for any* $x \in A$ *there exists an interval* $I_{j,k}^{(n)}$ *which contains* x *by the density of* D, *so* $A_n = A$:

$$A = \bigcup_{j,k} \left(A \cap I_{j,k}^{(n)} \right). \tag{2}$$

As a Borel and hence Lebesgue measurable set, Proposition 2.42 implies that for every n *there is a disjoint collection of right semi-closed intervals* $\{I_i^{(n)}\}$ *with* $A \subset \bigcup_i I_i^{(n)}$, *and:*

$$m\left(\bigcup_i I_i^{(n)} - A \right) \leq 1/n.$$

This follows since the open sets of Proposition 2.42 equal a disjoint union of open intervals by Proposition 2.12, and these can be replaced by a disjoint union of right semi-closed intervals noting that $m((a, b)) = m((a, b])$ *for Lebesgue measure.*

Then since $\bigcup_i I_i^{(n)} = A \bigcup \left(\bigcup_i I_i^{(n)} - A \right)$, *this and countable additivity obtain:*

$$m\left(\bigcup_i I_i^{(n)} \right) = \sum_i m\left(\bigcup_i I_i^{(n)} \right) \leq m(A) + 1/n. \tag{3}$$

We claim that this collection $\{I_i^{(n)}\}$ *can be replaced by a subcollection of* $\{I_{j,k}^{(n)}\}$ *defined above, by allowing for an increase from* $1/n$ *to* $2/n$ *in this inequality.*

To prove this, first note that $\{I_{j,k}^{(n)}\}$ *covers* A *as noted in* (2) *above. Now as* D *is dense, we can increase each of the original disjoint* $I_i^{(n)}$*-intervals to have endpoints in* D. *For example, if* $I_i^{(n)} = (a, b]$ *say, then* $m\left[(b_j, a]\right]$ *and* $m\left[(b, b_k]\right]$ *can be made arbitrarily small for* $b_j, b_k \in D$. *Thus* $I_i^{(n)}$ *can be replaced by* $(b_j, b_k]$ *with* $m\left[(b_j, b_k]\right] \leq m\left[I_i^{(n)}\right] + 2^{-i}/n$. *Of course it is possible that initially two sets in* $\{I_i^{(n)}\}$ *are* $(a, b]$ *and* $(b, c]$ *which would allow no such disjoint expansion, but in these cases we first combine to* $(a, c]$. *Thus* $\{I_i^{(n)}\}$ *is expanded to a disjoint union of such* $(b_j, b_k]$

with the maximum cost of $1/n$ in the inequality in (3). Then each such $(b_j, b_k]$ can be replaced by a disjoint union of $I_{j,k}^{(n)}$ by definition.

For notational simplicity we relabel this countable disjoint $\{I_{j,k}^{(n)}\}$ collection so obtained as $\{I_i^{(n)}\}$.

Now $A \subset \bigcup_i I_i^{(n)}$ by construction and $\mu_F\left(I_i^{(n)}\right) \leq (K_A + \epsilon) \, m\left(I_i^{(n)}\right)$ by (1). Then monotonicity, countable additivity and (3) modified to reflect $2/n$ obtains:

$$\mu_F(A) \leq \sum_i \mu_F\left(I_i^{(n)}\right)$$
$$\leq (K_A + \epsilon) \sum_i m\left(I_i^{(n)}\right)$$
$$\leq (K_A + \epsilon)\,(m(A) + 2/n).$$

Letting $n \to \infty$ it follows that

$$\mu_F(A) \leq (K_A + \epsilon)\, m(A),$$

and the upper bound result in (5.24) follows since ϵ is arbitrary.

  2. To prove that:

$$k_A m(A) \leq \mu_F(A),$$

the analysis above obtains that for $x \in A$, if $a < x < b$ and $m((a, b]) < \delta_x$ then because $F'(x) \geq k_A$:

$$\mu_F((a, b]) \geq (k_A - \epsilon)\, m((a, b]). \tag{4}$$

By Proposition 5.26 and item 1 of Example 2.4, for every n there is a disjoint collection of right semi-closed intervals $\{I_i^{(n)}\}$ with $A \subset \bigcup_i I_i^{(n)}$, and using countable additivity:

$$\sum_i \mu_F\left(I_i^{(n)}\right) \leq \mu_F(A) + 1/n.$$

As above, at a cost of $1/n$ in the upper bound of this inequality, each $I_i^{(n)}$-set can be replaced by a set $(b_j, b_k]$ with $\mu_F\left[(b_j, b_k]\right] \leq \mu_F\left[I_i^{(n)}\right] + 2^{-i}/n$. Each such $(b_j, b_k]$ is then replaced by a disjoint union of $I_{j,k}^{(n)}$-sets, which is possible by definition. Then relabelling this countable disjoint $\{I_{j,k}^{(n)}\}$ collection as $\{I_i^{(n)}\}$ obtains:

$$\mu_F(A) + 2/n \geq \sum_i \mu_F\left(I_i^{(n)}\right)$$
$$\geq (k_A - \epsilon) \sum_i m\left(I_i^{(n)}\right)$$
$$\geq (k_A - \epsilon)\, m(A).$$

Letting $n \to \infty$ the result follows as ϵ is arbitrary.   ∎

Corollary 5.31 (When $\mu_F = m$ on $\mathcal{B}(\mathbb{R})$) *Let F be an increasing, right continuous function, μ_F the Borel measure induced by F, and m Lebesgue measure. Then:*

$$\mu_F(A) = m(A), \tag{5.25}$$

for all $A \in \mathcal{B}(\mathbb{R})$ if and only if $F(x) = x + c$ with $c \in \mathbb{R}$.

Proof. *If $F(x) = x + c$, then in the above notation $k_A = K_A = 1$ for all $A \in \mathcal{B}(\mathbb{R})$. Thus (5.25) follows from (5.24).*

On the other hand, assume that $\mu_F(A) = m(A)$ for all $A \in \mathcal{B}(\mathbb{R})$, and let $A = (a,b]$. By Proposition 2.28 and Definition 2.40:

$$m(a,b] = m^*(a,b] = b - a,$$

while Propositions 5.23 and 5.20 and (5.6) obtain:

$$\mu_F(a,b] = \mu^*_A(a,b] = F(b) - F(a).$$

Thus (5.25) assures that $F(x) - x$ is constant, and the result follows. ∎

Corollary 5.32 (Equivalence: When $\mu_F \sim m$ on $\mathcal{B}(\mathbb{R})$) *Let F be an increasing, right continuous function, μ_F the Borel measure induced by F, and m Lebesgue measure. If F is differentiable with bounded derivative on compact sets, then for all $A \in \mathcal{B}(\mathbb{R})$:*

$$\mu_F(A) = 0 \text{ if and only if } m(A) = 0. \tag{5.26}$$

*In other words, μ_F and m are **equivalent measures**.*

Proof. *If $A \in \mathcal{B}(\mathbb{R})$ is bounded, then $A \subset [a,b]$ for some compact interval, and then (5.24) applies. Thus (5.26) follows for all bounded Borel sets. If A is unbounded define $A_n = A \bigcap I_n$ with $I_n \equiv [-n, -n+1) \bigcup [n-1, n)$. The prior result assures that $\mu_F(A_n) = 0$ if and only if $m(A_n) = 0$. The proof is complete since by countable additivity, $\mu_F(A) = 0$ if and only if $\mu_F(A_n) = 0$ for all n, and similarly for m.* ∎

Remark 5.33 (Equivalence of measures; Radon-Nikodým theorem) *In Book V we will return to a general study of properties of measures. One such property is **absolute continuity of measures**. In that terminology, μ_F **is absolutely continuous with respect to** m on $\mathcal{B}(\mathbb{R})$, denoted $\mu_F \ll m$, if $m(A) = 0$ implies that $\mu_F(A) = 0$. Thus this notation implies that $m(A) = 0$ forces $\mu_F(A) = 0$. But this notation does **not** imply that $\mu_F(A) \leq m(A)$ in general, or for any A in particular.*

*When both $\mu_F \ll m$ and $m \ll \mu_F$, we say that μ_F and m are **equivalent measures**, denoted $\mu_F \sim m$.*

*Absolute continuity of measures has profound implications in the forthcoming studies beginning in Book V, and continuing to later books on stochastic processes. As one of the most important examples of this, we cite the **Radon-Nikodým theorem**. This result is named for **Johann Radon** (1887–1956) who proved it on $\mathbb{R}^n$, and **Otto Nikodým** (1887–1974) who generalized Radon's result to σ-**finite measure spaces** $(X, \sigma(X), \mu)$, meaning that the measure μ is σ-**finite**.*

Definition 5.34 (σ-finite measure space) *The measure space $(X, \sigma(X), \mu)$ is said to be* **sigma finite**, *or* σ-finite, *if there exists a countable collection $\{B_j\} \subset \sigma(X)$ with $\mu(B_j) < \infty$ for all j, and $X = \bigcup_{j=1}^{\infty} B_j$. In this case it is also said that the measure μ is σ-finite.*

Proposition 5.35 (Radon-Nikodým theorem) *Let $(X, \sigma(X), \mu)$ be a σ-finite measure space, and υ a σ-finite measure on $\sigma(X)$ which is absolutely continuous with respect to μ, so $\upsilon \ll \mu$.*

Then there exists a nonnegative measurable function $f : X \to \mathbb{R}$, also denoted $f \equiv \frac{\partial \upsilon}{\partial \mu}$, so that for all $A \in \sigma(X)$:

$$\upsilon(A) = \int_A f d\mu. \tag{5.27}$$

Further, f is unique μ-a.e., meaning if g is a measurable function so that (5.27) *is true with g, then $g = f$, μ-a.e.*

All Borel measures are σ-finite by item 3 of Definition 5.1, as is Lebesgue measure. Hence this result can be applied to one conclusion of Corollary 5.32, that $\mu_F \ll m$ on $\mathcal{B}(\mathbb{R})$ when F is differentiable with bounded derivative on compact sets.

Then there exists a Borel measurable function f, unique m-a.e., so that the Borel measure can be defined on $\mathcal{B}(\mathbb{R})$ by a Lebesgue integral of Book III:

$$\mu_F(A) = \int_A f dm.$$

It will also follow from the integration theory in Book III that the assumption of differentiability of F in Corollary 5.32 assures that $f = F'$, m-a.e.

This theorem can also be applied to the other conclusion of Corollary 5.32 that $m \ll \mu_F$ on $\mathcal{B}(\mathbb{R})$, again assuming F differentiable.

Then there exists a Borel measurable function $\widetilde{f}$, unique μ_F-a.e. (and thus also m-a.e.), so that Lebesgue measure can be defined on $\mathcal{B}(\mathbb{R})$ by a Lebesgue-Stieltjes integral of Book V:

$$m(A) = \int_A \widetilde{f} d\mu_F.$$

Further, it is then the case that $\widetilde{f} = f^{-1}$, m-a.e.

6

Measures by Extension

6.1 Recap of Lebesgue and Borel Constructions

It may well have occurred to the reader that there was a strong similarity to the procedures by which the Lebesgue and general Borel measures, and associated measure spaces, were constructed.

1. In each case we began with a rudimentary notion of "measure" as applied to a special collection of simple sets:

    (a) In the Lebesgue theory, the collection of simple sets was the collection of **open intervals**, and the measure of an interval (a, b) was defined as its standard interval length:

    $$|(a, b)| = b - a,$$

    whether the interval was finite or infinite.

    (b) In the Borel theory, the collection of simple sets was the collection of finite or infinite **right semi-closed intervals**, $\{(a, b]\}$, and the measure of a given interval was defined by its F-length:

    $$|(a, b]|_F = F(b) - F(a),$$

    again whether the interval was finite or infinite. Here $F(x)$ was a given increasing, right continuous function.

2. In each case, this definition of interval measure was extended to a set function defined on all sets A of the power sigma algebra, $\sigma(P(\mathbb{R}))$. This extension was defined as the infimum of the total measures of all collections of simple sets which cover A. This was done in (2.4) and (5.9), respectively, and the set function extensions were called **outer measures**.

3. In the Lebesgue case, it was shown that Lebesgue outer measure could not be a Lebesgue measure due to the existence of a collection of highly irregular sets. Though not proved, this phenomenon is true in more general contexts. This implied that in order to eliminate such irregular sets, the power sigma algebra to which it was applied needed to be restricted.

DOI: 10.1201/9781003257745-6

4. A notion of measurable, which was formalized as **Carathéodory measurable** and named for **Constantin Carathéodory** (1873–1950), was introduced which provided such a restriction by requiring of measurable sets a certain type of regularity as defined in (2.14) and (5.10).

5. In both cases it was shown that the collection of Carathéodory measurable sets formed a sigma algebra which included the original collection of simple sets in item 1, as well as all sets with respective outer measure 0.

6. In both cases it was shown that when restricted to this sigma algebra, outer measure was a true measure, creating Lebesgue measure and general Borel measures, respectively. Further, these measures reproduced the values of the original measures on the collections of simple sets in item 1.

It would be a good guess at this point that the basic constructions used above can be generalized further. In other words, these constructions are special cases of a **general framework** which starts with an identification of the simple collection of sets and a rudimentary notion of their measure. This collection of sets would need to have enough structure to allow basic manipulations of sets without leaving the collection. This framework would then identify necessary conditions that would ensure that all of the above steps can be implemented to produce a complete measure space which extends the rudimentary measure from the collection of simple sets to a sigma algebra which includes this collection.

The goal of this chapter is to develop this framework.

6.2 Extension Theorems

There are several important "extension" theorems which underlie the general development noted above. We present these in "reverse" order to make it clear how we arrive at the ultimate goal. Had the theorems below been addressed at the beginning of this book, the development of Lebesgue measure and the Borel measures could have been presented as examples. A very advanced text might take this approach, and then identify the Lebesgue and Borel results as simple corollaries, or with details relegated to exercises. More commonly, authors develop the Lebesgue theory in detail, then the general theory.

The approach taken here reflects the author's view that a development of the details of these two specific applications of this general result helps in the understanding and appreciation of the general result.

While the Lebesgue development with open intervals seemed natural in that context, right semi-closed intervals were more natural for the Borel generalization to accommodate discontinuities in the function F. Finite unions from this latter collection also created the structure of an algebra, and hence provided a convenient first step to the more general sigma algebra results. The collection of open intervals cannot be so manipulated into an algebra structure.

The general theory will ultimately replicate this approach, starting with a **semi-algebra** of sets in place of the collection of right semi-closed intervals. This will again give rise to an algebra of sets by finite unions, and ultimately to a sigma algebra.

Thus the proofs in the Borel measure development will be seen to provide templates of the proofs of these more general, and indeed more abstract results. Consequently, some of the proofs below will be abbreviated.

6.2.1 From Outer Measure to Complete Measure

The first extension theorem we address was introduced by **Constantin Carathéodory** and generalizes the constructions in steps 4–6 above, once an "outer measure" has been defined. Carathéodory's result states that given an outer measure, which is defined next, the collection of Carathéodory measurable sets forms a complete sigma algebra. Further, the restriction of this outer measure to this sigma algebra is a measure, and together they form a complete measure space.

We begin with a definition. Note that the examples of outer measures given in (2.4) and (5.9) were proved to have these properties in Propositions 2.28 and 2.29, and Proposition 5.20, respectively.

Definition 6.1 (Outer measure) *Given a set X, a set function μ^* defined on the power sigma algebra $\sigma(P(X))$ of all subsets of X is an **outer measure** if:*

1. $\mu^*(\emptyset) = 0.$
2. *Monotonicity: For sets $A \subset B$:*

$$\mu^*(A) \le \mu^*(B).$$

3. *Countable Subadditivity: Given a countable collection $\{A_n\}$:*

$$\mu^*\left(\bigcup_n A_n\right) \le \sum_n \mu^*(A_n).$$

Carathéodory's result is that any outer measure obtains a measure on the sigma algebra of Carathéodory measurable sets.

Proposition 6.2 (Carathéodory Extension Theorem 1) *Let μ^* be an outer measure defined on a set X. Denote by $\mathcal{C}(X)$ the collection of all subsets of $\sigma(P(X))$ that are **Carathéodory measurable** with respect to μ^*. That is, $A \in \mathcal{C}(X)$ if for all $E \in \sigma(P(X))$:*

$$\mu^*(E) = \mu^*(A \bigcap E) + \mu^*(\tilde{A} \bigcap E). \tag{6.1}$$

Then $\mathcal{C}(X)$ is a complete sigma algebra.
If μ denotes the restriction of μ^* to $\mathcal{C}(X)$, then μ is a measure, and thus $(X, \mathcal{C}(X), \mu)$ is a complete measure space.

Remark 6.3 (On the proof) *The proof below is a generalization of the Lebesgue and Borel measure developments. For example, this requires the proofs of items 2 and 4 in the Borel development of Proposition 5.23 to be repeated in this general context, using only the properties of outer measure identified above. This will be less onerous than it first appears. As noted above,* *these outer measure properties were proved early in both the Lebesgue and Borel settings. It was then these properties, rather than the particular forms of the respective outer measures in (2.4) and (5.9), that were instrumental in the final proofs.*

*Also note that $\emptyset \in \mathcal{C}(X)$ by (6.1), and thus by properties 1 and 3 of Definition 6.1, outer measures also satisfy **finite subadditivity**. This implies that to prove $A \in \mathcal{C}(X)$ it is enough to prove that for all $E \in \sigma(P(X))$:*

$$\mu^*(E) \geq \mu^*(A \bigcap E) + \mu^*(\tilde{A} \bigcap E). \tag{6.2}$$

Proof. 1. $\mathcal{C}(X)$ *is a sigma algebra:* $\mathcal{C}(X)$ *is closed under complementation by symmetry in Definition 6.1, that $A \in \mathcal{C}(X)$ if and only if $\tilde{A} \in \mathcal{C}(X)$.*

To demonstrate that $\mathcal{C}(X)$ is closed under finite unions, it is sufficient by induction to show that if $A_1, A_2 \in \mathcal{C}(X)$, then $A_1 \bigcup A_2 \in \mathcal{C}(X)$. By the above Remark 6.3 we prove that for $E \subset \mathbb{R}$:

$$\mu^*(E) \geq \mu^*(E \bigcap (A_1 \bigcup A_2)) + \mu^*(E \bigcap \tilde{A}_1 \bigcap \tilde{A}_2),$$

since $\widetilde{(A_1 \bigcup A_2)} = \tilde{A}_1 \bigcap \tilde{A}_2$ by De Morgan's laws. Now since A_1 is measurable:

$$\mu^*(E) = \mu^*(E \bigcap A_1) + \mu^*(E \bigcap \tilde{A}_1).$$

Applying (6.1) with $E' \equiv E \bigcap \tilde{A}_1$ and measurable A_2 obtains:

$$\mu^*(E \bigcap \tilde{A}_1) = \mu^*(E \bigcap \tilde{A}_1 \bigcap A_2) + \mu^*(E \bigcap \tilde{A}_1 \bigcap \tilde{A}_2).$$

Combining:

$$\mu^*(E) = \mu^*(E \bigcap A_1) + \mu^*(E \bigcap \tilde{A}_1 \bigcap A_2) + \mu^*(E \bigcap \tilde{A}_1 \bigcap \tilde{A}_2).$$

But since $E \bigcap (A_1 \bigcup A_2) = (E \bigcap A_1) \bigcup (E \bigcap \tilde{A}_1 \bigcap A_2)$, by finite subadditivity:

$$\mu^*(E \bigcap A_1) + \mu^*(E \bigcap \tilde{A}_1 \bigcap A_2) \geq \mu^*(E \bigcap (A_1 \bigcup A_2)).$$

Combining results obtains (6.2) and so $A_1 \bigcup A_2 \in \mathcal{C}(X)$, as are all finite unions by induction.

To prove that $\mathcal{C}(X)$ is closed under countable unions, let $\{A_n\}_{n=1}^{\infty} \subset \mathcal{C}(X)$ and $A \equiv \bigcup_n A_n$. By Proposition 2.20 it can be assumed that the collection $\{A_n\}$ is disjoint, noting that the proof of this earlier result only requires that $\mathcal{C}(X)$ is closed under finite operations, a result already proved. Define $B_n = \bigcup_{j \leq n} A_j$ and note that $B_n \in \mathcal{C}(X)$ by the previous result. Since $B_n \subset A$ implies $\tilde{A} \subset \tilde{B}_n$, we have by monotonicity of μ^* that for any $E \subset \mathbb{R}$:

$$\begin{aligned} \mu^*(E) &= \mu^*(E \bigcap B_n) + \mu^*(E \bigcap \tilde{B}_n) \\ &\geq \mu^*(E \bigcap B_n) + \mu^*(E \bigcap \tilde{A}). \end{aligned} \tag{1}$$

By measurability of A_n applied to $E' \equiv E \cap B_n$:

$$\mu^*(E \cap B_n) = \mu^*(E \cap B_n \cap A_n) + \mu^*(E \cap B_n \cap \tilde{A}_n)$$

$$= \mu^*(E \cap A_n) + \mu^*(E \cap B_{n-1}),$$

nothing that $B_n \cap A_n = A_n$ and $B_n \cap \tilde{A}_n = B_{n-1}$. Then by induction:

$$\mu^*(E \cap B_n) = \sum_{j=1}^{n} \mu^*(E \cap A_j).$$

Combining with (1), this obtains for all n:

$$\mu^*(E) \geq \sum_{j=1}^{n} \mu^*(E \cap A_j) + \mu^*(E \cap \tilde{A}).$$

Since $E \cap A = \bigcup_j (E \cap A_j)$, subadditivity of μ^* yields:

$$\mu^*(E) \geq \sum_{j=1}^{\infty} \mu^*(E \cap A_j) + \mu^*(E \cap \tilde{A})$$

$$\geq \mu^*(E \cap A) + \mu^*(E \cap \tilde{A}).$$

Hence A is measurable and $\mathcal{C}(X)$ is a sigma algebra.

   2. $\mathcal{C}(X)$ is complete: *To show that $\mu^*(A) = 0$ implies that $A \in \mathcal{C}(X)$, we repeat the proofs in the Lebesgue and Borel measure cases. If $\mu^*(A) = 0$. then for any $E \subset \mathbb{R}$ it follows from $E \cap A \subset A$ and monotonicity that $\mu^*(E \cap A) = 0$. Also $\mu^*(E \cap \tilde{A}) \leq \mu^*(E)$ by monotonicity since $E \cap \tilde{A} \subset E$. Combining yields (6.2) and hence $A \in \mathcal{C}(X)$.*

   3. μ is a measure on $\mathcal{C}(X)$: *The final step is to show that $\mu \equiv \mu^*$ is a measure on $\mathcal{C}(X)$. Now $\mu(\emptyset) = \mu^*(\emptyset) = 0$ by outer measure property 1. For finite additivity, which again follows by induction if it is true for two disjoint sets, let $A_1, A_2 \in \mathcal{C}(X)$ be disjoint. Then by definition of μ and the measurability of A_1:*

$$\mu(A_1 \bigcup A_2) \equiv \mu^*(A_1 \bigcup A_2)$$

$$= \mu^*(A_1 \bigcup A_2 \cap A_1) + \mu^*(A_1 \bigcup A_2 \cap \tilde{A}_1)$$

$$= \mu^*(A_1) + \mu^*(A_2)$$

$$\equiv \mu(A_1) + \mu(A_2).$$

For countable additivity, let $A = \bigcup_n A_n \in \mathcal{C}(X)$ be a disjoint union of measurable sets. Then for any n, monotonicity and finite additivity assure:

$$\mu(A) \geq \mu \left(\bigcup_{j \leq n} A_j \right) = \sum_{j \leq n} \mu(A_j),$$

and so

$$\mu(A) \geq \sum_{j=1}^{\infty} \mu(A_j).$$

But the reverse inequality follows from countable subadditivity of μ^*, and the result follows.

   Thus μ is a measure on $\mathcal{C}(X)$.   ∎

6.2.2 Measure on an Algebra to a Complete Measure

The next theorem generalizes the proof of Proposition 5.20, that starting with a measure on an algebra obtained in Proposition 5.13, an outer measure could be defined (Definition 5.16) and proved to have the properties identified in Definition 6.1. This theorem is known as the Hahn-Kolmogorov extension theorem, named for **Hans Hahn** (1879–1934) and **Andrey Kolmogorov** (1903–1987). This theorem requires an algebra $\mathcal{A}$ and a measure $\mu_{\mathcal{A}}$ on this algebra in the sense of Definition 5.9.

If this construction is implemented, as the outer measure μ^* of Proposition 6.2, the sigma algebra $\mathcal{C}(X)$ of the Carathéodory theorem will contain the original algebra of sets, and the measure μ of the Carathéodory theorem will equal $\mu_{\mathcal{A}}$ on this algebra. In other words, μ and $\mathcal{C}(X)$ will truly extend the original measure $\mu_{\mathcal{A}}$ and algebra $\mathcal{A}$.

Proposition 6.4 (Hahn-Kolmogorov Extension theorem) *Let $\mathcal{A}$ be an algebra of sets on X and $\mu_{\mathcal{A}}$ a measure on $\mathcal{A}$ in the sense of Definition 5.9. Then $\mu_{\mathcal{A}}$ gives rise to an outer measure $\mu_{\mathcal{A}}^*$ on $\sigma(P(X))$ such that $\mu_{\mathcal{A}}^*(A) = \mu_{\mathcal{A}}(A)$ for all $A \in \mathcal{A}$.*

   In addition, there exists a complete sigma algebra $\mathcal{C}(X)$ with $\mathcal{A} \subset \mathcal{C}(X)$, and $\mu \equiv \mu_{\mathcal{A}}^*$ is a measure on $\mathcal{C}(X)$.

   Thus $(X, \mathcal{C}(X), \mu)$ is a complete measure space and $\mu(A) = \mu_{\mathcal{A}}(A)$ for all $A \in \mathcal{A}$.

Proof. *The first step is to define a candidate for the outer measure $\mu_{\mathcal{A}}^*$ using the measure $\mu_{\mathcal{A}}$ and the algebra $\mathcal{A}$. As in (5.8), for any set $A \subset X$, define the **outer measure of** A, or the **outer measure of** A **induced by** $\mu_{\mathcal{A}}$, denoted $\mu_{\mathcal{A}}^*(A)$, by:*

$$\mu_{\mathcal{A}}^*(A) = \inf \left\{ \sum_n \mu_{\mathcal{A}}(A_n) \mid A \subset \bigcup_n A_n \right\}, \tag{6.3}$$

where $A_n \in \mathcal{A}$.

   *1. $\mu_{\mathcal{A}}^*$ **is an outer measure**: Since $\emptyset \in \mathcal{A}$ and $\mu_{\mathcal{A}}$ is a measure on $\mathcal{A}$, it follows that $\mu_{\mathcal{A}}(\emptyset) = 0$. Hence $\mu_{\mathcal{A}}^*(\emptyset) = 0$ since by definition, $\mu_{\mathcal{A}}^*(A) \leq \mu_{\mathcal{A}}(A)$ for all $A \in \mathcal{A}$.*

   If $A \subset \bigcup_n A_n$ and $\mu_{\mathcal{A}}^*(A_n) = \infty$ for any n, then countable subadditivity is satisfied. So assume $\mu_{\mathcal{A}}^*(A_n) < \infty$ for all n. Given $\epsilon > 0$, for each n let $\{A_{n_m}\}_{m=1}^{\infty} \subset \mathcal{A}$ with $A_n \subset \bigcup_m A_{n_m}$ and:

$$\sum_m \mu_{\mathcal{A}}(A_{n_m}) < \mu_{\mathcal{A}}^*(A_n) + \epsilon/2^n.$$

This is possible because $\mu^*_{\mathcal{A}}(A_n)$ is equal to the infimum of all such summations by definition. Then since $A \subset \bigcup_{n,m} A_{n_m}$ and $\mu^*_{\mathcal{A}}(A)$ is defined as an infimum:

$$\mu^*_{\mathcal{A}}(A) \leq \sum_{n,m} \mu_{\mathcal{A}}(A_{n_m}) < \sum_n \mu^*_{\mathcal{A}}(A_n) + \epsilon.$$

Thus $\mu^*_{\mathcal{A}}$ is countable subadditive since ϵ is arbitrary.

Finally, monotonicity follows from countable subadditivity and $\mu^*_{A}(\emptyset) = 0$, by choosing $A_1 = B$ and all other $A_n = \emptyset$.

2. $\mu^*_{\mathcal{A}}(A) = \mu_{\mathcal{A}}(A)$ *for all* $A \in \mathcal{A}$: As a measure on an algebra, $\mu_{\mathcal{A}}$ is countably subadditive on $\mathcal{A}$ as noted in Remark 5.15. Hence if $A \in \mathcal{A}$ and $A \subset \bigcup_n A_n$ with $A_n \in \mathcal{A}$, then:

$$\mu_{\mathcal{A}}(A) \leq \sum_n \mu_{\mathcal{A}}(A_n).$$

As this is true for all such collections $\{A_n\}$:

$$\mu_{\mathcal{A}}(A) \leq \inf \sum \mu_{\mathcal{A}}(A_n) \equiv \mu^*_{\mathcal{A}}(A).$$

But for $A \in \mathcal{A}$ choose all $A_1 = A$ and all other $A_n = \emptyset$ as a cover in (6.3), and it follows that $\mu^*_{\mathcal{A}}(A) \leq \mu_{\mathcal{A}}(A)$.

3. **Complete sigma algebra and measure**: The Carathéodory extension theorem now applies to produce a complete sigma algebra $\mathcal{C}(X)$ and measure μ on $\mathcal{C}(X)$, where $\mu \equiv \mu^*_{\mathcal{A}}$ on $\mathcal{C}(X)$.

4. $\mathcal{A} \subset \mathcal{C}(X)$: To prove that $A \in \mathcal{A}$ is Carathéodory measurable, it is enough to prove (6.2) as noted in Remark 6.3. Let $E \subset X$ be given. Then as above, for any $\epsilon > 0$ there is a collection: $\{A_n\} \subset \mathcal{A}$ so that $E \subset \bigcup_n A_n$ and:

$$\sum_n \mu_{\mathcal{A}}(A_n) < \mu^*_{\mathcal{A}}(E) + \epsilon.$$

As $\mu_{\mathcal{A}}$ is a measure on $\mathcal{A}$ and thus finitely additive, it follows that for any n:

$$\mu_{\mathcal{A}}(A_n) = \mu_{\mathcal{A}}(A_n \bigcap A) + \mu_{\mathcal{A}}(A_n \bigcap \tilde{A})$$

$$= \mu^*_{\mathcal{A}}(A_n \bigcap A) + \mu^*_{\mathcal{A}}(A_n \bigcap \tilde{A}),$$

applying step 2.

Now $E \subset \bigcup_n A_n$ obtains:

$$E \bigcap A \subset \bigcup_n (A_n \bigcap A),$$

and

$$E \bigcap \tilde{A} \subset \bigcup_n (A_n \bigcap \tilde{A}).$$

Then by monotonicity and countable subadditivity of $\mu^*_{\mathcal{A}}$ by Definition 6.1:

$$\mu^*_{\mathcal{A}}(E) + \epsilon > \sum_n \mu^*_{\mathcal{A}}(A_n \bigcap A) + \sum_n \mu^*_{\mathcal{A}}(A_n \bigcap \tilde{A})$$

$$\geq \mu^*_{\mathcal{A}}\left(\bigcup_n (A_n \bigcap A)\right) + \mu^*_{\mathcal{A}}\left(\bigcup_n (A_n \bigcap \tilde{A})\right)$$

$$\geq \mu^*_{\mathcal{A}}(E \bigcap A) + \mu^*_{\mathcal{A}}(E \bigcap \tilde{A}).$$

This obtains (6.2) since $\epsilon > 0$ is arbitrary.   ∎

6.2.3 Approximating Carathéodory Measurable Sets

A very useful by-product of defining an outer measure in terms of a measure on an algebra $\mathcal{A}$, is that one obtains various approximations of measurable sets $B \in \mathcal{C}(X)$ with sets derived from the $\mathcal{A}$-sets. For example, if $\mu(B) < \infty$ we have ϵ-approximations by supersets $A \in \mathcal{A}_\sigma$, the collection of countable unions of sets in the algebra $\mathcal{A}$, and subsets $C \in \mathcal{A}_\delta$, the collection of countable intersections of sets in the algebra $\mathcal{A}$. Then using $\mathcal{A}_{\sigma\delta}$ supersets and $\mathcal{A}_{\delta\sigma}$ subsets, we can approximate to a μ-measure of 0. The proof of this result is virtually identical with that of Proposition 5.26 which addressed general Borel measures.

Looking to that earlier proof, these approximations also applied when $\mu(B) = \infty$, but to prove that utilized that $\mathbb{R}$ could be expressed as a countable union of disjoint sets of finite measure. For the measure space $(\mathbb{R}, \mathcal{M}_{\mu_F}(\mathbb{R}), \mu_F)$, this decomposition of $\mathbb{R}$ was achieved with $\{I_n\}$ defined by $I_n = [-n, -n+1)\bigcup[n-1, n)$, and then defining $B_n = B \bigcap I_n$. That these sets had finite measure was assured by monotonicity and property 3 of the definition of a Borel measure, that the measure of a compact set is finite. Then since $B = \bigcup_n B_n$ a disjoint union, the results for $\mu(B) < \infty$ could be generalized to $\mu(B) = \infty$ by countable additivity.

In the case of a general measure space $(X, \mathcal{C}(X), \mu)$, to approximate sets B with $\mu(B) = \infty$ it will be necessary to explicitly assume that $\mu_{\mathcal{A}}$ is a **sigma finite measure**, or σ-**finite measure** as defined in Definition 5.34. Then X can be expressed as a union, and hence by Proposition 2.20 a disjoint union, of sets of finite measure. This definition includes **finite measures** for which $\mu(X) < \infty$, and in particular probability measures for which $\mu(X) = 1$.

Thus for sigma finite measure spaces, the approximations noted above are possible for all $B \in \mathcal{C}(X)$.

Proposition 6.5 (Approximations with $\mathcal{A}$-derived sub/supersets) *Let $\mathcal{A}$ be an algebra of sets on X and $\mu_{\mathcal{A}}$ a σ-finite measure on $\mathcal{A}$. Let the measure μ and the complete sigma algebra $\mathcal{C}(X)$ be as given by Proposition 6.4.*

If $B \in \mathcal{C}(X)$, then given $\epsilon > 0$:

1. *There is a set $A \in \mathcal{A}_\sigma$, the collection of countable unions of sets in the algebra $\mathcal{A}$, so that $B \subset A$ and:*

$$\mu(A) \leq \mu(B) + \epsilon, \qquad \mu(A - B) \leq \epsilon. \tag{6.4}$$

2. *There is a set $C \in \mathcal{A}_\delta$, the collection of countable intersections of sets in the algebra $\mathcal{A}$, so that $C \subset B$ and:*

$$\mu(B) \leq \mu(C) + \epsilon, \qquad \mu_F(B - C) \leq \epsilon. \tag{6.5}$$

3. *There is a set $A' \in \mathcal{A}_{\sigma\delta}$, the collection of countable intersections of sets in $\mathcal{A}_\sigma$, and $C' \in \mathcal{A}_{\delta\sigma}$, the collection of countable unions of sets in $\mathcal{A}_\delta$, so that $C' \subset B \subset A'$ and:*

$$\mu(A' - B) = \mu(B - C') = 0. \tag{6.6}$$

If $\mu_\mathcal{A}$ is a general measure on $\mathcal{A}$, then the above results remain true for all $B \in \mathcal{C}(X)$ with $\mu(B) < \infty$.

Proof. *We address these statements in turn.*
1. *For $B \in \mathcal{C}(X)$, Proposition 6.4 obtains that $\mu(B) \equiv \mu^*_\mathcal{A}(B)$, and so:*

$$\mu(B) = \inf\{\textstyle\sum_j \mu(A_j) | B \subset \bigcup_j A_j\},$$

with $\{A_j\} \subset \mathcal{A}$. This reformulation of the definition of $\mu^*_\mathcal{A}(B)$ reflects that $\mu_\mathcal{A}(A_j) = \mu^*_\mathcal{A}(A_j) = \mu(A_j)$ for such A_j by this proposition. By Proposition 2.20 these $\mathcal{A}$-sets can be assumed to be disjoint, and thus with $A \equiv \bigcup_j A_j$, it follows that $A \in \mathcal{A}_\sigma \subset \mathcal{C}(X)$ and $\mu(A) = \sum_j \mu(A_j)$ for any such collection.

If $\mu(B) < \infty$, then by definition of infimum, for any $\epsilon > 0$ there is a finite or countable disjoint collection $\{A_j\} \subset \mathcal{A}$ with $\mu(A) \leq \mu(B) + \epsilon$, where $A \equiv \bigcup_j A_j \in \mathcal{A}_\sigma$. This proves the first inequality in (6.4). In this case, finite additivity yields:

$$\mu(A) = \mu(B) + \mu(A - B),$$

and so $\mu(A - B) \leq \epsilon$ and the second inequality follows.

If $\mu(B) = \infty$, the first inequality of (6.4) is true by monotonicity of μ with any collection $\{A_j\}$ with $B \subset A \equiv \bigcup_j A_j$, independent of ϵ. For the second inequality, let $\{I_n\} \subset X$ be disjoint measurable sets with $\bigcup_n I_n = X$ and $\mu(I_n) < \infty$ for all n, and define $B_n \equiv B \bigcap I_n$. Then $B_n \in \mathcal{C}(X)$, $\mu(B_n) < \mu(I_n) < \infty$, and so by the result just proved there exists $A_n \in \mathcal{A}_\sigma$ with $B_n \subset A_n$, and $\mu_F(A_n - B_n) \leq \epsilon/2^n$.

Letting $A \equiv \bigcup_n A_n \in \mathcal{A}_\sigma$ obtains $B = \bigcup_n B_n \subset A$, and using De Morgan's laws:

$$A - B \equiv \left(\bigcup_n A_n\right) \bigcap \left(\widetilde{\bigcup_n B_n}\right)$$

$$= \bigcup_n \left(A_n \bigcap \left(\bigcap_n \widetilde{B_n}\right)\right)$$

$$\subset \bigcup_n \left(A_n \bigcap \widetilde{B_n}\right).$$

From the monotonicity and subadditivity of μ:

$$\mu(A - B) \leq \mu\left[\bigcup_n \left(A_n \bigcap \widetilde{B_n}\right)\right]$$

$$\leq \sum \mu(A_n - B_n) \leq \epsilon,$$

which is the second inequality in (6.4) when $\mu(B) = \infty$.

2. Applying the second conclusion in part 1 to $\widetilde{B} \in \mathcal{C}(X)$ assures that there exists $A \in \mathcal{A}_\sigma$ with $\widetilde{B} \subset A$ and $\mu(A - \widetilde{B}) \leq \epsilon$. Since A equals a countable union of elements of $\mathcal{A}$, defining $C = \widetilde{A}$ obtains that C is a countable intersection of complements of elements of $\mathcal{A}$. Because an algebra is closed under complementation, C is also a countable intersection of elements of $\mathcal{A}$ and so $C \in \mathcal{A}_\delta$. Now $C \subset B$ by construction, and since:

$$B - C = B \bigcap \widetilde{C} = A - \widetilde{B},$$

it follows that $\mu(B - C) \leq \epsilon$, which is the second inequality in (6.5).

The first inequality then follows from finite additivity since $B = C \bigcup (B - C)$.

3. For (6.6), define $A_n \in \mathcal{A}_\sigma$ and $C_n \in \mathcal{A}_\delta$ so that the second inequalities in (6.4) and (6.5) are true with $\epsilon = 1/n$. In other words, $C_n \subset B \subset A_n$ and:

$$\mu(A_n - B) \leq 1/n, \quad \mu(B - C_n) \leq 1/n. \tag{1}$$

Letting $A' \equiv \bigcap_n A_n$ and $C' \equiv \bigcup_n C_n$, note that $A' \in \mathcal{A}_{\sigma\delta}$, $C' \in \mathcal{A}_{\delta\sigma}$, and $C' \subset B \subset A'$.

To prove (6.6), define $A'_n \equiv \bigcap_{j \leq n} A_j$. Then $A' \subset A'_n \subset A_n$ implies:

$$A' - B \subset A'_n - B \subset A_n - B,$$

and hence $\mu(A' - B) = 0$ by monotonicity and (1).

The same argument applies to C', defining $C'_n \equiv \bigcup_{j \leq n} C_j$. Then $C_n \subset C'_n \subset C'$ implies:

$$B - C' \subset B - C'_n \subset A - C_n. \qquad \blacksquare$$

6.2.4 Pre-Measure on Semi-Algebra to Measure on Algebra

The final step in the process is another **Carathéodory** result. It generalizes the extension of an initial notion of "measure" on a collection of simple sets, to a true measure on an algebra generated by those sets. In the Borel development, the collection of simple sets was the collection of right semi-closed intervals $\{(a, b]\}$, and the initial notion of measure was F-length. In the Lebesgue development, the sets were the open intervals and initial measure was ordinary interval length.

For the current result, the Borel development is the better model for generalization. The Lebesgue development is more the exception, but justified by the fact that ordinary interval length is the same whether the interval is open, closed or semi-open. As seen in Exercise 5.22 for Borel measures, this property does not generalize.

In general, the collection of simple sets must have enough structure, and the initial notion of measure enough "measure-like" properties, that these can be extended to a true measure on an algebra. The basic properties needed for the initial measure defined on a collection of sets is often formalized by the notion of a **pre-measure**, defined as follows.

This definition generalizes Definition 5.9 of pre-measure or measure on an algebra.

Definition 6.6 (Pre-measure on a collection of sets) *Given a collection of sets, $\mathcal{D}$, a set function $\mu_0 : \mathcal{D} \to [0, \infty]$ is a **pre-measure** if:*

1. $\emptyset \in \mathcal{D}$ and $\mu_0(\emptyset) = 0$, **or:**

1′. **Finite additivity:** *If $\{D_j\}_{j=1}^n \subset \mathcal{D}$ is a disjoint finite collection of sets and $\bigcup_{j=1}^n D_j \in \mathcal{D}$, then:*

$$\mu_0\left(\bigcup_{j=1}^n D_n\right) = \sum_{j=1}^n \mu_0(D_n);$$

2. **Countable additivity:** *If $\{D_n\}_{j=1}^\infty \subset \mathcal{D}$ is a disjoint countable collection of sets and $\bigcup_{j=1}^\infty D_n \in \mathcal{D}$, then:*

$$\mu_0\left(\bigcup_{j=1}^\infty D_n\right) = \sum_{j=1}^\infty \mu_0(D_n).$$

Remark 6.7 (On pre-measure definition)

1. **Either/or:** *Note that items 1 and 2 imply finite additivity in item 1′ since all but finitely many of the D_j could be taken as empty sets. In the absence of item 1, finite additivity must be explicitly assumed in item 1′. Then given item 1′, if it happens that $\emptyset \in \mathcal{D}$, it must be the case that $\mu_0(\emptyset) = 0$ since $\emptyset \bigcup \emptyset = \emptyset$.*

2. **Monotonicity:** *Because μ_0 is nonnegative valued, monotonicity is assured by finite additivity but only in the following **limited sense**. If D, D' are elements of $\mathcal{D}$ with $D \subset D'$, then it follows that $\mu_0(D) \leq \mu_0(D')$ as long as $D' - D \in \mathcal{D}$ as well.*

3. *If the collection $\mathcal{D}$ is endowed with more structure, for example assuming $\{D_n\} \subset \mathcal{D}$ implies $\bigcup_n D_n \in \mathcal{D}$, or $\{D, D'\} \subset \mathcal{D}$ with $D \subset D'$ implies $D' - D \in \mathcal{D}$, a pre-measure will then have the requisite properties of a measure. In other words, a pre-measure would be a measure if only $\mathcal{D}$ had enough structure to allow all the basic set manipulations needed for the measure definition.*

4. *If $\mathcal{D} = \mathcal{A}$: When the collection $\mathcal{D}$ is an **algebra**, it is common to refer to μ_0 as a **measure** and sometimes as a **pre-measure** as noted in Definition 5.9. When $\mathcal{D}$ has less structure than an algebra, such as that of a semi-algebra defined next, μ_0 is virtually always referred to as a pre-measure.*

As for the basic collection of sets, the key requirement is that it must have enough structure so that it is possible to construct an algebra which contains this collection, as well as possible to extend the pre-measure defined on this collection to a measure on that algebra. There are several such structures that have been investigated, and **Carathéodory** worked with **rings** and **semi-rings**, while others since have used **algebras** and **semi-algebras**, the approach we follow.

Recall that algebra is also defined in Definition 2.1. See item 3 of Remark 6.9.

Definition 6.8 (Semi-algebra of sets) *A collection of sets $\mathcal{A}'$ is a **semi-algebra** on a space X:*

1. *If $A_1', A_2' \in \mathcal{A}'$, then $A_1' \bigcap A_2' \in \mathcal{A}'$, and thus this holds for all finite intersections by induction.*

2. *If $A' \in \mathcal{A}'$, then there exists disjoint $\{A_j'\}_{j=1}^n \subset \mathcal{A}'$ so that $\widetilde{A}' = \bigcup_{j=1}^n A_j'$.*

The collection $\mathcal{A}' = \mathcal{A}$, *an* **algebra**, *if in place of item 2 we have:*

2'. *If* $A \in \mathcal{A}$ *then* $\widetilde{A} \in \mathcal{A}$.

Remark 6.9 (On semi-algebra definition)

1. *Note that for a semi-algebra it need not be the case that* $\emptyset$, $X \in \mathcal{A}'$. *The reason for this is that while item 2 assures existence of disjoint* $\{A'_j\}_{j=1}^n \subset \mathcal{A}'$ *so that* $\widetilde{A}' = \bigcup_j A'_j$, *we cannot in general conclude that* $\widetilde{A}' \in \mathcal{A}$ *since a semi-algebra is not required to be closed under finite unions. However, if* $X \in \mathcal{A}'$ *then item 2 assures that* $\emptyset \in \mathcal{A}'$. *If* $\emptyset \in \mathcal{A}'$ *we can only conclude that* $X = \bigcup_{j=1}^n A'_j$ *with disjoint* $\{A'_j\}_{j=1}^n \subset \mathcal{A}'$.

2. *If* $\widetilde{A}' \in \mathcal{A}'$ *for even one* $A' \in \mathcal{A}'$, *then item 1 assures that* $\emptyset \in \mathcal{A}'$. *But we can not then conclude that* $A' \bigcup \widetilde{A}' = X \in \mathcal{A}'$ *since a semi-algebra is not required to be closed under finite unions.*

3. *In contrast, once item 2 is replaced by item 2', then* $\emptyset$, $X \in \mathcal{A}$ *and the resulting algebra is closed under finite unions. In detail,* $\emptyset \in \mathcal{A}$ *since for any* $A \in \mathcal{A}$, $A \bigcap \widetilde{A} = \emptyset \in \mathcal{A}$ *by 1, and then* $\widetilde{\emptyset} = X \in \mathcal{A}$ *by item 2'. The conclusion on finite unions follows from item 2' and item 1, recalling De Morgan's laws, that given* $A_1, A_2 \in \mathcal{A}$:

$$A_1 \bigcup A_2 = \widetilde{\widetilde{A_1} \bigcap \widetilde{A_2}},$$

which generalizes to all finite unions by induction. Thus this definition and Definition 2.1 are equivalent for algebras.

Exercise 6.10 (Algebra generated by a semi-algebra) *Prove that given a semi-algebra* $\mathcal{A}'$, *the collection consisting of the empty set and* **all finite disjoint unions** *of elements of the semi-algebra forms an algebra* $\mathcal{A}$. *If the semi-algebra already contains* $\emptyset$, *the inclusion of this set is unnecessary. Hint: Use the above version of the definition of algebra, reflecting complementation and finite intersections, and De Morgan's laws. Then, to prove that a given set is in* $\mathcal{A}$ *is to prove that it can be represented as a finite disjoint union of* $\mathcal{A}'$-sets.

Notation 6.11 ($\mathcal{A}'$ and $\mathcal{A}$) *As implied in Definition 6.8, we will denote semi-algebras by* $\mathcal{A}'$, *and algebras by* $\mathcal{A}$. *Often* $\mathcal{A}$ *will be the algebra generated by* $\mathcal{A}'$ *as in Exercise 6.10.*

Example 6.12 (Borel semi-algebra) *In the Borel development, the collection of right semi-closed intervals* $\{(a, b]\}$ *of Example 2.4 forms a semi-algebra* $\mathcal{A}'$ *that contains* $\emptyset$ *and* $\mathbb{R}$, *recalling the convention that* $(a, \infty] \equiv (a, \infty)$.

Firstly, the intersection of two such intervals is of the same type or empty. The empty set, $\emptyset = (a, a]$ *for any* a, *and is hence a member of this collection. Also, if* $-\infty < a, b < \infty$:

$$\widetilde{(a, b]} = (-\infty, a] \bigcup (b, \infty),$$

while for the other cases of $(-\infty, b]$ *or* (a, ∞):

$$\widetilde{(-\infty, b]} = (b, \infty),$$
$$\widetilde{(a, \infty)} = (-\infty, a].$$

The collection of all finite unions of elements of this semi-algebra forms an algebra $\mathcal{A}$ *as demonstrated in Example 2.4. This earlier example may appear at odds with Exercise 6.10, which used finite **disjoint** unions of elements. But recall that in the earlier example it was demonstrated that every* $A \in \mathcal{A}$ *had a representation with disjoint* $\mathcal{A}'$-*sets.*

For a general semi-algebra, if the associated algebra of Exercise 6.10 was defined with arbitrary finite unions, it would not in general be possible to restate such unions in terms of disjoint sets. The reason for this is that the construction in Proposition 2.20 would not in general produce $\mathcal{A}'$-*sets.*

The final extension result implicit in the Borel development is again attributable to **Constantin Carathéodory**. It addresses the question of when a pre-measure on a semi-algebra can be extended to a measure on the associated algebra of all finite disjoint unions of sets (recall Exercise 6.10). See Proposition 6.18 for a modest generalization.

This proof is long because we must carefully translate the desired statements in $\mathcal{A}$ to statements in $\mathcal{A}'$, so that the pre-measure properties can be applied.

Proposition 6.13 (Carathéodory Extension Theorem 2) *Let* $\mathcal{A}'$ *be a semi-algebra and* μ_0 *a pre-measure on* $\mathcal{A}'$. *Then* μ_0 *can be extended to a measure* $\mu_{\mathcal{A}}$ *on the algebra* $\mathcal{A}$, *defined as the collection of all finite disjoint unions of sets in* $\mathcal{A}'$, *including* $\emptyset$, *if necessary.*

Proof. 1. Definition and well-definedness of $\mu_{\mathcal{A}}$: *For* $A \in \mathcal{A}$ *given by* $A = \bigcup_{k=1}^{m} A_k'$ *with disjoint* $A_k' \in \mathcal{A}'$, *define:*

$$\mu_{\mathcal{A}}(A) \equiv \sum_{k=1}^{m} \mu_0(A_k'). \tag{1}$$

Thus in particular, $\mu_{\mathcal{A}} = \mu_0$ *on* $\mathcal{A}'$.

To demonstrate that this definition is independent of the collection of disjoint sets used to define A, *assume also that* $A = \bigcup_{j=1}^{n} B_j'$ *with disjoint* $B_j' \in \mathcal{A}'$. *Define* $A_{jk}' = B_j' \bigcap A_k'$ *for* $1 \le j \le n$ *and* $1 \le k \le m$, *and note that* $\{A_{jk}'\} \subset \mathcal{A}'$, *by Definition 6.8 and that these sets are disjoint with potentially many empty. Also:*

$$\bigcup_j A_{jk}' = A_k', \quad \bigcup_k A_{jk}' = B_j'.$$

Because $A_k', B_j' \in \mathcal{A}'$ *and* μ_0 *is finitely additive on* $\mathcal{A}'$ *by Definition 6.6:*

$$\mu_0(A_k') = \sum_{j=1}^{n} \mu_0(A_{jk}'), \quad \mu_0(B_j') = \sum_{k=1}^{m} \mu_0(A_{jk}').$$

It then follows by summation that:

$$\sum_{k=1}^{m} \mu_0(A'_k) = \sum_{j=1}^{n} \mu_0(B'_j) = \sum_{k=1}^{m} \sum_{j=1}^{n} \mu_0(A'_{jk}).$$

Note that this double sum can be ordered even if infinite since $\mu_0(A'_{jk}) \geq 0$ for all j, k. Thus $\mu_{\mathcal{A}}$ is well defined.

   2. **Finite additivity of** $\mu_{\mathcal{A}}$: *Let disjoint $\{A_j\}_{j=1}^{n} \subset \mathcal{A}$, and denote $A \equiv \bigcup_{j=1}^{n} A_j \in \mathcal{A}$. We must show that:*

$$\mu_{\mathcal{A}}(A) = \sum_{j=1}^{n} \mu_{\mathcal{A}}(A_j). \tag{2}$$

By definition of $\mathcal{A}$, each $A_j = \bigcup_{k=1}^{n_j} A'_{jk}$ with disjoint $\{A'_{jk}\}_{k=1}^{n_j} \subset \mathcal{A}'$, and so $\{A'_{jk}\}_{j=1,k=1}^{n,n_j} \subset \mathcal{A}'$ are disjoint. Because $A \in \mathcal{A}$, it also follows by definition that $A = \bigcup_{l=1}^{m} B'_l$ with disjoint $\{B'_l\}_{l=1}^{m} \subset \mathcal{A}'$. Summarizing:

$$A = \bigcup_{l=1}^{m} B'_l = \bigcup_{j=1}^{n} \bigcup_{k=1}^{n_j} A'_{jk},$$

with disjoint $\mathcal{A}'$-sets in both union expressions.

   Now define $A'_{jkl} = B'_l \bigcap A'_{jk}$. Then $A'_{jkl} \in \mathcal{A}'$ for all indexes by Definition 6.8, these sets are disjoint, and:

$$B'_l = \bigcup_{j=1}^{n} \bigcup_{k=1}^{n_j} A'_{jkl}.$$

Finite additivity of μ_0 on $\mathcal{A}'$ obtains:

$$\mu_0\left(B'_l\right) = \sum_{j=1}^{n} \sum_{k=1}^{n_j} \mu_0\left(A'_{jkl}\right),$$

and so by (1):

$$\mu_{\mathcal{A}}(A) \equiv \sum_{l=1}^{m} \mu_0\left(B'_l\right) = \sum_{l=1}^{m} \sum_{j=1}^{n} \sum_{k=1}^{n_j} \mu_0\left(A'_{jkl}\right). \tag{3}$$

To prove (2), observe that:

$$A'_{jk} = \bigcup_{l=1}^{m} A'_{jkl},$$

so finite additivity again provides:

$$\mu_0\left(A'_{jk}\right) = \sum_{l=1}^{m}\mu_0\left(A'_{jkl}\right).$$

Thus:

$$\mu_{\mathcal{A}}(A_j) \equiv \sum_{k=1}^{n_j}\mu_0\left(A'_{jk}\right) = \sum_{k=1}^{n_j}\sum_{l=1}^{m}\mu_0\left(A'_{jkl}\right).$$

Summing over j and comparing with (3) proves (2) and finite additivity. As noted above, this triple sum can be reordered even if infinite.

*3. **Countable additivity of $\mu_{\mathcal{A}}$:** For countable additivity, the proof is identical to that for finite additivity with $n = \infty$, though we must verify the needed manipulations. Let $\{A_j\}_{j=1}^{\infty} \subset \mathcal{A}$ be a disjoint countable collection of sets with $\bigcup_j A_j = A \in \mathcal{A}$. Note that it is an assumption that such $A \in \mathcal{A}$ since in general algebras are not closed under countable unions. For countable additivity, we must show that:*

$$\mu_{\mathcal{A}}(A) = \sum_{j=1}^{\infty}\mu_{\mathcal{A}}\left(A_j\right). \tag{4}$$

By definition of $\mathcal{A}$, each $A_j = \bigcup_{k=1}^{n_j}A'_{jk}$ as a finite union of disjoint $\mathcal{A}'$-sets, and so $\{A'_{jk}\}_{j=1,k=1}^{\infty,n_j} \subset \mathcal{A}'$ are disjoint. Also, $A \in \mathcal{A}$ is again a finite disjoint union of $\mathcal{A}'$-sets, say $A = \bigcup_{l=1}^{m}B'_l$. Now define $A'_{jkl} = A'_{jk}\bigcap B'_l$ for $k \leq n_j, l \leq m$, and all j. Then $\{A'_{jkl}\}_{jkl} \subset \mathcal{A}'$ are disjoint and:

$$B'_l = \bigcup_{j=1}^{\infty}\bigcup_{k=1}^{n_j}A'_{jkl}.$$

Since $B'_l \in \mathcal{A}'$, countable additivity of μ_0 on $\mathcal{A}'$ obtains:

$$\mu_0\left(B'_l\right) = \sum_{j=1}^{\infty}\sum_{k=1}^{n_j}\mu_0\left(A'_{jkl}\right),$$

and so by (1):

$$\mu_{\mathcal{A}}(A) \equiv \sum_{l=1}^{m}\mu_0\left(B'_l\right) = \sum_{l=1}^{m}\sum_{j=1}^{\infty}\sum_{k=1}^{n_j}\mu_0\left(A'_{jkl}\right). \tag{5}$$

To prove (4), observe that:

$$A'_{jk} = \bigcup_{l=1}^{m}A'_{jkl},$$

so finite additivity again provides:

$$\mu_0\left(A'_{jk}\right) = \sum_{l=1}^{m} \mu_0\left(A'_{jkl}\right).$$

Thus:

$$\mu_{\mathcal{A}}(A_j) \equiv \sum_{k=1}^{n_j} \mu_0\left(A'_{jk}\right) = \sum_{k=1}^{n_j}\sum_{l=1}^{m} \mu_0\left(A'_{jkl}\right).$$

Summing over j and comparing with (5) proves (4) and countable additivity, since this triple sum can be reordered as noted above, even if infinite. ∎

6.2.5 Uniqueness of Extensions 1

In this section we prove that if $\mu_{\mathcal{A}}$ defined on the algebra $\mathcal{A}$ is **sigma finite**, then the extension to μ in Proposition 6.4 is unique on the **smallest sigma algebra** that contains $\mathcal{A}$. By Definition 5.34, that $\mu_{\mathcal{A}}$ is sigma finite or σ-finite means that the measure space X can be expressed as a countable union of measurable sets of finite measure. That is, there exists $\{X_j\} \subset \sigma(X)$ so that:

$$X = \bigcup_{j=1}^{\infty} X_j, \quad \mu_{\mathcal{A}}(X_j) < \infty, \text{ all } j. \tag{6.7}$$

By Proposition 2.20, there is no loss of generality in assuming that the collection $\{X_j\}$ is disjoint.

Considering the measures and measure spaces of greatest interest in these books, **all Lebesgue and Borel measures** on $\mathbb{R}$ (and $\mathbb{R}^n$ defined below) are sigma finite, as are **all probability measures** and more generally **all finite measures** $\mu_{\mathcal{A}}$, meaning $\mu_{\mathcal{A}}(X) < \infty$.

It is common to see this assumption stated either as the measure μ on X is sigma finite, or, the measure space $(X, \sigma(X), \mu)$ is sigma finite. Recalling Proposition 6.13 which extended the pre-measure μ_0 from the semi-algebra $\mathcal{A}'$ to the measure $\mu_{\mathcal{A}}$ on the associated algebra $\mathcal{A}$, and Proposition 6.4 which extended $\mu_{\mathcal{A}}$ on the algebra $\mathcal{A}$ to the measure μ on the complete sigma algebra $\mathcal{C}(X)$, note that:

1. If μ_0 is **finite** or σ-**finite** on $\mathcal{A}'$, which of necessity is now assumed to contain X, then $\mu_{\mathcal{A}}$ will be **finite** or σ-**finite** on $\mathcal{A}$ since $\mu_{\mathcal{A}} = \mu_0$ on $\mathcal{A}'$.

2. If $\mu_{\mathcal{A}}$ is **finite** or σ-**finite** on $\mathcal{A}$, the extension to μ via the outer measure $\mu_{\mathcal{A}}^*$ will have the same property since $\mathcal{A} \subset \mathcal{C}(X)$ and $\mu = \mu_{\mathcal{A}}$ on $\mathcal{A}$.

For the statement of the following result, note that while μ is defined as the extension of $\mu_{\mathcal{A}}$ induced by the outer measure $\mu_{\mathcal{A}}^*$, the notion that μ' is another extension of $\mu_{\mathcal{A}}$ is meant quite generally. In other words, there is no implied construction by which μ' is so created. That μ' is an extension of $\mu_{\mathcal{A}}$ simply means that μ' is defined on some sigma algebra $\sigma(X)$, that $\mathcal{A} \subset \sigma(X)$, and for all $A \in \mathcal{A}$ that $\mu'(A) = \mu_{\mathcal{A}}(A)$. Any such measure is automatically defined on $\sigma(\mathcal{A})$, the smallest sigma algebra that contains $\mathcal{A}$ since by definition $\sigma(\mathcal{A}) \subset \sigma(X)$. The uniqueness result states that in the sigma finite case, every such extension of $\mu_{\mathcal{A}}$ must agree on this smallest sigma algebra.

In the examples below this result will be applied to the Lebesgue and Borel constructions.

See also Proposition 6.24 for an extension of this result.

Proposition 6.14 (Uniqueness of Extensions to $\sigma(\mathcal{A})$) *Let $\mu_{\mathcal{A}}$ be a σ-finite measure on an algebra $\mathcal{A}$, and μ the extension of $\mu_{\mathcal{A}}$ to the sigma algebra $\mathcal{C}(X)$, induced by the outer measure $\mu_{\mathcal{A}}^*$. Let μ' be any other extension of $\mu_{\mathcal{A}}$ from $\mathcal{A}$ to a sigma algebra $\sigma(X)$.*
Then for all $B \in \sigma(\mathcal{A})$:

$$\mu(B) = \mu'(B),$$

where $\sigma(\mathcal{A})$ denotes the smallest sigma algebra that contains $\mathcal{A}$.

More generally, if μ, μ' are σ-finite measures on an sigma algebra $\sigma(X)$, and $\mu(B) = \mu'(B)$ for all $B \in \mathcal{A}$ for some algebra $\mathcal{A} \subset \sigma(X)$, then $\mu = \mu'$ on $\sigma(\mathcal{A}) \subset \sigma(X)$.

Proof. *Note that $\mathcal{A} \subset \sigma(\mathcal{A})$ by definition, and as extensions of $\mu_{\mathcal{A}}$, it follows that $\mu = \mu'$ on $\mathcal{A}$. But then by countable additivity, $\mu = \mu'$ on $\mathcal{A}_{\sigma} \subset \sigma(\mathcal{A})$, the collection of countable unions of members of $\mathcal{A}$. Recalling Proposition 2.20, any such union can be expressed as a countable union of disjoint sets, and thus countable additivity applies.*

We first prove that if $B \in \sigma(\mathcal{A})$ with $\mu(B) < \infty$, then $\mu'(B) = \mu(B)$. Such B is automatically μ-measurable and μ'-measurable since as the smallest such sigma algebra that contains $\mathcal{A}$, both $\sigma(\mathcal{A}) \subset \mathcal{C}(X)$ and $\sigma(\mathcal{A}) \subset \sigma(X)$. An application of Proposition 6.5 provides that for any $\epsilon > 0$ there is an $A \in \mathcal{A}_{\sigma}$ with $B \subset A$ and:

$$\mu(A) \le \mu(B) + \epsilon. \tag{1}$$

Now $B \subset A$ implies my monotonicity that $\mu'(B) \le \mu'(A)$, while $\mu'(A) = \mu(A)$ since $A \in \mathcal{A}_{\sigma}$ as proved above. Combining, $\mu'(B) \le \mu(B) + \epsilon$ for all $\epsilon > 0$ and thus:

$$\mu'(B) \le \mu(B). \tag{2}$$

By μ-measurability of $B \in \sigma(\mathcal{A})$ and since $A \in \mathcal{A}_{\sigma}$:

$$\mu(A) = \mu(B) + \mu(A - B).$$

As $\mu(B) < \infty$ by assumption, subtraction and (1) obtain that $\mu(A-B) < \epsilon$, and then $\mu'(A-B) < \epsilon$ by (2).
Then, since $B \subset A$ and $A \in \mathcal{A}_{\sigma}$:

$$\begin{aligned}
\mu(B) &\le \mu(A) \\
&= \mu'(A) \\
&= \mu'(B) + \mu'(A - B) \\
&\le \mu'(B) + \epsilon,
\end{aligned}$$

and so $\mu_{\mathcal{A}}(B) \leq \mu'(B)$ as above. Combining with (2) it follows that if $B \in \sigma(\mathcal{A})$ with $\mu(B) < \infty$, then:

$$\mu'(B) = \mu(B).$$

Now for arbitrary $B \in \sigma(\mathcal{A})$, let $\{X_i\} \subset \mathcal{A}$ be a countable collection of disjoint sets with $\bigcup_i X_i = X$ and $\mu(X_i) < \infty$ for all i. Such a collection exists by the sigma finiteness of $\mu_{\mathcal{A}}$. Then:

$$B = \bigcup_i (B \bigcap X_i),$$

is a disjoint union, and so by countable additivity both $\mu(B)$ and $\mu'(B)$ equal summations of the respective measures of these sets. But since $B \bigcap X_i \in \sigma(\mathcal{A})$ and $\mu(B \bigcap X_i) < \infty$, the above result applies and $\mu'(B \bigcap X_i) = \mu(B \bigcap X_i)$ for all i.

Hence $\mu'(B) = \mu(B)$ for all $B \in \sigma(\mathcal{A})$.

More generally, given μ, μ' on $\sigma(X)$, then $\mu_{\mathcal{A}} \equiv \mu$ is a measure on $\mathcal{A}$ and thus has an extension $\widetilde{\mu}$ to the sigma algebra $\mathcal{C}(X)$, induced by the outer measure $\mu^*_{\mathcal{A}}$. But then by the above result $\widetilde{\mu} = \mu$ and $\widetilde{\mu} = \mu'$ on $\sigma(\mathcal{A})$. ∎

Example 6.15 (Uniqueness of Borel Measure) *In the Borel measure development, $\mathcal{A}'$ was defined as the semi-algebra of right semi-closed intervals, and the pre-measure μ_0 was defined by F-length in (5.4) as induced by a right continuous, increasing function $F(x)$:*

$$\mu_0\{(a,b]\} \equiv F(b) - F(a).$$

This pre-measure is sigma finite in general since

$$\mu_0(\mathbb{R}) = \sum_{n \in \mathbb{Z}} \mu_0\{(n, n+1]\},$$

and $\mu_0\{(n, n+1]\}$ is finite for all n since $F(x)$ is real-valued. In the case of bounded $F(x)$, μ_0 is in fact a finite pre-measure. The associated $\mu_{\mathcal{A}}$ and μ, denoted $\mu_{\mathcal{A}}$ and μ_F respectively in Chapter 5, are similarly sigma finite or finite.

Thus the extension to $\mu \equiv \mu_F$ is unique on the smallest sigma algebra that contains the algebra $\mathcal{A}$. Since this smallest sigma algebra contains the intervals, it follows that it must be the Borel sigma algebra, $\mathcal{B}(\mathbb{R})$. Thus μ_F is the unique extension of $\mu_{\mathcal{A}}$ from $\mathcal{A}$ to $\mathcal{B}(\mathbb{R})$. Recalling Section 5.3 on consistency of Borel constructions, this affirmatively answers the second question posited there.

These results do not assure that μ_F is the unique extension of $\mu_{\mathcal{A}}$ from $\mathcal{A}$ to the larger complete sigma algebra $\mathcal{M}_{\mu_F}(\mathbb{R})$. But as will be noted in Section 6.5, uniqueness extends to $\mathcal{M}_{\mu_F}(\mathbb{R})$.

Example 6.16 (Uniqueness of Lebesgue Measure) *As another application of this uniqueness result consider the Borel development with $F(x) = x + c$. It was confirmed in Corollary 5.31 that the resulting Borel measure μ_F is identical to Lebesgue measure m, but we investigate this anew from an alternative perspective.*

First, the Lebesgue pre-measure was defined on the collection $\mathcal{D}$ of all open intervals, which is not a semi-algebra. In other words, in the Lebesgue case, we began with:

$$\mu_0\{(a,b)\} = b - a \text{ for } (a,b) \in \mathcal{D}.$$

In the Borel development, we would have begun with:

$$\mu_0\{(a,b]\} = b - a \text{ for } (a,b] \in \mathcal{A}',$$

where $\mathcal{A}'$ was defined as the semi-algebra of right semi-closed intervals.

The associated outer measures, $\mu^*_{\mathcal{A}}$ and m^*, were then defined in terms of these pre-measures and associated collections of intervals.

In Proposition 2.28 it was shown that for any interval, the Lebesgue outer measure m^* reproduced interval length. In other words, $m^*\{(a,b]\} = b - a$ for $(a,b] \in \mathcal{A}'$. Hence, once both developments reach the outer measure step, m^* and $\mu^*_{\mathcal{A}}$ agree on the semi-algebra $\mathcal{A}'$ and hence agree on the algebra $\mathcal{A}$ of all finite unions of right semi-closed intervals.

So by the above result, the extensions of m^* and $\mu^*_{\mathcal{A}}$ to the smallest sigma algebra that contains $\mathcal{A}$ is unique. As this sigma algebra is $\mathcal{B}(\mathbb{R})$ as noted above, we conclude that on the Borel sigma algebra:

$$m(A) = \mu_F(A) \text{ for } A \in \mathcal{B}(\mathbb{R}).$$

Here μ_F is the Borel measure associated with $F(x) = x + c$.

Again, we can not conclude from the above result that this identity extends to the complete sigma algebras $\mathcal{M}_{\mu_F}(\mathbb{R})$ and $\mathcal{M}_L(\mathbb{R})$, nor even that these sigma algebras agree. But this will be resolved affirmatively in Section 6.5 below.

6.3 Summary of Construction Process

The above extension and uniqueness propositions provide a road map to the construction of a complete measure space $(X, \mathcal{C}(X), \mu)$, which generalizes the development of general Borel measures.

The construction process is as follows:

1. **Required Step (Choose** (a) or (b)**):**

   (a) Define a set function μ_0 on a semi-algebra $\mathcal{A}'$, which can be shown to be a pre-measure.

   (b) Define a set function $\mu_{\mathcal{A}}$ on an algebra $\mathcal{A}$, which can be shown to be a measure.

   The algebra $\mathcal{A}$ in 1(b) can be the algebra generated by a given subalgebra $\mathcal{A}'$ by Exercise 6.10, or defined independently of a semi-algebra.

   Remark 6.17 (On measure/pre-measure verification) *For this required step, it is only necessary to demonstrate countable additivity in the cases where the disjoint union produces a set in the semi-algebra $\mathcal{A}'$, or in the algebra $\mathcal{A}$. Similarly, the demonstration of finite additivity on $\mathcal{A}'$ can again be restricted to these special cases. These demonstrations* *can be challenging enough as will be seen, but still immeasurably easier than such a demonstration on the sigma algebra generated by these classes.*

This will be easier because if a set A is a member of a semi-algebra $\mathcal{A}'$ or algebra $\mathcal{A}$, we will typically know a great deal more about the properties of this set than simply knowing that A is a member of the associated sigma algebras. And it is precisely these properties which motivate the potential approaches to the desired verifications.

2. **"Free" Steps**:

   (a) From 1(a), if μ_0 is a pre-measure on a semi-algebra $\mathcal{A}'$, then by the Carathéodory Extension theorem II of Proposition 6.13, μ_0 can be extended to a measure $\mu_{\mathcal{A}}$ on $\mathcal{A}$, the algebra generated by $\mathcal{A}'$ as in Exercise 6.10. Alternatively from 1(b), we have a measure $\mu_{\mathcal{A}}$ on an algebra $\mathcal{A}$.

   (b) In either case from step 2(a), $\mu_{\mathcal{A}}$ and $\mathcal{A}$ then generate an outer measure $\mu_{\mathcal{A}}^*$ on $\sigma(P(X))$, the power set sigma algebra on X. This follows from the Hahn-Kolmogorov Extension theorem of Proposition 6.4 with $\mu_{\mathcal{A}}^*$ is defined as in (6.3), and this outer measure then satisfies the conditions specified in the Carathéodory Extension theorem I of Proposition 6.2.

   (c) The collection of sets that are Carathéodory measurable with respect to $\mu_{\mathcal{A}}^*$ as defined in (6.1) then forms a complete sigma algebra $\mathcal{C}(X)$, and the restriction of $\mu_{\mathcal{A}}^*$ to $\mathcal{C}(X)$ is a measure μ, and hence $(X, \mathcal{C}(X), \mu)$ is a complete measure space. Moreover, $\mathcal{A} \subset \mathcal{C}(X)$ and μ extends $\mu_{\mathcal{A}}$ in that $\mu(A) = \mu_A(A)$ for all $A \in \mathcal{A}$.

   (d) If μ_0 from 1(a) is a finite or σ-finite pre-measure on $\mathcal{A}'$, then so too is $\mu_{\mathcal{A}}$ constructed from μ_0, and μ constructed from $\mu_{\mathcal{A}}$. The same applies if $\mu_{\mathcal{A}}$ is a finite or σ-finite measure on $\mathcal{A}$. In these cases μ will be the unique extension of $\mu_{\mathcal{A}}$ to the smallest sigma algebra containing $\mathcal{A}$ by Proposition 6.14.

   (e) In the σ-finite case, given $B \in \mathcal{C}(X)$ and $\epsilon > 0$, there is by Proposition 6.5 a superset $A \in \mathcal{A}_\sigma$ and subset $C \in \mathcal{A}_\delta$ so that:

   $$\mu(A - B) < \epsilon \text{ and } \mu(B - C) < \epsilon.$$

   In addition, there is a superset $A' \in \mathcal{A}_{\sigma\delta}$ and subset $C' \subset \mathcal{A}_{\delta\sigma}$ so that:

   $$\mu(A' - B) = \mu(B - C') = 0.$$

   (f) In general constructions that are not σ-finite, conclusion 2(e) applies to sets $B \in \mathcal{C}(X)$ with $\mu(B) < \infty$.

The above road map makes it clear that to produce a complete measure space we need to focus on the **Required Step**. In general, this step begins with a plausible definition of a set function μ_0, defined on a class of sets $\mathcal{D}$. From this starting point, there are usually two challenges to completing this required step:

1. The collection $\mathcal{D}$ must be expanded to a semi-algebra $\mathcal{A}'$ or an algebra $\mathcal{A}$, and the definition of μ_0 extended to this collection.

2. It must be proved that the set function μ_0 is a pre-measure on the semi-algebra $\mathcal{A}'$, or that the set function $\mu_{\mathcal{A}}$ is a measure on the algebra $\mathcal{A}$. In the former case, 2(a) automatically yields a measure $\mu_{\mathcal{A}}$ on the algebra $\mathcal{A}$ generated by $\mathcal{A}'$.

In most applications it is the demonstration of the pre-measure or measure properties that is the greatest challenge, and demonstrating countable additivity is usually the most challenging part of this step.

The next section provides some useful results for such a demonstration.

6.4 Approaches to Countable Additivity

6.4.1 Countable Additivity on a Semi-Algebra

The following result provides an approach which circumvents the need to directly prove countable additivity on the given semi-algebra $\mathcal{A}'$. It states that finite additivity and countable **subadditivity** on $\mathcal{A}'$ are sufficient to ensure a unique extension to a measure $\mu_{\mathcal{A}}$ on $\mathcal{A}$, the algebra generated by $\mathcal{A}'$ identified in Exercise 6.10.

Thus this result represents a modest generalization of the **Carathéodory Extension Theorem** 2 of Proposition 6.13.

Proposition 6.18 (Set function μ_0 on $\mathcal{A}'$ to $\mu_{\mathcal{A}}$ on $\mathcal{A}$) *Let μ_0 be a non-negative set function defined on a semi-algebra $\mathcal{A}'$ where $\emptyset \in \mathcal{A}'$. Assume that μ_0 satisfies:*

1. *Finite Additivity: If $\{A_i'\}_{i=1}^n \subset \mathcal{A}'$ are disjoint and $A' \equiv \bigcup_{i=1}^n A_i' \in \mathcal{A}'$:*

$$\mu_0\left(A'\right) = \sum_{i=1}^n \mu_0(A_i').$$

2. *Countable Subadditivity: If $\{A_i'\}_{i=1}^\infty \subset \mathcal{A}'$ are disjoint with $A' \equiv \bigcup_{i=1}^\infty A_i' \in \mathcal{A}'$:*

$$\mu_0\left(A'\right) \leq \sum_{i=1}^\infty \mu_0(A_i').$$

Then μ_0 has a unique extension to a measure $\mu_{\mathcal{A}}$ on $\mathcal{A}$, the algebra generated by $\mathcal{A}'$.

Proof. *As $\mathcal{A}$ is the collection of all finite disjoint unions of sets in $\mathcal{A}'$, it follows that $\emptyset \in \mathcal{A}$ and $\mu_0(\emptyset) = 0$ by finite additivity.*

1. Well-definedness and finite additivity of $\mu_{\mathcal{A}}$: Recalling the proof of Proposition 6.13, we again define $\mu_{\mathcal{A}}$ on $A \equiv \bigcup_{i=1}^n A_i \in \mathcal{A}$ by:

$$\mu_{\mathcal{A}}(A) \equiv \sum_{i=1}^n \mu_0(A_i). \tag{1}$$

So in particular, $\mu_{\mathcal{A}} = \mu_0$ on $\mathcal{A}'$. The proof that $\mu_{\mathcal{A}}$ is a well-defined and finitely additive set function on $\mathcal{A}$ is identical in every detail with the former proof, since all that was used was finite additivity of μ_0 on $\mathcal{A}'$.

2. Monotonicity of $\mu_{\mathcal{A}}$: *We first prove that finite additivity implies monotonicity on $\mathcal{A}$. To see this, if $A, A' \in \mathcal{A}$ with $A' \subset A$, then as an algebra $A - A' = A \bigcap \tilde{A}' \in \mathcal{A}$. Thus $A = A' \bigcup (A - A')$ is a disjoint union of $\mathcal{A}$-sets, and by finite additivity of $\mu_{\mathcal{A}}$:*

$$\mu_{\mathcal{A}}(A') \leq \mu_{\mathcal{A}}(A),$$

since $\mu_{\mathcal{A}}(A - A') \geq 0$.

3. Countable additivity of $\mu_{\mathcal{A}}$: *Let $A = \bigcup_{j=1}^{\infty} A_j \in \mathcal{A}$, a disjoint union of $\mathcal{A}$-sets. It is an assumption that such $A \in \mathcal{A}$ since in general, algebras are not closed under countable unions. However, $\bigcup_{j=1}^{n} A_j \in \mathcal{A}$ as an algebra, and $\bigcup_{j=1}^{n} A_j \subset A$. Applying finite additivity and monotonicity of $\mu_{\mathcal{A}}$:*

$$\sum_{j=1}^{n} \mu_{\mathcal{A}}(A_j) = \mu_{\mathcal{A}}\left(\bigcup_{j=1}^{n} A_j\right) \leq \mu_{\mathcal{A}}(A).$$

As this is true for all n, this implies that:

$$\sum_{j=1}^{\infty} \mu_{\mathcal{A}}(A_j) \leq \mu_{\mathcal{A}}(A). \tag{2}$$

For countable additivity of $\mu_{\mathcal{A}}$, we are left to show:

$$\mu_{\mathcal{A}}(A) \leq \sum_{j=1}^{\infty} \mu_{\mathcal{A}}(A_j). \tag{3}$$

By definition of $\mathcal{A}$, each $A_j = \bigcup_{k=1}^{n_j} A'_{jk}$ as a finite union of disjoint $\mathcal{A}'$-sets, and so $\{A'_{jk}\}_{j=1,k=1}^{\infty,n_j} \subset \mathcal{A}'$ are disjoint. Also, $A \in \mathcal{A}$ is again a finite disjoint union of $\mathcal{A}'$-sets, say, $A = \bigcup_{l=1}^{m} B'_l$. Define $A'_{jkl} = A'_{jk} \bigcap B'_l$ for $k \leq n_j, l \leq m$, and all j. Then $\{A'_{jkl}\}_{jkl} \subset \mathcal{A}'$ are disjoint and:

$$B'_l = \bigcup_{j=1}^{\infty} \bigcup_{k=1}^{n_j} A'_{jkl}.$$

Since $B'_l \in \mathcal{A}'$, countable subadditivity of μ_0 on $\mathcal{A}'$ obtains:

$$\mu_0\left(B'_l\right) \leq \sum_{j=1}^{\infty} \sum_{k=1}^{n_j} \mu_0\left(A'_{jkl}\right),$$

and so by (1):

$$\mu_{\mathcal{A}}(A) \equiv \sum_{l=1}^{m} \mu_0\left(B'_l\right) \leq \sum_{l=1}^{m} \sum_{j=1}^{\infty} \sum_{k=1}^{n_j} \mu_0\left(A'_{jkl}\right). \tag{4}$$

To prove (3), observe that:

$$A'_{jk} = \bigcup_{l=1}^{m} A'_{jkl},$$

so finite additivity provides:

$$\mu_0\left(A'_{jk}\right) = \sum_{l=1}^{m} \mu_0\left(A'_{jkl}\right).$$

Thus:

$$\mu_{\mathcal{A}}(A_j) \equiv \sum_{k=1}^{n_j} \mu_0\left(A'_{jk}\right) = \sum_{k=1}^{n_j}\sum_{l=1}^{m} \mu_0\left(A'_{jkl}\right).$$

Summing over j and comparing with (4) proves (3) and countable additivity. For this, note that this triple sum can be reordered even if infinite, since $\mu_0\left(A'_{jkl}\right) \geq 0$ *for all indexes.* ∎

6.4.2 Countable Additivity on an Algebra

The next result is useful in providing an alternative means of demonstrating countable additivity for a finitely additive set function μ defined on an algebra $\mathcal{A}$. To set the stage, recall Proposition 2.45 where it was stated that all measures μ on a measure space $(X, \sigma(X), \mu)$ had the following continuity properties. Given $\{A_i\} \subset \sigma(X)$:

1. **Continuity from Below:** If $A_i \subset A_{i+1}$ for all i:

$$\mu\left(\bigcup_{i=1}^{\infty} A_i\right) = \lim_{i \to \infty} \mu(A_i),$$

   where this limit may be finite or infinite.

2. **Continuity from Above:** If $A_{i+1} \subset A_i$ for all i and $\mu(A_1) < \infty$:

$$\mu\left(\bigcap_{i=1}^{\infty} A_i\right) = \lim_{i \to \infty} \mu(A_i).$$

The following proposition states that a finitely additive set function μ on an algebra $\mathcal{A}$ is in fact countably additive if either of the following applies. The second result will be of immediate use in the investigations on product spaces.

1. μ is continuous from below, or,

2. μ is finite, and is continuous from above when $\bigcap_{i=1}^{\infty} A_i = \emptyset$.

By finite it is meant that $\mu(X) < \infty$. Although this assumption appears quite restrictive, we will see that this result is applicable in the more general context of σ-finite spaces, where σ-finite is defined in Definition 5.34.

As an exercise, the reader is invited to identify where this proposition fails if the algebra $\mathcal{A}$ is replaced by a semi-algebra $\mathcal{A}'$.

Proposition 6.19 (Set function μ_0 on $\mathcal{A}$ to $\mu_\mathcal{A}$ on $\mathcal{A}$) *Let μ be a non-negative, finitely additive set function on an algebra $\mathcal{A}$.*

*If **either** of the following is satisfied, then μ is countably additive and hence a measure on $\mathcal{A}$:*

1. ***Continuity from Below:*** *Given $\{A_i\}_{i=1}^\infty$ with $A_i \subset A_{i+1}$ for all i, and $\bigcup_{i=1}^\infty A_i \in \mathcal{A}$, then:*

$$\lim_{i\to\infty} \mu(A_i) = \mu\left(\bigcup_{i=1}^\infty A_i\right);\tag{6.8}$$

2. ***Continuity from Above and Finite:*** *If $\mu(X) < \infty$, given $\{A_i\}_{i=1}^\infty$ with $A_{i+1} \subset A_i$ for all i, and $\bigcap_{i=1}^\infty A_i = \emptyset$, then:*

$$\lim_{i\to\infty} \mu(A_i) = 0.\tag{6.9}$$

Proof. *Let $\{B_j\} \subset \mathcal{A}$ be a disjoint collection with $B \equiv \bigcup_{j=1}^\infty B_j$ and assume that $B \in \mathcal{A}$. Recall that algebras need not be closed under countable unions, and countable additivity only needs to be demonstrated in cases where it is so closed.*

For item 1, define $A_i = \bigcup_{j \leq i} B_j$. Then $\{A_i\} \subset \mathcal{A}$ and so by finite additivity:

$$\mu(A_i) = \sum_{j\leq i} \mu(B_j).$$

But since $B = \bigcup_{i=1}^\infty A_i$ and $A_i \subset A_{i+1}$ for all i, it follows from (6.8) that:

$$\mu(B) = \lim_{i\to\infty} \sum_{j\leq i} \mu(B_j).$$

This is equivalent to countable additivity since these partial sums are increasing and thus have a well defined limit of $\sum_j \mu(B_j)$.

For item 2, define $A_i = \bigcup_{j\geq i} B_j$, and so $B = A_i \bigcup \left(\bigcup_{j<i} B_j\right)$. Then since $A_i = B \cap \left(\widetilde{\bigcup_{j<i} B_j}\right)$, with $\widetilde{D}$ denotes the complement of D, this obtains that $\{A_i\} \subset \mathcal{A}$. Again by finite additivity applied to $B = \left(\bigcup_{j<i} B_j\right) \bigcup A_i$:

$$\mu(B) = \mu(A_i) + \sum_{j<i} \mu(B_j).$$

Now $\mu(A_1) = \mu(B) < \infty$ since μ is finite, and $\bigcap_{i=1}^{\infty} A_i = \emptyset$, so (6.9) obtains:

$$\mu(B) = \lim_{i \to \infty} \sum_{j < i} \mu(B_j).$$

∎

6.5 Completion of a Measure Space

There is one final extension theorem which has not yet been needed, but will be used in the next chapter. Although the measure spaces created with outer measures and the Carathéodory measurability criterion have always been complete by Definition 2.48, many measure spaces such as $(\mathbb{R}, \mathcal{B}(\mathbb{R}), m)$ are not complete. The question then arises as to the possibility of extending such a measure space to a complete measure space. This extension is called the **completion of a measure space**.

It turns out that every measure space can be completed, meaning that the associated sigma algebra can be expanded to include all subsets of sets of measure 0. Further, this extension can be accomplished without changing the value of the measure on the sets of the original sigma algebra.

Proposition 6.20 (Completion of a measure space) *Given a measure space $(X, \sigma(X), \mu)$, there is a smallest sigma algebra $\sigma^C(X)$ and unique measure μ^C on $\sigma^C(X)$ so that:*

1. *$\sigma(X) \subset \sigma^C(X)$;*
2. *$\mu^C(A) = \mu(A)$ for all $A \in \sigma(X)$;*
3. *$(X, \sigma^C(X), \mu^C)$ is a complete measure space, called the **completion** of $(X, \sigma(X), \mu)$.*

Proof. 1. Definition of $\sigma^C(X)$ and μ^C: *Define the collection of sets:*

$$\sigma^C(X) = \{A | B \subset A \subset C, \text{ where } B, C \in \sigma(X) \text{ and } \mu(C - B) = 0\}.$$

For $A \in \sigma^C(X)$, define:

$$\mu^C(A) = \mu(B). \tag{6.10}$$

Choosing $B = C$ obtains that $\sigma(X) \subset \sigma^C(X)$.

We must show that $\sigma^C(X)$ is a sigma algebra and that the definition of $\mu^C(A)$ is well defined and independent of the pair of sets (B, C) used to identify A. It will then follow that $\mu^C(A) = \mu(A)$ for all $A \in \sigma(X)$ by taking $B = C = A$.

2. Well-definedness of μ^C: *Let (B, C) and (B', C') be pairs of sets in $\sigma(X)$ with $B \subset A \subset C$, $B' \subset A \subset C'$, and:*

$$\mu(C - B) = \mu(C' - B') = 0.$$

Then since μ is finitely additive it follows that $\mu(C) = \mu(B)$ and $\mu(C') = \mu(B')$. Also, $B \subset A \subset C'$ implies that $\widetilde{C}' \subset \widetilde{B}$, and so $\widetilde{C}' \cap C \subset \widetilde{B} \cap C$. In other words, $C - C' \subset C - B$, and so $\mu(C - C') = 0$ by monotonicity. Thus $\mu(C) = \mu(C')$ and then $\mu(B) = \mu(B')$ by finite additivity, and so $\mu^C(A)$ is well defined.

3. $\sigma^C(X)$ is a sigma algebra: Note that $B \subset A \subset C$ implies that $\widetilde{C} \subset \widetilde{A} \subset \widetilde{B}$. But then $\mu(\widetilde{B} - \widetilde{C}) = 0$ since $\widetilde{B} - \widetilde{C} = C - B$, so if $A \in \sigma^C(X)$ then $\widetilde{A} \in \sigma^C(X)$.

Now let $\{A_j\}_j \subset \sigma^C(X)$. Then by definition there exists $\{B_j, C_j\}_j \subset \sigma(X)$ with $B_j \subset A_j \subset C_j$ and $\mu(C_j - B_j) = 0$ for all j. Define $B = \bigcup B_j$, $A = \bigcup A_j$, and $C = \bigcup C_j$. Then $B \subset A \subset C$ and by De Morgan's laws:

$$C - B \subset \bigcup_j (C_j - B_j).$$

As $\{C_j - B_j\}$ need not be disjoint, countable subadditivity applies and yields $\mu(C - B) = 0$. Thus $\bigcup A_j \in \sigma^C(X)$ and $\sigma^C(X)$ is a sigma algebra.

4. $\sigma^C(X)$ is complete: To prove completeness by Definition 2.48, we show that the above characterization is equivalent to the earlier definition. If $C \in \sigma^C(X)$ has measure 0, then $\sigma^C(X)$ contains all subsets of C by choosing $B = \emptyset$ above, so $\mu^C(A)$ is complete by Definition 2.47. Conversely, any set A above would be in the complete sigma algebra of Definition 2.47 because $\mu(C - B) = 0$ implies that every subset of $C - B$ is $\sigma^C(X)$-measurable. As $A - B \subset C - B$, this implies that $A - B$ as well as $A = B \bigcup (A - B)$ are in this sigma algebra.

Note: *The advantage of defining $\sigma^C(X)$ in terms of pairs of sets is that this definition produces a sigma algebra directly as was demonstrated, whereas simply adding sets of measure 0 to $\sigma(X)$ would not produce a sigma algebra until such sets were unioned, intersected, etc. with all other sets.*

5. μ^C is a measure and is unique: By definition, μ^C is nonnegative and $\mu^C(\emptyset) = 0$. Next, let $\{A_j\} \subset \sigma^C(X)$ be disjoint sets, and $\{B_j\} \subset \sigma(X)$ the associated subsets in the definition above. Then $\{B_j\}$ is also a disjoint collection since $B_j \subset A_j$. Defining $B \equiv \bigcup_j B_j$ obtains $B \subset A$ and $\mu(A - B) = 0$ as in item 3 above. Thus, by the countable additivity of μ on $\sigma(X)$:

$$\mu^C \left(\bigcup_j A_j \right) \equiv \mu \left(\bigcup_j B_j \right) = \sum_j \mu(B_j) = \sum_j \mu^C(A_j).$$

Uniqueness follows from finite additivity of measures and (6.10).

6. $\sigma^C(X)$ is the smallest such sigma algebra: To prove that $\sigma^C(X)$ is the smallest sigma algebra which completes $\sigma(X)$, we show that any other such sigma algebra must contain the A-sets of the above definition, and have measure as defined above. To see this let $(X, \sigma'(X), \mu')$ be another complete measure space for which $\sigma(X) \subset \sigma'(X)$ and $\mu'(D) = \mu(D)$ for all $D \in \sigma(X)$. Assume that $B, C \in \sigma(X)$ with $\mu(C - B) = 0$. If A is any set with $B \subset A \subset C$, then $A - B \subset C - B$ obtains $\mu'(C - B) = \mu(C - B) = 0$ by monotonicity. As $\sigma'(X)$ is assumed complete, $A - B \in \sigma'(X)$ and $\mu'(A - B) = 0$. But then $B \subset A$ assures that $A = (A - B) \bigcup B$ and thus $A \in \sigma'(X)$ and so $\sigma^C(X) \subset \sigma'(X)$.

Finally, by finite additivity $\mu'(A) = \mu'(B) \equiv \mu(B)$ and thus $\mu'(A) = \mu^C(A)$ for all such A. ∎

Remark 6.21 (On completion) *To emphasize what is perhaps obvious, it is important to appreciate that the completion of a measure space depends both on the starting sigma algebra, and on the measure defined on that sigma algebra.*

1. Two measures on $\sigma(X)$: *Given a space X and two measures μ_1 and μ_2 defined on the same sigma algebra $\sigma(X)$, different complete sigma algebras, $\sigma_1^C(X)$ and $\sigma_2^C(X)$ can result. If $A \in \sigma_1^C(X)$ say, this implies that there exists $B_1, C_1 \in \sigma(X)$ with $B_1 \subset A \subset C_1$ and $\mu_1(C_1 - B_1) = 0$. But it can happen that $\mu_2(C_1 - B_1) \neq 0$, and that $\mu_2(C_2 - B_2) \neq 0$ for any other $\sigma(X)$-sets with $B_2 \subset A \subset C_2$, and thus $A \notin \sigma_2^C(X)$.*

For example, let $X = \mathbb{R}$, $\sigma(X) = \mathcal{B}(\mathbb{R})$, the Borel sigma algebra, $\mu_1 = m$, and μ_2 the rationals counting measure of Example 2.24. Then $\sigma_1^C(X) = \mathcal{M}_L(\mathbb{R})$ by Example 6.23, and we claim that $\sigma_2^C(X) = \sigma(P(\mathbb{R}))$, the power sigma algebra of Example 2.7. If A is any set, define $B = \{r \in \mathbb{Q} | r \in A\}$, and $C = \mathbb{R} - \{r \in \mathbb{Q} | r \notin A\}$. Then both B and C are Borel measurable, B as a subset of rationals, and C as $\mathbb{R}$ less a subset of rationals. Also $B \subset A \subset C$ by construction, and $\mu_2(C - B) = 0$ since $C - B$ contains no rationals. Thus $A \in \sigma_2^C(X)$.

As an explicit example of $A \in \sigma(P(\mathbb{R})) - \mathcal{M}_L(\mathbb{R})$ let $A = A_0 \subset [0,1]$, the set constructed in Proposition 2.31. If there existed $B, C \in \mathcal{B}(\mathbb{R})$ with $B \subset A_0 \subset C$ and $m(C - B) = 0$, this would imply that $m(A_0 - B) = 0$ and thus $A_0 \in \mathcal{M}_L(\mathbb{R})$ since $A_0 = B + (A_0 - B)$. This is a contradiction.

2. Two sigma algebras and μ: *Given a space X with two sigma algebras $\sigma_1(X)$ and $\sigma_2(X)$ and one measure μ, the complete sigma algebras $\sigma_1^C(X)$ and $\sigma_2^C(X)$ need not agree. If $A \in \sigma_1^C(X)$ say, this implies that there exists $B_1, C_1 \in \sigma_1(X)$ with $B_1 \subset A \subset C_1$ and $\mu(C_1 - B_1) = 0$. But it can happen that $B_1, C_1 \notin \sigma_2(X)$, and that $\mu(C_2 - B_2) \neq 0$ for any other $\sigma_2(X)$-sets with $B_2 \subset A \subset C_2$, and so $A \notin \sigma_2^C(X)$.*

An extreme example here is $X = \mathbb{R}$, $\sigma_1(X) = \sigma(P(\mathbb{R}))$, the power sigma algebra, and $\sigma_1(X) = \sigma(\emptyset, \mathbb{R})$, the trivial sigma algebra of Example 2.7. Also, let μ denote the rationals counting measure of Example 2.24. Then $\sigma_1^C(X) = \sigma_1(X)$, since $\sigma_1(X)$ already contains all sets, and $\sigma_2^C(X) = \sigma_2(X)$, since $\sigma_2(X)$ has no set of measure 0 other than $\emptyset$.

6.5.1 Uniqueness of Extensions 2

In Section 6.2.5 on Uniqueness of Extensions, two questions were left unanswered regarding Borel measures and Lebesgue measures, which we address here.

Example 6.22 (Uniqueness of Borel measure on $\mathcal{M}_{\mu_F}(\mathbb{R})$) *Using Proposition 6.14 on the uniqueness of extensions from an algebra to the smallest sigma algebra containing that algebra, it was earlier concluded that μ_F was the unique extension of the measure $\mu_{\mathcal{A}}$ from the algebra $\mathcal{A}$ to $\sigma(\mathcal{A})$, the smallest sigma algebra that contains $\mathcal{A}$. Now, $\mathcal{A} \subset \mathcal{B}(\mathbb{R})$ implies that $\sigma(\mathcal{A}) \subset \mathcal{B}(\mathbb{R})$. Conversely, $\sigma(\mathcal{A})$ contains the open intervals and so $\mathcal{B}(\mathbb{R}) \subset \sigma(\mathcal{A})$. Thus, $\sigma(\mathcal{A}) = \mathcal{B}(\mathbb{R})$.*

By (5.19) of Proposition 5.26 but changing notation, for any $A \in \mathcal{M}_{\mu_F}(\mathbb{R})$ there is a set $C \in \mathcal{A}_{\sigma\delta}$, the collection of countable intersections of sets in $\mathcal{A}_\sigma$, and $B \in \mathcal{A}_{\delta\sigma}$, the collection of countable unions of sets in $\mathcal{A}_\delta$, so that $B \subset A \subset C$ and $\mu_F(C-A) = \mu_F(A-B) = 0$. This implies that $\mu_F(C - B) = 0$ and also $\mu_F(A) = \mu_F(B)$. Since $\mathcal{A}_{\sigma\delta} \subset \mathcal{B}(\mathbb{R})$ and $\mathcal{A}_{\delta\sigma} \subset \mathcal{B}(\mathbb{R})$, it follows that every set in $\mathcal{M}_{\mu_F}(\mathbb{R})$ is bounded by Borel sets with the same measure. Thus, with a small variation on the notation of the above proposition, $\mathcal{M}_{\mu_F}(\mathbb{R}) \subset B(\mathbb{R})^C$, where this completion is defined relative to μ_F.

Conversely, given $B, C \in \mathcal{B}(\mathbb{R})$ with $B \subset C$ and $\mu_F(C - B) = 0$, then given a set A with $B \subset A \subset C$ it must be the case that $A \in \mathcal{M}_{\mu_F}(\mathbb{R})$. In detail, since $A - B \subset C - B$ and $\mu_F(C-B) \equiv \mu_{\mathcal{A}}^*(C - B) = 0$, it follows by monotonicity from Proposition 5.20 that $\mu_{\mathcal{A}}^*(A - B) = 0$, and by Proposition 5.23 that $A - B \in \mathcal{M}_{\mu_F}(\mathbb{R})$. But then $A = B \bigcup (A - B) \in \mathcal{M}_{\mu_F}(\mathbb{R})$. In other words, $\mathcal{B}(\mathbb{R})^C \subset \mathcal{M}_{\mu_F}(\mathbb{R})$.

Combining with the above result,

$$\mathcal{B}(\mathbb{R})^C = \mathcal{M}_{\mu_F}(\mathbb{R}). \tag{6.11}$$

Hence $\mathcal{M}_{\mu_F}(\mathbb{R})$ is the smallest completion of $\mathcal{B}(\mathbb{R})$ in the μ_F measure, or more precisely $(\mathbb{R}, \mathcal{M}_{\mu_F}(\mathbb{R}), \mu_F)$ is the smallest completion of $(\mathbb{R}, \mathcal{B}(\mathbb{R}), \mu_F)$ given by the above proposition.

Thus μ_F is the unique extension of the measure $\mu_{\mathcal{A}}$ from algebra $\mathcal{A}$ to $\mathcal{B}(\mathbb{R})$, and by the above proof, the measure of every set in $\mathcal{M}_{\mu_F}(\mathbb{R})$ is defined in terms of the measure of $\mathcal{B}(\mathbb{R})$ sets by (6.10). Thus we can conclude that μ_F is also the unique extension of the measure $\mu_{\mathcal{A}}$ from algebra $\mathcal{A}$ to the complete sigma algebra $\mathcal{M}_{\mu_F}(\mathbb{R})$.

Example 6.23 (Lebesgue Equals Borel Measure with $F(x) = x + c$) *In Section 6.2.5 it was concluded that on the Borel sigma algebra $\mathcal{B}(\mathbb{R})$:*

$$m(A) = \mu_F(A) \text{ for } A \in \mathcal{B}(\mathbb{R}),$$

where μ_F is the Borel measure induced by $F(x) = x + c$. However, by the same argument as above, $\mathcal{M}_L(\mathbb{R})$ is the smallest completion of $\mathcal{B}(\mathbb{R})$ in the m measure, while $\mathcal{M}_{\mu_F}(\mathbb{R})$ is is the smallest completion of $\mathcal{B}(\mathbb{R})$ in the μ_F measure.

But as these measures agree on $\mathcal{B}(\mathbb{R})$, we conclude by (6.10) that for $F(x) = x + c$:

$$\mathcal{M}_{\mu_F}(\mathbb{R}) = M_L(\mathbb{R}), \tag{6.12}$$

and that

$$m(A) = \mu_F(A) \text{ for } A \in \mathcal{M}_{\mu_F}(\mathbb{R}).$$

In other words, μ_F is the unique extension of the measure $\mu_{\mathcal{A}}$ from algebra $\mathcal{A}$ to the complete sigma algebra $\mathcal{M}_{\mu_F}(\mathbb{R})$. Further, this measure and sigma algebra agree with Lebesgue measure m and the sigma algebra of Lebesgue measurable sets $\mathcal{M}_L(\mathbb{R})$.

Combining this with the conclusion of Example 6.22 obtains that under m:

$$\mathcal{B}(\mathbb{R})^C = \mathcal{M}_L(\mathbb{R}). \tag{6.13}$$

The following result generalizes the discussion in these examples. The summary conclusion is that a σ-finite measure on an algebra $\mathcal{A}$ extends uniquely to the Proposition 6.20 completion of $\sigma(\mathcal{A})$, the smallest sigma algebra that contains $\mathcal{A}$. Further, by Proposition 6.5, this complete sigma algebra agrees with the complete sigma algebra of Carathéodory measurable sets given by the Hahn-Kolmogorov Extension theorem of Proposition 6.4.

Proposition 6.24 (Uniqueness of Extensions to $\mathcal{C}(X)$) *Let $\mathcal{A}$ be an algebra of sets on X, $\mu_{\mathcal{A}}$ a σ-finite measure on $\mathcal{A}$, and $(X, \mathcal{C}(X), \mu)$ the complete measure space of Carathéodory measurable sets given by the Hahn-Kolmogorov Extension theorem. Thus $\mathcal{A} \subset \mathcal{C}(X)$ and $\mu(A) = \mu_{\mathcal{A}}(A)$ for all $A \in \mathcal{A}$.*

Let $\sigma(\mathcal{A})$ denote the smallest sigma algebra that contains $\mathcal{A}$, and v a measure on $\sigma(\mathcal{A})$ with $v(A) = \mu(A)$ for all $A \in \mathcal{A}$.

Then $v = \mu$ on $\sigma(\mathcal{A})$, and with superscript C_v denoting the Proposition 6.20 completion of $\sigma(\mathcal{A})$ with respect to v:

$$\sigma^{C_v}(\mathcal{A}) = \mathcal{C}(X), \qquad v^C = \mu. \tag{6.14}$$

In particular:

$$\sigma^{C_\mu}(\mathcal{A}) = \mathcal{C}(X), \qquad \mu^C = \mu.$$

More generally, if v, v' are σ-finite measures on $\sigma(\mathcal{A})$ with $v(A) = v'(A)$ for all $A \in \mathcal{A}$, then $v = v'$ on $\sigma(\mathcal{A})$, and with C_v and C'_v denoting the Proposition 6.20 completions of $\sigma(\mathcal{A})$ with respect to v and v':

$$\sigma^{C_v}(\mathcal{A}) = \sigma^{C'_v}(\mathcal{A}), \qquad v^C = \left(v'\right)^C.$$

Proof. *That $v = \mu$ on $\sigma(\mathcal{A})$ follows from Proposition 6.14 by σ-finiteness of $\mu_{\mathcal{A}}$. Then by Proposition 6.20, $\sigma^{C_v}(\mathcal{A})$ is the smallest complete sigma algebra relative to v that contains $\sigma(\mathcal{A})$. Now $\mathcal{A} \subset \mathcal{C}(X)$ and thus $\sigma(\mathcal{A}) \subset \mathcal{C}(X)$, so it follows that since $\mathcal{C}(X)$ is complete that $\sigma^{C_v}(\mathcal{A}) \subset \mathcal{C}(X)$. On the other hand, if $A \in \mathcal{C}(X)$, then by Proposition 6.5 there exists $A' \in \mathcal{A}_{\delta\sigma}$ and $A'' \in \mathcal{A}_{\sigma\delta}$ with $A' \subset A \subset A''$ and $\mu(A'' - A') = 0$. But as $\sigma(\mathcal{A})$ is the smallest sigma algebra that contains $\mathcal{A}$, it follows that $A', A'' \in \sigma(\mathcal{A})$ and hence $A \in \sigma^{C_v}(\mathcal{A})$ by definition of $\sigma^{C_v}(\mathcal{A})$. Combining obtains $\sigma^C(\mathcal{A}) = \mathcal{C}(X)$.*

Given $A \in \sigma^{C_v}(\mathcal{A})$, there exists $A', A'' \in \sigma(\mathcal{A})$ with $A' \subset A \subset A''$ and $v(A'' - A') = 0$, and so $v^C(A) \equiv v(A')$ by (6.10). But $v(A') = \mu(A')$ since $A' \in \sigma(\mathcal{A})$ and $v = \mu$ on $\sigma(\mathcal{A})$. Similarly, $A'' - A' \in \sigma(\mathcal{A})$ obtains that $\mu(A'' - A') = 0$, and since $A - A' \subset A'' - A'$ it follows that $\mu(A - A') = 0$ by completeness of $\mathcal{C}(X)$. Combining results:

$$\mu(A) = \mu(A') + \mu(A - A') = v^C(A),$$

and thus $v^C = \mu$.

The second result then follows as a change of notation, letting v equal the restriction of μ to $\sigma(\mathcal{A})$.

Finally, given such v, v' defined on $\sigma(\mathcal{A})$ with $v(A) = v'(A)$ for all $A \in \mathcal{A}$, then $v = v'$ on $\sigma(\mathcal{A})$ by Proposition 6.14 and σ-finiteness. Now, if $A \in \sigma^{C_v}(\mathcal{A})$, then there exists $A', A'' \in \sigma(\mathcal{A})$ with $A' \subset A \subset A''$ and $v(A'' - A') = 0$. But then $v'(A'' - A') = 0$ and so $A \in \sigma^{C'_v}(\mathcal{A})$. And for such A by (6.10):

$$v^C(A) \equiv v(A') = v'(A') \equiv \left(v'\right)^C(A).$$

Thus $\sigma^{C_\nu}(\mathcal{A}) \subset \sigma^{C'_\nu}(\mathcal{A})$ and $\nu^C = (\nu')^C$ on $\sigma^{C_\nu}(\mathcal{A})$. The proof of the reverse conclusion is identical. ∎

The next result addresses uniqueness of extensions of a measure from different algebras in the space. In particular, it provides a criterion which assures that the Proposition 6.4 extensions agree. See Example 6.26 for what can happen when this criterion fails.

For this result, we continue the notational convention of Proposition 6.24, that $\sigma^{C_\mu}(\mathcal{A})$ for example, denotes the Proposition 6.20 completion of $\sigma(\mathcal{A})$ relative to μ.

Corollary 6.25 (Uniqueness of extensions from different algebras) *Let $\mathcal{A}^+$ be an algebra of sets on X, $\mu_{\mathcal{A}}$ a σ-finite measure on $\mathcal{A}^+$, and $(X, \mathcal{C}(X), \mu)$ the complete measure space of Carathéodory measurable sets given by the Hahn-Kolmogorov Extension theorem. Let $\mathcal{A} \subset \mathcal{A}^+$ be an algebra of sets and again using $\mu_{\mathcal{A}}$, let $(X, \mathcal{C}'(X), \mu')$ be the associated complete measure space.*

If $\mathcal{A}^+ \subset \mathcal{C}'(X)$, then:

$$\mathcal{C}(X) = \mathcal{C}'(X), \quad \mu = \mu'.$$

Proof. *1. $\mu = \mu'$ on $\mathcal{C}'(X)$:*

Given $(X, \mathcal{C}'(X), \mu')$, $\mathcal{A} \subset \mathcal{A}^+$ assures that $\mu = \mu_{\mathcal{A}} = \mu'$ on $\mathcal{A}$. Proposition 6.24 then obtains $\mu = \mu'$ on $\sigma(\mathcal{A})$, and also that $\mu^C = \mu'$ and $\sigma^{C_\mu}(\mathcal{A}) = \mathcal{C}'(X)$. Now $\sigma^{C_\mu}(\mathcal{A}) \subset \mathcal{C}(X)$ as the smallest complete sigma algebra extending $\sigma(\mathcal{A})$ relative to μ. It then follows that $\mu^C = \mu$ on $\sigma^{C_\mu}(\mathcal{A})$ since μ is a complete measure on $\mathcal{C}(X)$. Thus $\mu = \mu'$ on $\mathcal{C}'(X)$.

2. $\mathcal{C}(X) = \mathcal{C}'(X)$:

From item 1 and that $\mu = \mu'$ on $\sigma(\mathcal{A})$ obtains by the construction of Proposition 6.20 that:

$$\sigma^{C_\mu}(\mathcal{A}) = \sigma^{C_{\mu'}}(\mathcal{A}) = \mathcal{C}'(X). \tag{1}$$

Then given $(X, \mathcal{C}(X), \mu)$, since $\mathcal{A}^+ \subset \mathcal{C}'(X)$ we have from item 1 that $\mu' = \mu$ on $\mathcal{A}^+$. Using the above argument, $\mu' = \mu$ on $\sigma(\mathcal{A}^+)$, and also $(\mu')^C = \mu$ and $\sigma^{C_{\mu'}}(\mathcal{A}^+) = \mathcal{C}(X)$. Further, $\mu' = \mu$ on $\sigma(\mathcal{A}^+)$ obtains by the construction of Proposition 6.20 that:

$$\sigma^{C_{\mu'}}(\mathcal{A}^+) = \sigma^{C_\mu}(\mathcal{A}^+) = \mathcal{C}(X). \tag{2}$$

Now $\sigma(\mathcal{A}) \subset \sigma(\mathcal{A}^+)$ implies that $\sigma^{C_\mu}(\mathcal{A}) \subset \sigma^{C_\mu}(\mathcal{A}^+)$ and thus by (1) and (2):

$$\mathcal{C}'(X) \subset \mathcal{C}(X).$$

On the other hand, $\mathcal{A}^+ \subset \mathcal{C}'(X)$ obtains $\sigma(\mathcal{A}^+) \subset \mathcal{C}'(X)$, and thus $\sigma^{C_{\mu'}}(\mathcal{A}^+) \subset \mathcal{C}'(X)$ by the construction of Proposition 6.20 and the completeness of $\mathcal{C}'(X)$ relative to μ'. Hence by (2):

$$\mathcal{C}(X) \subset \mathcal{C}'(X),$$

and the proof is complete. ∎

Given the set-up in Corollary 6.25, that $\mathcal{A} \subset \mathcal{C}'(X)$ and $\mathcal{A} \subset \mathcal{A}^+ \subset \mathcal{C}(X)$, one might wonder if the assumption $\mathcal{A}^+ \subset \mathcal{C}'(X)$ automatically follows, and is thus always true. The next example addresses this question.

Example 6.26 (Failure of $\mathcal{A}^+ \subset \mathcal{C}'(X)$) *On* $\mathbb{R}$ *define two classes of sets:*

$$\mathcal{A}' \equiv \{(n,m] | n, m \in \mathbb{Z} \bigcup \{-\infty, \infty\}\},$$

$$(\mathcal{A}^+)' \equiv \{(a,b] | a, b \in \mathbb{R} \bigcup \{-\infty, \infty\}\},$$

where we define $(c, \infty] = (c, \infty)$ *for any c. The collection* $(\mathcal{A}^+)'$ *is the semi-algebra of right semi-closed intervals of Example 6.12 underlying the Borel measure development of Chapter 5, and it is an exercise to check that* $\mathcal{A}'$ *is also a semi-algebra. Now let* $\mathcal{A}$ *and* $\mathcal{A}^+$ *denote the associated algebras of finite disjoint unions of sets, recalling Exercise 6.10.*

For specificity, let $F(x)$ *be a continuous distribution function with* $F(-\infty) = 0$ *and* $F(\infty) = 1$, *for example the normal distribution function:*

$$F(x) \equiv \frac{1}{\sqrt{2\pi}} \int_{-\infty}^{x} e^{-y^2/2} dy.$$

On $\mathcal{A}'$ *and* $(\mathcal{A}^+)'$ *define the set function* $\mu_{\mathcal{A}}$ *by:*

$$\mu_{\mathcal{A}}[(c,d]] = F(d) - F(c),$$

and then extend to $\mathcal{A}$ *and* $\mathcal{A}^+$ *additively. Then* $\mu_{\mathcal{A}}$ *is a measure on* $\mathcal{A}^+$ *by Proposition 5.13, and then also a measure on* $\mathcal{A}$ *since* $\mathcal{A} \subset \mathcal{A}^+$.

The Proposition 6.4 extension of $\mu_{\mathcal{A}}$ *and* $\mathcal{A}^+$ *obtains the complete measure space* $(\mathbb{R}, \mathcal{M}_F(\mathbb{R}), \mu_F)$ *of Proposition 5.23. Let* $(\mathbb{R}, \mathcal{C}'(\mathbb{R}), \mu_F')$ *denote the Proposition 6.4 extension of* $\mu_{\mathcal{A}}$ *and* $\mathcal{A}$, *which also obtained that* $\mathcal{A} \subset \mathcal{C}'(\mathbb{R})$. *We now show that* $\mathcal{C}'(\mathbb{R})$ *does not contain* $\mathcal{A}^+$, *and we do this by showing that:*

$$\mathcal{C}'(\mathbb{R}) = \mathcal{A}. \tag{1}$$

Let $\mu_{\mathcal{A}}^*$ *denote the outer measure associated with* $\mu_{\mathcal{A}}$ *and* $\mathcal{A}$, *defined on all* $A \subset \mathbb{R}$ *as in (5.8) of Definition 5.16. By the argument there this can be restated as in (5.9):*

$$\mu_{\mathcal{A}}^*(A) = \inf \left\{ \sum_n \mu_{\mathcal{A}}(A_n') \mid A \subset \bigcup_n A_n' \right\},$$

where $\{A_n'\} \subset \mathcal{A}'$. *It then follows that for any set A:*

$$\mu_{\mathcal{A}}^*(A) = \sum_j \mu_{\mathcal{A}}((m_j, m_j + 1]), \tag{2}$$

where $A \subset \bigcup_j (m_j, m_j + 1]$ *and* $A \bigcap (m_j, m_j + 1] \neq \emptyset$ *for all j. From this we obtain that* $\mu_{\mathcal{A}}^*(A) = \mu_{\mathcal{A}}(A)$ *for all* $A \in \mathcal{A}'$ *as expected.*

To prove that $\mathcal{C}'(\mathbb{R}) = \mathcal{A}$, first note that if $A \in \mathcal{C}'(\mathbb{R})$, then $A \bigcap (n, n+1] \in \mathcal{C}'(\mathbb{R})$ for all $(n, n+1] \in \mathcal{A}'$ since $\mathcal{A} \subset \mathcal{C}'(\mathbb{R})$. Thus it is sufficient to prove that if $A \bigcap (n, n+1] \equiv A_n \in \mathcal{C}'(\mathbb{R})$ and $A_n \neq \emptyset$, then $A_n = (n, n+1]$.

Assume to the contrary that $A_n \subsetneqq (n, n+1]$. Letting $E = (n, n+1]$ in (5.10) obtains:

$$\mu_{\mathcal{A}}^*((n, n+1]) = \mu_{\mathcal{A}}^*(A_n \bigcap (n, n+1]) + \mu_{\mathcal{A}}^*(\widetilde{A_n} \bigcap (n, n+1]).$$

Since both sets on the right are nonempty by the assumption that $A_n \subsetneqq (n, n+1]$, this implies by (2) that:

$$\mu_{\mathcal{A}}((n, n+1]) = 2\mu_{\mathcal{A}}((n, n+1]),$$

which is a contradiction since $\mu_{\mathcal{A}}((n, n+1]) \neq 0$. Thus $A_n = (n, n+1]$, and this implies (1).

7

Finite Products of Measure Spaces

7.1 Product Space Semi-Algebras

In this chapter we provide the first application of the extension theorems of Chapter 6. Other results of this type will be developed in the later chapters. The application here is to "product spaces" which are in effect, multi-dimensional versions of the measures spaces addressed so far. For example, the 2-dimensional Euclidean space $\mathbb{R}^2$ is a product space $\mathbb{R} \times \mathbb{R}$. Not surprisingly, the development above for Lebesgue or Borel measures on $\mathbb{R}$ can be generalized to produce Lebesgue, Borel measures or general measure spaces on $\mathbb{R}^2$ or $\mathbb{R}^n$.

The following question is addressed in this chapter. Given "n copies" of the Lebesgue measure space, or n Borel or general measure spaces, how can we generate a measure space on $\mathbb{R}^n$ or X^n that is consistent with these component measure spaces? A more general development for measures on $\mathbb{R}^n$ will be undertaken in the next chapter.

For example, define a 2-dimensional **right semi-closed rectangle**:

$$(a, b] \times (c, d] = \{(x, y) | a < x \leq b, c < y \leq d\}.$$

A logical starting point for a measure on $\mathbb{R}^2$ is to define a set function $\mu \times \mu$ on these rectangles by:

$$\mu \times \mu((a, b] \times (c, d]) \equiv \mu((a, b]) \mu((c, d]),$$

where μ denotes m in the Lebesgue case, or μ_F in the general Borel case.

The collection of all such rectangles, $\{(a, b] \times (c, d]\}$, is not a sigma algebra as is easily demonstrated by considering a union of two elements or a complement. Thus, while this potential definition of a measure on $\mathbb{R}^2$ seems logical, if not compelling, there remains the question: Can this definition be extended to a sigma algebra $\sigma(\mathbb{R}^2)$ on $\mathbb{R}^2$ which contains these rectangles, and thus yields a measure space $(\mathbb{R}^2, \sigma(\mathbb{R}^2), \mu \times \mu)$?

The development in this section is quite general and not specifically applicable only to Lebesgue and Borel measure spaces, nor only to products of 1-dimensional measure spaces. So we resort to general notation and assume that for $i = 1, ..., n$ that $(X_i, \sigma(X_i), \mu_i)$ is a measure space. We assume also that

$$\mathcal{A}_i' \subset \mathcal{A}_i \subset \sigma(X_i),$$

DOI: 10.1201/9781003257745-7

denoting that the algebra $\mathcal{A}_i$ is generated in the usual way (Exercise 6.10) by the semi-algebra $\mathcal{A}_i'$, and $\sigma(X_i)$ is a sigma algebra that contains $\mathcal{A}_i$. For example, it could be the smallest sigma algebra that contains $\mathcal{A}_i$.

For a Borel example, X_i could be taken as $\mathbb{R}$, $\mathcal{A}_i'$ the semi-algebra of right semi-closed intervals, $\mathcal{A}_i$ the algebra of finite disjoint unions of right semi-closed intervals, and $\sigma(X_i)$ the sigma algebra of Borel measurable sets.

Definition 7.1 (Product space and set function) *Given measure spaces $\{(X_i, \sigma(X_i), \mu_i)|i = 1, ..., n\}$, the product space:*

$$X = \prod_{i=1}^{n} X_i,$$

is defined:

$$X = \{x \equiv (x_1, x_2, ..., x_n)|x_i \in X_i\}. \tag{7.1}$$

A measurable rectangle in X is a set A:

$$A = \prod_{i=1}^{n} A_i = \{x \in X|x_i \in A_i\}, \tag{7.2}$$

where $A_i \in \sigma(X_i)$. We denote the collection of measurable rectangles in X by $\mathcal{A}'$.

For general and not necessarily measurable sets $A_i \subset X_i$, the set $A = \prod_{i=1}^{n} A_i$ is called a **rectangle** *in X.*

The **product set function** *μ_0 is defined on $A = \prod_{i=1}^{n} A_i \in \mathcal{A}'$ by:*

$$\mu_0(A) = \prod_{i=1}^{n} \mu_i(A_i), \tag{7.3}$$

where we explicitly define $0 \cdot \infty = 0$.

The goal of this chapter is to utilize the extension theorems in the prior chapter to show that the product set function defined in (7.3) has a unique extension to a measure on the sigma algebra generated by the measurable rectangles. To this end, the first result is that as the notation implies, $\mathcal{A}'$ is a semi-algebra.

Then after investigating properties of such semi-algebras, we will turn to μ_0 defined by (7.3), and show that it is a pre-measure on $\mathcal{A}'$ in the sense of Definition 6.6.

Proposition 7.2 ($\mathcal{A}'$ is a semi-algebra) *With the notation above, let $\mathcal{A}'$ denote the collection of measurable rectangles in X. Then $\mathcal{A}'$ is a semi-algebra.*

Proof. *If $A = \prod_{i=1}^{n} A_i$ and $B = \prod_{i=1}^{n} B_i$ where $A_i, B_i \in \sigma(X_i)$, then:*

$$A \bigcap B = \prod_{i=1}^{n} (A_i \bigcap B_i).$$

Since $A_i \bigcap B_i \in \sigma(X_i)$ it follows that $A \bigcap B \in \mathcal{A}'$.

Consider next $\widetilde{A} = \widetilde{\prod_{i=1}^{n} A_i}$. *Express X as:*

$$X = \prod_{i=1}^{n} (A_i \bigcup \widetilde{A_i}).$$

It is an exercise to show that X can be expressed as a disjoint union:

$$X = \bigcup_I \prod_{i=1}^{n} D_i.$$

There are 2^n terms in the union associated with the 2^n possible n-tuples of 0s and 1s. For each such n-tuple I, the associated measurable rectangle $\prod_{i=1}^{n} D_i$ is defined by $D_i = A_i$ if the ith component of I is 1, and $D_i = \widetilde{A_i}$ otherwise.
   One of these n-tuples is all 1s and so has all $D_i = A_i$. Hence:

$$\widetilde{A} = X - A = \bigcup_{I'} \prod_{i=1}^{n} D_i,$$

where in each of the $2^n - 1$ terms in the union at least one $D_i = \widetilde{A_i}$.
   Hence, $\widetilde{A}$ is a finite union of disjoint elements of $\mathcal{A}'$, and thus $\mathcal{A}'$ is a semi-algebra. ∎

To motivate the notation of the next result note that this semi-algebra of measurable rectangles, $\mathcal{A}'$, could have been denoted by $\mathcal{A}'(\sigma(X_i))$ to designate that the component sets in the rectangles came from the respective sigma algebras. This raises the question: What if these component sets came from the respective semi-algebras or algebras defined on X_i?

Corollary 7.3 (Other semi-algebras) *Let $\mathcal{A}'(\mathcal{A}'_i)$, respectively $\mathcal{A}'(\mathcal{A}_i)$, denote the collection of measurable rectangles in X, defined by $A = \prod_{i=1}^{n} A_i$, with, respectively:*

1. *$A_i \in \mathcal{A}'_i$, where $\mathcal{A}'_i \subset \sigma(X_i)$ is a semi-algebra;*
2. *$A_i \in \mathcal{A}_i$, where $\mathcal{A}_i \subset \sigma(X_i)$ is an algebra.*

   Then $\mathcal{A}'(\mathcal{A}'_i)$, respectively $\mathcal{A}'(\mathcal{A}_i)$, is a semi-algebra.

Proof. *As above:*

$$A \bigcap B = \prod_{i=1}^{n} (A_i \bigcap B_i).$$

If $A, B \in \mathcal{A}'(\mathcal{A}'_i)$, then $A_i, B_i \in \mathcal{A}'_i$ and $A_i \bigcap B_i \in \mathcal{A}'_i$ as a semi-algebra, and so $A \bigcap B \in \mathcal{A}'(\mathcal{A}'_i)$. The same applies if $A, B \in \mathcal{A}'(\mathcal{A}_i)$. So both $\mathcal{A}'(\mathcal{A}'_i)$ and $\mathcal{A}'(\mathcal{A}_i)$ are closed under finite intersections.
   Also as above:

$$\widetilde{A} = X - A = \bigcup_{I'} \prod_{i=1}^{n} D_i,$$

where in each of the $2^n - 1$ terms in the union, at least one $D_i = \widetilde{A_i}$.

If each $A_i \in \mathcal{A}'_i$ then $\widetilde{A}_i$ is a finite disjoint union of elements of $\mathcal{A}'_i$:

$$\widetilde{A}_i = \bigcup\nolimits_{j \leq n(i)} B_j,$$

and so $\widetilde{A}$ can be expressed as a finite disjoint union of rectangles from $\mathcal{A}'\left(\mathcal{A}'_i\right)$. For example, if $D_1 = \widetilde{A}_1$ and all other $D_i = A_i$, then

$$\prod\nolimits_{i=1}^{n} D_i = \left(\bigcup\nolimits_{j \leq n(1)} B_j \right) \times \prod\nolimits_{i=2}^{n} A_i$$
$$= \bigcup\nolimits_{j \leq n(1)} \left(B_j \times \prod\nolimits_{i=2}^{n} A_i \right),$$

a disjoint union of measurable rectangles in $\mathcal{A}'\left(\mathcal{A}'_i\right)$. This extends by induction to all other examples.

If each $A_i \in \mathcal{A}_i$, then $\widetilde{A}_i \in \mathcal{A}_i$ as in the sigma algebra case, and so again $\widetilde{A}$ is a finite disjoint union of rectangles from $\mathcal{A}'\left(\mathcal{A}_i\right)$.

Hence, both $\mathcal{A}'\left(\mathcal{A}'_i\right)$ and $\mathcal{A}'\left(\mathcal{A}_i\right)$ are semi-algebras. ∎

Example 7.4 (Semi-algebras for Lebesgue/Borel measure) *For the application to multi-dimensional Lebesgue measure, if we begin with $(X_i, \sigma(X_i), \mu_i) = (\mathbb{R}, \mathcal{M}_L, m)$ for all i, then the above results imply that the collection of measurable rectangles in $\mathbb{R}^n$:*

$$\mathcal{A}' = \left\{ A = \prod\nolimits_{i=1}^{n} A_i \,\Big|\, A_i \in \mathcal{M}_L \right\},$$

is a semi-algebra if each component set is Lebesgue measurable. This also true if $(X_i, \sigma(X_i), \mu_i) = (\mathbb{R}, \mathcal{B}(\mathbb{R}), m)$ for all i.

In addition, $\mathcal{A}\left(\mathcal{A}'_i\right) = \{A = \prod\nolimits_{i=1}^{n} A_i\}$ is a semi-algebra if the A_i-sets are taken from the semi-algebra of right semi-closed intervals $\mathcal{A}'_i = \{(a, b]\}$, as is $\mathcal{A}\left(\mathcal{A}_i\right)$ where the A_i-sets are taken from the associated algebra $\mathcal{A}_i$ of all finite unions of elements of $\mathcal{A}'_i$.

The same statements apply to the multi-dimensional Borel measure space where each $(X_i, \sigma(X_i), \mu_i) = (\mathbb{R}, \mathcal{M}_{\mu_F}(\mathbb{R}), \mu_F)$. Again, $\mathcal{A}' = \{A = \prod\nolimits_{i=1}^{n} A_i\}$ is a semi-algebra if either $A_i \in \mathcal{M}_{\mu_F}(\mathbb{R})$ or $A_i \in \mathcal{B}(\mathbb{R})$, or if the A_i-sets are restricted to the semi-algebra of right semi-closed rectangles or the associated algebra.

This generalizes further to the case where $(X_i, \sigma(X_i), \mu_i) = (\mathbb{R}, \mathcal{M}_{\mu_{F_i}}(\mathbb{R}), \mu_{F_i})$, where the component measures differ.

In theory we could begin the extension process of Chapter 6 with any one of the four semi algebras defined on these product spaces, and it is natural to wonder if in general these different starting points will obtain different final sigma algebras. For example, in the Lebesgue measure case:

$$\mathcal{A}'_i \subsetneqq \mathcal{A}_i \subsetneqq \mathcal{B}(\mathbb{R}) \subsetneqq \mathcal{M}_L,$$

and these inclusions are strict since not all Lebesgue measurable sets are Borel sets, not all Borel sets are a finite union of right semi-closed intervals, and of course not all finite unions of such intervals form such an interval. Consequently, in this case the product space semi-algebras satisfy:

$$\mathcal{A}'(\mathcal{A}'_i) \subsetneqq \mathcal{A}'(\mathcal{A}_i) \subsetneqq \mathcal{A}'(\mathcal{B}(\mathbb{R})) \subsetneqq \mathcal{A}'(\mathcal{M}_L).$$

But note that the four algebras associated with these semi-algebras, each defined as in Exercise 6.10, reduce to three distinct algebras. With the apparent notational convention:

$$\mathcal{A}(\mathcal{A}'_i) = \mathcal{A}(\mathcal{A}_i) \subsetneqq \mathcal{A}(\mathcal{B}(\mathbb{R})) \subsetneqq \mathcal{A}(\mathcal{M}_L).$$

That $\mathcal{A}(\mathcal{A}'_i) = \mathcal{A}(\mathcal{A}_i)$ follows from the fact that an algebra is the collection of finite disjoint unions of sets from the associated semi-algebra. But the collections of finite disjoint unions of sets from $\mathcal{A}'(\mathcal{A}'_i)$ agrees with the collection of such unions from $\mathcal{A}'(\mathcal{A}_i)$, since each $\mathcal{A}_i$-set is a finite union of $\mathcal{A}'_i$-sets. In contrast, $\mathcal{A}(\mathcal{A}_i) \subsetneqq \mathcal{A}(\mathcal{B}(\mathbb{R}))$ since $\mathcal{A}_i \subsetneqq \mathcal{B}(\mathbb{R})$, and so the basic measurable rectangles can differ. Moreover, the component Borel sets in general require countable many operations with $\mathcal{A}_i$-sets, while $\mathcal{A}(\mathcal{A}_i)$ only permits finitely many. Though not proved, it was noted in Section 2.8 that $\mathcal{B}(\mathbb{R}) \subsetneqq \mathcal{M}_L$, and then the argument is similar.

For the above application of Lebesgue or Borel measure, there would seem to be an advantage to beginning the extension process of Proposition 6.13 with a pre-measure μ_0 on one of the semi-algebras $\mathcal{A}'(\mathcal{A}'_i)$ or $\mathcal{A}'(\mathcal{A}_i)$. This would produce a relatively simple measure on the associated algebra $\mathcal{A}(\mathcal{A}'_i) = \mathcal{A}(\mathcal{A}_i)$, and in turn a more accessible definition of "outer measure" in (6.3). Specifically, the outer measure of a set would be defined in terms of sums of the μ_0-measures of simple measurable rectangles defined in terms of right semi-closed intervals. And this would also be handy for the approximation results of Proposition 6.5. For example, all measurable sets could then be approximated within ϵ by supersets in $\mathcal{A}_\sigma$, and within measure 0 by supersets in $\mathcal{A}_{\sigma\delta}$.

However, in the more general case developed in this chapter, the semi-algebra $\mathcal{A}'_i$ underlying $(X_i, \sigma(X_i), \mu_i)$ need not have such a convenient or simple structure as that of right semi-closed intervals. In this case, one simply has $\mathcal{A}'_i \subset \mathcal{A}_i \subset \sigma(X_i)$. For this reason we will assume that the semi-algebra on the product space, $\mathcal{A}'$, is defined as in Definition 7.1, in terms of measurable rectangles using sets from the respective sigma algebras $\sigma(X_i)$.

7.2 Properties of the Semi-Algebra $\mathcal{A}'$

For the pre-measure results below, we will need to work with a finite or countable collection of disjoint measurable rectangles, the union of which is another measurable rectangle. So we begin with an investigation into what this constraint – the union of which is another measurable rectangle – implies. Consistent with Definition 7.1, we assume that measurable rectangles are defined in terms of sigma algebras, so $A \in \mathcal{A}'$ implies $A = \prod_{i=1}^{n} A_i$ with $A_i \in \sigma(X_i)$. Naturally, any result demonstrated below on this semi-algebra would need to be reevaluated in terms of other semi-algebras if such results are needed. We will make note of this below.

Assume that a finite or countable collection of measurable rectangles as $B_j \in \mathcal{A}'$ is given, and define:

$$A = \bigcup_j B_j.$$

Writing each measurable rectangle as $B_j = \prod_{i=1}^n A_{ji}$ with $A_{ji} \in \sigma(X_i)$ for all j, we have:

$$A = \bigcup_j \prod_{i=1}^n A_{ji}.$$

As semi-algebras are in general not closed under unions, a union of measurable rectangles need not be a measurable rectangle.

Example 7.5 (Unions of measurable rectangles) *In $\mathbb{R}^2$, if $B_1 = (0,1] \times (1,2]$ and $B_2 = (1,2] \times (2,4]$, then $B_1 \bigcup B_2$ is not a measurable rectangle. The problem is the asymmetry in the union set. Specifically, if $(x,y) \in B_1 \bigcup B_2$ then the y-component is a function of x, $y = y(x)$, and the y-options for $x \in (0,1]$ differ from those available when $x \in (1,2]$. To be a measurable rectangle by (7.2) requires that the y-components available are independent of x.*

If the sets $B_3 = (0,1] \times (2,4]$ and $B_4 = (1,2] \times (1,2]$ are included in the union, then $\bigcup_{j \le 4} B_j = (0,2] \times (1,4]$, a measurable rectangle. Each factor of the product rectangle is a union of the respective factors of the component rectangles. Expressed in order, with some redundancy:

$$(0,2] = (0,1] \bigcup (1,2] \bigcup (0,1] \bigcup (1,2],$$

$$(1,4] = (1,2] \bigcup (2,4] \bigcup (2,4] \bigcup (1,2].$$

The next result generalizes this example. It shows that if $\{B_j\}_{j=1}^M \subset \mathcal{A}'$ with $B_j = \prod_{i=1}^n A_{ji}$, and $A \equiv \bigcup_{j=1}^M B_j \in \mathcal{A}'$ with $A = \prod_{i=1}^n A_i$, then $A_i = \bigcup_{j=1}^M A_{ji}$ for every i. And this is true whether there are finitely many, or countably many B_j-sets.

Proposition 7.6 (When a union is a measurable rectangle 1) *Let $\{B_j\}_{j=1}^M \subset \mathcal{A}'$ with $B_j = \prod_{i=1}^n A_{ji}$ be a finite ($M < \infty$) or countable ($M = \infty$) collection of measurable rectangles. If $A = \bigcup_{j=1}^M B_j \in \mathcal{A}'$ with $A = \prod_{i=1}^n A_i$, then for each i:*

$$A_i = \bigcup_{j=1}^M A_{ji}. \tag{7.4}$$

Further:

$$A = \prod_{i=1}^n \left(\bigcup_{j=1}^M A_{ji} \right) = \bigcup_N \left[\prod_{i=1}^n A_{N(i),i} \right], \tag{7.5}$$

*where the set $N = \{(N(1), N(2), ..., N(n)) | N(i) \in \{1, 2, 3, ..., M\}\}$. In words, this latter union is over **all** finitely or countably many n-tuples of j-indexes.*

Proof. *It follows by definition that $\bigcup_{j=1}^{M} A_{ji} \subset A_i$, and we prove equality by contradiction. Assume that for some k that $C_k \equiv A_k - \bigcup_{j=1}^{M} A_{jk} \neq \emptyset$. Then since $\bigcup_{j=1}^{M} A_{jk} \subset A_k$ it follows that $A_k = C_k \bigcup \left(\bigcup_{j=1}^{M} A_{jk} \right)$, a disjoint union of two nonempty sets. Thus:*

$$A = \prod_{i=1}^{n} A_i$$
$$= \left(\prod_{i=1}^{k-1} A_i \times C_k \times \prod_{i=k+1}^{n} A_i \right) \bigcup \left(\prod_{i=1}^{k-1} A_i \times \bigcup_{j=1}^{M} A_{jk} \times \prod_{i=k+1}^{n} A_i \right).$$

But the rectangle $\prod_{i=1}^{k-1} A_i \times C_k \times \prod_{i=k+1}^{n} A_i$ cannot be a subset of $A = \bigcup_{j=1}^{M} \left(\prod_{i=1}^{n} A_{ji} \right)$, since $C_k \bigcap \{ \bigcup_{j=1}^{M} A_{jk} \} = \emptyset$. Hence $A_i = \bigcup_{j=1}^{M} A_{ji}$ for all i, which is (7.4).

For the representations in (7.5), the first expression is a result of the identity in (7.4), while the second is a combinatorial exercise. If $M < \infty$ and there are a finite number of B_j-rectangles whose union is A, there will be M^n terms in the right-most union in (7.5) since for all i, $N(i) \in \{1, 2, 3..., M\}$.

When there are countably many B_j-rectangles whose union is A, this union will have countably many terms. To see this, note that between any two integers N_1 and N_2, there are only finitely many terms in the union for which the sum of the indexes satisfies:

$$N_1 \leq \sum_{i=1}^{n} N(i) < N_2.$$

Choosing $N_j = 10^j$ for example, the above union is equivalent to a countable union of finite sets, and this is countable (indeed, a countable union of countable sets is countable). ∎

Example 7.7 ($n = 2$) *When $M = \infty$ and $n = 2$, the cumbersome notation in (7.5) simplifies to:*

$$A = A_1 \times A_2 = \bigcup_{j,k=1}^{\infty} (A_{j1} \times A_{k2}).$$

where $A_i = \bigcup_{j=1}^{\infty} A_{ji}$. Thus the index 2-tuples are the collection of all index pairs (j, k), with $j, k = 1, 2, 3...$

Example 7.8 (On Example 7.5) *The last union in (7.5) need not be a disjoint union even when the B_j-sets are disjoint. As will be seen below, this happens precisely because the representations $A_i = \bigcup_j A_{ji}$ cannot all be disjoint unions even in this case of disjoint B_j-sets.*

Returning to Example 7.5 above, where the four B_j-sets were disjoint and unioned to a measurable rectangle, the second expression for $A = \bigcup_j B_j$ is formally:

$$A = \left\{ (0,1] \bigcup (1,2] \bigcup (0,1] \bigcup (1,2] \right\} \times \left\{ (1,2] \bigcup (2,4] \bigcup (2,4] \bigcup (1,2] \right\}.$$

For this example, the union of measurable rectangles on the right in (7.5) contains $M^n = 16$ terms, with many redundancies since there are only four distinct measurable rectangles.

In this example neither of the unions, $A_i = \bigcup_{j \leq 4} A_{ji}$ for $i = 1, 2$, is a disjoint union.

The next result states that when $A \equiv \bigcup_j B_j$ is a measurable rectangle, at most one $A_i = \bigcup_j A_{ji}$ can be a disjoint union.

Proposition 7.9 (When a union is a measurable rectangle 2) *Let* $\{B_j\}_{j=1}^M = \{\prod_{i=1}^n A_{ji}\}_{j=1}^M$ $\subset \mathcal{A}'$ *be a collection of finitely ($M < \infty$) or countably ($M = \infty$) many measurable rectangles, and* $A \equiv \bigcup_{j=1}^M B_j \in \mathcal{A}'$.

Then $A_i = \bigcup_{j=1}^M A_{ji}$ *is a disjoint union for **at most one** i.*

Proof. *If* $A = \prod_{i=1}^n A_i$, *then by Proposition 7.6,* $A_i = \bigcup_j A_{ji}$ *for all i. Assume that* A_1 *is a disjoint union. We claim that for* $i > 1$, *that* $A_{ji} = A_i$ *for all j and so:*

$$B_j = A_{j1} \times \prod_{i=2}^n A_i.$$

Thus the associated unions $A_i = \bigcup_j A_{ji}$ *are not disjoint for* $i > 1$, *and in fact are unions of identical sets.*

To see this, note that A has two representations:

$$A = \bigcup_{j=1}^M \left[A_{j1} \times \prod_{i=2}^n A_{ji} \right] = \bigcup_{j=1}^M \left[A_{j1} \times \prod_{i=2}^n A_i \right].$$

The expression on the left is $\bigcup_{j=1}^M B_j$, *while that on the right derives from* $A_1 \times \prod_{i=2}^n A_i$ *and the union representation of* A_1.

Since $\{A_{j1}\}$ *are disjoint it follows that for all j:*

$$A_{j1} \times \prod_{i=2}^n A_{ji} = A_{j1} \times \prod_{i=2}^n A_i,$$

and the conclusion follows. ∎

7.3 Measure on the Algebra $\mathcal{A}$

The goal of this section is to prove that the set function μ_0 of (7.3) induces a measure on the algebra $\mathcal{A}$, which is the algebra generated by $\mathcal{A}'$ as in Exercise 6.10. To this end, we first develop the finite additivity result of μ_0 on the semi-algebra $\mathcal{A}'$, which will generalize to the associated algebra $\mathcal{A}$, and then pursue countable additivity on $\mathcal{A}$.

For certain types of sets in $\mathcal{A}'$, finite and countable additivity are easy to prove.

Proposition 7.10 (Additivity of μ_0 is sometimes apparent) *For* $1 \leq i \leq n$ *let* $\{A_{ji}\}_{j=1}^M \subset \sigma(X_i)$ *be a finite ($M < \infty$) or countable ($M = \infty$) collection of disjoint measurable sets,* $A_i \equiv \bigcup_{j=1}^M A_{ji}$ *and* $A \equiv \prod_{i=1}^n A_i$.

Then A is a disjoint union of measurable rectangles:

$$A = \bigcup_N \left[\prod_{i=1}^n A_{N(i),i} \right],$$

and:

$$\mu_0(A) = \sum_N \mu_0 \left(\prod_{i=1}^n A_{N(i),i} \right).$$

This union and summation are over N, the set of all finitely or countably many n-tuples $(N(1), N(2), ..., N(n))$ *such that* $N(i) \in \{1, 2, 3, ..., M\}$ *for all i.*

Proof. *The representation for A as a union over the set N follows as in Proposition 7.6.*

To see that $\{\prod_{i=1}^n A_{N(i),i}\}_N$ *is a disjoint collection, let* $x \in \prod_{i=1}^n A_{N(i),i} \cap \prod_{i=1}^n A_{N'(i),i}$. *Thus* $A_{N(i),i} \cap A_{N'(i),i} \neq \emptyset$ *for each i, and it then follows that* $N(i) = N'(i)$ *for all i by disjointness of* $\{A_{ji}\}_{j=1}^M$.

To prove that μ_0 *is additive over this union of sets, first assume that* $\mu_i(A_i) < \infty$ *for all i. With* $\mu_0(A)$ *defined as in (7.3), finite* $(M < \infty)$ *or countable* $(M = \infty)$ *additivity each* μ_i *and disjointness of* $\{A_{ji}\}_{j=1}^M \subset \sigma(X_i)$ *obtains that* $\mu_i(A_{ji}) < \infty$ *for all i, j. Now by definition of* μ_0:

$$\mu_0(A) \equiv \prod_{i=1}^n \mu_i(A_i)$$

$$= \prod_{i=1}^n \left(\sum_{j=1}^M \mu_i(A_{ji}) \right)$$

$$= \sum_N \prod_{i=1}^n \mu_i(A_{N(i),i})$$

$$\equiv \sum_N \mu_0 \left(\prod_{i=1}^n A_{N(i),i} \right).$$

There is no ambiguity in these manipulations, since even if $M = \infty$, *the* $\mu_i(A_{ji})$-*summations are absolutely convergent and thus they can be multiplied componentwise.*

If $\mu_i(A_i) = \infty$ *for* $i = 1$, *say, and all other* $\mu_j(A_j) > 0$, *then* $\mu_0(A) = \infty$ *by (7.3). Defining* $N' = \{(k, 1, 1, ..., 1) | 1 \leq k \leq M\}$, *then by (7.3) and since* μ_1 *is a measure:*

$$\sum_N \mu_0 \left(\prod_{i=1}^n A_{N(i),i} \right) \geq \sum_{N'} \mu_0 \left(\prod_{i=1}^n A_{N(i),i} \right)$$

$$= \sum_k \mu_0 \left(A_{k,1} \times \prod_{i=2}^n A_{1,i} \right)$$

$$= \mu_1(A_1) \times \prod_{i=2}^n \mu_i(A_{1,i})$$

$$= \infty.$$

Finally, in the case where $\mu_i(A_i) = \infty$ *for* $i = 1$, *say, and* $\mu_j(A_j) = 0$ *for* $j = 2$ *say, then* $\mu_0(A) = 0$ *by (7.3). That each term in the summation is also zero follows as in the previous derivation, noting that* $\mu_2(A_{N(2),2}) = 0$ *for all* $N(2)$ *by monotonicity of* μ_2. ∎

Thus for such special collections of disjoint measurable rectangles, finite and countable additivity of μ_0 is relatively easy to prove. Example 7.5 above is special in this way in that:

$$\bigcup_{j \leq 4} B_j = (0, 2] \times (1, 4] = \left[(0, 1] \bigcup (1, 2] \right] \times \left[(1, 2] \bigcup (2, 4] \right].$$

Further, the products on the right reproduce the original B_j-sets.

In the general development below, $\bigcup_{j \leq M} B_j$ will be assumed to be a measurable rectangle $\prod_{i=1}^{n} A_i$. It will not be difficult to represent each A_i as a disjoint union, and thus represent $\bigcup_{j \leq M} B_j$ as a product of disjoint unions. But what makes the Proposition 7.10 derivation work is that this product of disjoint unions reproduces the original B_j-sets.

It is not difficult to produce a collection of disjoint measurable rectangles which unions to a measurable rectangle, yet cannot be expressed this way. For example, consider a small change to Example 7.5:

$$B_1 = (0, 1] \times (1, 2], \ B_2 = (1, 2] \times (3, 4],$$
$$B_3 = (0, 1] \times (2, 4], \ B_4 = (1, 2] \times (1, 3].$$

Now $\bigcup_{j \leq M} B_j$ has many representations as a product of disjoint unions, but none that reproduces these four sets. The key to the next section's development is the observation that $\bigcup_{j \leq M} B_j$ can always be decomposed as the product of disjoint unions such that each B_j-set is a disjoint union of one or more of these product sets.

Before proceeding, the reader is encouraged to develop the application of this idea to the above example.

7.3.1 Finite Additivity on the Semi-Algebra $\mathcal{A}'$

In order to apply the idea in Proposition 7.10 to prove that μ_0 is finitely additive on $\mathcal{A}'$, assume that we are given **finitely** many disjoint measurable rectangles $\{B_j\}_{j=1}^{M} = \{\prod_{i=1}^{n} A_{ji}\}_{j=1}^{M}$, and that:

$$A \equiv \bigcup_{j=1}^{M} B_j = \prod_{i=1}^{n} A_i,$$

a measurable rectangle.

We then claim that there exists disjoint collections of sets $\{A'_{ki}\}_{k=1}^{M'_i} \subset \sigma(X_i)$, where $M'_i \leq 2^M - 1$, so that:

1. For each i:

$$A_i = \bigcup_{j=1}^{M} A_{ji} = \bigcup_{k=1}^{M'_i} A'_{ki}.$$

Hence, as in Proposition 7.10:

$$A = \bigcup_N \left[\prod_{i=1}^{n} A'_{N(i),i} \right],$$

where the union is over the set $N = \{(N(1), N(2), ..., N(n)) |\ 1 \leq N(i) \leq M'_i\}$. Thus N contains up to $\prod_{i=1}^{n} M'_i \leq (2^M - 1)^n$ such n-tuples.

2. For each i there is a collection of subsets $\{I_{ji}\}_{j=1}^{M}$, with $\bigcup_{j=1}^{M} I_{ji} = I \equiv \{1, 2, \ldots, M'\}$, so that:

$$A_{ji} = \bigcup_{k \in I_{ji}} A'_{ki}.$$

3. There exists a partition $\{N_j\}_{j=1}^{M}$ of N defined as in item 1, so that for all j:

$$B_j = \bigcup_{N_j} \left[\prod_{i=1}^{n} A'_{N(i),i} \right].$$

By a partition it is meant that $\{N_j\}_{j=1}^{M}$ is a disjoint collection of subsets of N with $\bigcup_{j=1}^{M} N_j = N$, and by disjoint it is meant as always that $N_j \bigcap N_k = \emptyset$ for $j \neq k$.

Once items 1-3 are implemented, the proof of finite additivity of μ_0 will follow as in Proposition 7.10 above.

Remark 7.11 (The simple idea) *What the precise yet cumbersome notation above is intended to reflect is relatively simple. If $A \equiv \bigcup_{j=1}^{M} B_j$ is a measurable rectangle, so $A = \prod_{i=1}^{n} A_i$, then each component of A, say $A_1 = \bigcup_{j=1}^{M} A_{j1}$, can be expressed as a disjoint union of sets, say $A_1 = \bigcup_{k=1}^{M_1} A'_{k1}$. That's the easier part, and one that is already addressed in Proposition 2.20. What makes this construction and resulting disjoint collection different and suitable for the forthcoming derivation, is that each set in the A_1-union, each A_{j1}, can also be expressed as a union of a subset of this disjoint collection. This latter property is not enjoyed by the construction in Proposition 2.20.*

Continuing with A_1, while each A_{j1} is the union of a subset of this collection of disjoint sets, in general the collection of subsets $\{I_{j1}\}_{j=1}^{M}$ is not disjoint, meaning that any one of the A'_{k1}-sets may be needed for one or more of the A_{j1}-sets, and hence will be needed in one or more of the B_j-sets. This will be observed in the example below. However, once these A'_{ki} are used in the representation of B_j as a union of rectangles, the n-tuple indexing sets, N_j, will be disjoint.

It will be seen below in Example 7.17 that this construction fails when $M = \infty$ because the collection $\{A'_{ji}\}_{j=1}^{2^M - 1}$ can then be uncountable. Hence, this approach cannot be used for countable additivity.

The next proposition accomplishes this construction, but first an example.

Example 7.12 (Proposition 7.13 illustrated) *In $\mathbb{R}^2$ let $B_j = A_{j1} \times A_{j2}$ be defined for $1 \leq j \leq 5$ as:*

$$B_1 = (1, 3] \times (0, 1],$$
$$B_2 = (1, 2] \times (1, 2],$$
$$B_3 = (1, 2] \times (2, 3],$$

$$B_4 = (2,3] \times (1,3],$$
$$B_5 = (3,4] \times (0,3].$$

Then $A = \bigcup_{j=1}^{5} B_j = (1,4] \times (0,3] \equiv A_1 \times A_2$ with:

$$A_1 = (1,3] \bigcup (1,2] \bigcup (1,2] \bigcup (2,3] \bigcup (3,4],$$

$$A_2 = (0,1] \bigcup (1,2] \bigcup (2,3] \bigcup (1,3] \bigcup (0,3].$$

Of course $A_i = \bigcup_{j=1}^{5} A_{ji}$ as proved in Proposition 7.6, and here $\{A_{ji}\}_{j=1}^{5}$ is not a disjoint collection for either $i = 1, 2$, consistent with Proposition 7.9.
  Define:

$$A'_{11} = (1,2], \ A'_{21} = (2,3], \ A'_{31} = (3,4],$$

$$A'_{12} = (0,1], \ A'_{22} = (1,2], \ A'_{32} = (2,3].$$

As noted in the introduction above, this example requires only $M'_i = 3$, A'_{ki}-sets for each A_i, compared with the upper bound of $2^5 - 1 = 31$ such sets.
  Then each $A_i = \bigcup_{k=1}^{3} A'_{ki}$ as disjoint unions. Also each A_{ji} equals a union of $\{A'_{ki}\}$-sets:

$$A_{ji} = \bigcup_{k \in I_{ji}} A'_{ki}.$$

For each i, $\{I_{ji}\}_{j=1}^{5}$ is a collection of subsets of $I \equiv \{1,2,3\}$ with $\bigcup_{j=1}^{5} I_{ji} = I$.
  Thus A and each B_j can be expressed as disjoint unions:

$$A = \left(\bigcup_{k=1}^{3} A'_{k1} \right) \times \left(\bigcup_{k=1}^{3} A'_{k2} \right),$$

$$B_j = \left(\bigcup_{k \in I_{j1}} A'_{k1} \right) \times \left(\bigcup_{k \in I_{j2}} A'_{k2} \right).$$

The set N in item 3 above is $I \times I$, while N_j associated with B_j is defined as the set $I_{j1} \times I_{j2}$.
  Note that in the representations for the B_j-sets that a given A'_{ki} may be used for more than one j. For example, $A'_{21} = (2,3]$ is used in the first component of B_1 and B_4, while $A'_{32} = (2,3]$ is used in the second component of B_3, B_4, and B_5.

This example is generalized in the following proposition, and will form the basis for the proof of the finite additivity of μ_0. Admittedly, the notation is cumbersome but the central idea of the conclusion is that the construction accomplished in the above example can be implemented in general.

Proposition 7.13 (Disjoint partitions of rectangles) *Let* $\{B_j\}_{j=1}^M \subset \mathcal{A}'$ *denote a finite disjoint collection of measurable rectangles,* $B_j = \prod_{i=1}^n A_{ji}$ *where* $\{A_{ji}\}_{j=1}^M \subset \sigma(X_i)$ *for each i. Assume that* $A \equiv \bigcup_{j=1}^M B_j \in \mathcal{A}'$, *so* $A = \prod_{i=1}^n A_i$ *with* $A_i \equiv \bigcup_{j=1}^M A_{ji} \in \sigma(X_i)$.

Then for each i there exists disjoint $\{A_{ki}'\}_{k=1}^{M_i'} \subset \sigma(X_i)$, *with* $M_i' \leq 2^M - 1$, *so that for* $i = 1, 2, ..., n$:

$$A_i \equiv \bigcup_{j=1}^M A_{ji} = \bigcup_{k=1}^{M_i'} A_{ki}'. \tag{7.6}$$

Thus:

$$A = \bigcup_N \left[\prod_{i=1}^n A_{N(i),i}' \right], \tag{7.7}$$

where the union is over the set $N = \{(N(1), N(2), ..., N(n)) | 1 \leq N(i) \leq M_i'\}$.

In addition, for each i there is a not necessarily disjoint collection of subsets $\{I_{ji}\}_{j=1}^M$ *with* $\bigcup_{j=1}^M I_{ji} = I_i \equiv \{1, 2, ..., M_i'\}$, *so that:*

$$A_{ji} = \bigcup_{k \in I_{ji}} A_{ki}', \tag{7.8}$$

and hence:

$$B_j = \prod_{i=1}^n \left(\bigcup_{k \in I_{ji}} A_{ki}' \right).$$

Finally, there is a partition of the set N into M disjoint subsets $\{N_j\}_{j=1}^M$ *so that:*

$$B_j = \bigcup_{N_j} \left[\prod_{i=1}^n A_{N_j(i),i}' \right], \tag{7.9}$$

where by partition it is meant that $\bigcup_{j=1}^M N_j = N$ *and* $N_j \cap N_k = \emptyset$ *for* $j \neq k$. *Specifically,* $N_j = \prod_{i=1}^n I_{ji}$, *the collection of all n-tuples with ith component from* I_{ji}.

Proof. *We implement the construction explicitly for* $i = 1$ *to simplify notation. Given* $\{A_{j1}\}_{j=1}^M \subset \sigma(X_1)$, *note that:*

$$X_1 = \bigcap_{j=1}^M \left(A_{j1} \bigcup \widetilde{A}_{j1} \right)$$
$$= \bigcup \left(\bigcap_{j=1}^M D_{j1} \right),$$

where the union is over all 2^M *possible intersections for which either* $D_{j1} = A_{j1}$ *or* $D_{j1} = \widetilde{A}_{j1}$. *This is a disjoint union by construction, since any two such intersection sets differ by at least one*

D_{j1}-*value. Further, for any such intersection set,* $\bigcap_{j=1}^{M} D_{j1} \in \sigma(X_1)$, *and thus this is a disjoint union of measurable sets.*

One of these sets is $\bigcap_{j=1}^{M} \widetilde{A_{j1}} = \left(\widetilde{\bigcup_{j=1}^{M} A_{j1}} \right)$ *by De Morgan's laws, and so:*

$$\bigcup_{j=1}^{M} A_{j1} = X_1 - \bigcap_{j=1}^{M} \widetilde{A_{j1}}$$
$$= \bigcup{}' \left(\bigcap_{j=1}^{M} D_{j1} \right).$$

This $\bigcup'$-*disjoint union is now over all* $2^M - 1$ *possible intersections for which **at least** one* D_{j1} *equals* A_{j1}.

Let $M'_1 \leq 2^M - 1$ *denote the number of nonempty intersection sets, and define* $\{A'_{k1}\}_{k=1}^{M'_1}$ *as these* M'_1 *sets in some order. With this construction, (7.6) now follows. If this construction is implemented for each i, then (7.7) follows since:*

$$A \equiv \prod_{i=1}^{n} \left[\bigcup_{k=1}^{M'_i} A'_{ki} \right] = \bigcup_{N} \left[\prod_{i=1}^{n} A'_{N(i),i} \right],$$

where N is defined above.

Continuing with $i = 1$, *we next define the* $\{I_{j1}\}_{j=1}^{M}$ *subsets of the index set* $I_1 \equiv \{1, 2, ..., M'_1\}$, *so that* $\bigcup_{j=1}^{M} I_{j1} = I_1$ *and (7.8) is validated. To this end, recall that each* A'_{k1} *is defined as one of the* M'_1 *nonempty intersection sets* $\bigcap_{i=1}^{M} D_{i1}$ *for which at least one* $D_{i1} = A_{i1}$, *so define:*

$$I_{j1} = \{k | A'_{k1} \subset A_{j1}\}.$$

Thus I_{j1} *is the set of subscripts k so that* $A'_{k1} = \bigcap_{i=1}^{M} D_{i1}$ *where* $D_{i1} = A_{j1}$ *for some i.*
  By definition:

$$\bigcup_{k \in I_{j1}} A'_{k1} \subset A_{j1}.$$

Then to prove (7.8), observe that for any j:

$$A_{j1} = \left[\bigcap_{k=1}^{j-1} \left(A_{k1} \cup \widetilde{A_{k1}} \right) \right] \cap A_{j1} \cap \left[\bigcap_{k=j+1}^{M} \left(A_{k1} \cup \widetilde{A_{k1}} \right) \right],$$

simply because $A_{k1} \cup \widetilde{A_{k1}} = X_1$ *for any k. The right-hand expression can be rewritten as above, as a union of intersections of sets. With the above notation, this union now contains every set* $\bigcap_{i=1}^{M} D_{i1}$ *for which some* $D_{i1} = A_{j1}$. *Said differently, this union now contains every* A'_{k1}-*set in the definition of* I_{j1}. *Thus* $\bigcup_{k \in I_{j1}} A'_{k1} = A_{j1}$. *Then* $\bigcup_{j=1}^{M} I_{j1} = I_1$ *since by definition of* A'_{k1}, *each is contained in at least one* A_{j1}.

Finally, we can rewrite the expression $B_j = \prod_{i=1}^{n} \left(\bigcup_{k \in I_{ji}} A'_{ki} \right)$ *as a union of measurable rectangles as in (7.9), and define* N_j *as the collection of n-tuples that arise. Notationally, this can be expressed* $N_j = \prod_{i=1}^{n} I_{ji}$. *Then because* $\{B_j\}$ *is a disjoint collection of rectangles with union equal to A, it follows that* $\bigcup_{j=1}^{M} N_j = N$ *and* $N_j \cap N_k \neq \emptyset$ *for* $j \neq k$. ∎

Remark 7.14 (Proposition 7.13 on $\mathcal{A}'(\mathcal{A}_i)$, but not $\mathcal{A}'(\mathcal{A}_i')$) *Recalling the notation introduced in Corollary 7.3 and its introductory paragraph, Proposition 7.13 was explicitly stated and proved relative to the semi-algebra $\mathcal{A}' \equiv \mathcal{A}'\left(\sigma\left(X_i\right)\right)$. This notation implies that the component sets for measurable rectangles are selected from the respective sigma algebras, $\{\sigma\left(X_i\right)\}$.*

This result is also true in $\mathcal{A}'(\mathcal{A}_i)$, the semi-algebra defined with rectangle components selected from the $\mathcal{A}_i$-algebras, where $\mathcal{A}_i \subset \sigma\left(X_i\right)$. The statement of assumptions now includes $A \equiv \bigcup_{j=1}^{M} B_j \in \mathcal{A}'(\mathcal{A}_i)$. This generalization follows since $A_{ji} \in \mathcal{A}_i$ implies that $\widetilde{A}_{ji} \in \mathcal{A}_i$, and then because an algebra is closed under finite intersections we derive that $\bigcap_{j=1}^{M} D_{ji} \in \mathcal{A}_i$. In other words, $\{A_{ji}'\}_{j=1}^{M_i'} \subset \mathcal{A}_i$, and thus $\prod_{j=1}^{n} A_{N(j),j}' \in \mathcal{A}'(\mathcal{A}_i)$ for any $N(j)$.

But this result is not true in $\mathcal{A}'(\mathcal{A}_i')$, the semi-algebra defined with rectangle components selected from the $\mathcal{A}_i'$-semi-algebras, where $\mathcal{A}_i' \subset \mathcal{A}_i$. Now $A_{ji} \in \mathcal{A}_i'$ implies only that $\widetilde{A}_{ji}$ equals a finite union of $\mathcal{A}_i'$-sets. Hence, $\bigcap_{j=1}^{M} D_{ji}$ is a finite union of finite intersections of $\mathcal{A}_i'$-sets. However, while each finite intersection set is in $\mathcal{A}_i'$, semi-algebras are not closed under finite unions, as the semi-algebra of right semi-closed intervals illustrates. Thus, we cannot conclude that $\bigcap_{j=1}^{M} D_{ji} \in \mathcal{A}_i'$, and so Proposition 7.13 does not generalize to the semi-algebra $\mathcal{A}'(\mathcal{A}_i')$.

With the construction of the above proposition, the proof that μ_0 is finitely additive on $\mathcal{A}'$ is now relatively straightforward. The proof of Proposition 6.13 will then obtain that μ_0 induces a well-defined and finitely additive set function on the algebra $\mathcal{A}$ generated by $\mathcal{A}'$. Recall that by Exercise 6.10 that $\mathcal{A}$ equals the collection of all finite disjoint unions of sets from $\mathcal{A}'$, plus the empty set when not already included in $\mathcal{A}'$.

Proposition 7.15 (μ_0 extends to finitely additive $\mu_{\mathcal{A}}$ on $\mathcal{A}$) *Let $\mathcal{A}'$ denote the collection of measurable rectangles in $X = \prod_{i=1}^{n} X_i$ defined in (7.2).*

Then the product set function μ_0, defined on $\mathcal{A}'$ by (7.3), is finitely additive on $\mathcal{A}'$, and has a well defined and finitely additive extension $\mu_{\mathcal{A}}$ on the algebra $\mathcal{A}$ generated by $\mathcal{A}'$.

Proof. *1. Finite additivity of μ_0 on $\mathcal{A}'$:* To show that μ_0 is finitely additive on $\mathcal{A}'$, let $\{B_j\}_{j=1}^{M} \subset \mathcal{A}'$ denote a finite disjoint collection of measurable rectangles, with $B_j = \prod_{i=1}^{n} A_{ji}$ $\{A_{ji}\}_{j=1}^{M} \subset \sigma\left(X_i\right)$ for each i. Assume that $A \equiv \bigcup_{j=1}^{M} B_j \in \mathcal{A}'$, so $A = \prod_{i=1}^{n} A_i$, and by Proposition 7.6, $A_i \equiv \bigcup_{j=1}^{M} A_{ji} \in \sigma\left(X_i\right)$. Then, by Proposition 7.13, each $A_i \equiv \bigcup_{k=1}^{M_i'} A_{ki}'$ as a disjoint union.

If $\mu_i(A_i) < \infty$ for all i, then monotonicity of μ_i implies that $\mu_i(A_{ki}') < \infty$ for all i, k. Then with $N(i)$, etc. as defined in Proposition 7.13:

$$\mu_0(A) \equiv \prod_{i=1}^{n} \mu_i(A_i)$$

$$= \prod_{i=1}^{n} \left(\sum_{k=1}^{M_i'} \mu_i(A_{ki}') \right)$$

$$= \sum_{N} \prod_{i=1}^{n} \mu_i(A_{N(i),i}')$$

$$\equiv \sum_{N} \mu_0 \left(\prod_{i=1}^{n} A_{N(i),i}' \right).$$

Similarly by (7.9):

$$\mu_0(B_j) = \sum_{N_j} \mu_0 \left(\prod_{i=1}^{n} A'_{N_j(i),i} \right).$$

Since $\bigcup_j N_j = N$, it follows that:

$$\mu_0(A) = \sum_{j=1}^{M} \mu_0(B_j). \tag{1}$$

If $\mu_i(A_i) = \infty$ for some i, then by Definition 7.1, $\mu_0(A) = 0$ if $\mu_k(A_k) = 0$ for some k, and $\mu_0(A) = \infty$ otherwise. Now if $\mu_i(A_i) = \infty$ and $\mu_k(A_k) = 0$, then by monotonicity $\mu_k(A_{jk}) = 0$ for all j, and then, $\mu_k(A'_{jk}) = 0$ for all j by (7.6). Thus, since each term in the union in (7.9) for B_j uses an A'_{jk}-set in its kth component, $\mu_0(B_j) = 0$ for all j. This proves (1) in this case.

For the final case, assume that $\mu_i(A_i) = \infty$ and all other $\mu_k(A_k) > 0$. To prove (1) is to prove that $\mu_0(B_j) = \infty$ for some j. To this end, first note that by finite subadditivity of μ_i that $\mu_i(A_{ji}) = \infty$ for some j, and similarly $\mu_i(A'_{li}) = \infty$ for some l with $A'_{li} \subset A_{ji}$. Consider now $N' \subset N$ defined as $N' \equiv \{(N(1), N(2), ..., N(n)) | 1 \leq N(j) \leq M'_j, N(i) = l\}$. In other words, N' is the collection of n-tuples in N with ith component equal to l, and thus every measurable rectangle $\prod_{j=1}^{n} A'_{N'(j),j}$ contains A'_{li}. Now define:

$$A' = \bigcup_{N'} \left[\prod_{j=1}^{n} A'_{N'(j),j} \right],$$

and note that A' is a measurable rectangle. In fact, expressing each A_k for $k \neq l$ as a union of $A'_{N'(k),k}$:

$$A' = \prod_{k=1}^{l-1} A_k \times A'_{li} \times \prod_{k=l+1}^{n} A_k.$$

By assumption $\mu_k(A_k) > 0$ for $k \neq l$, and since $A_k = \bigcup_{N'} A'_{N'(k),k}$ for each such k, at least one of these unioned sets has positive measure, which we denote by $A'_{N''(k),k}$. Now define:

$$A'' = \prod_{k=1}^{l-1} A'_{N''(k),k} \times A'_{li} \times \prod_{k=l+1}^{n} A'_{N''(k),k}.$$

Then A'' is a measurable rectangle which by Proposition 7.13 is contained in a unique set B_j, and $\mu_0(A'') = \infty$ by construction. By monotonicity of μ_k, $\mu_0(B_j) \geq \mu_0(A'') = \infty$, and this proves (1) in this case.

Combining the above, the proof of finite additivity is complete.

*2. **Extension to finitely additive $\mu_{\mathcal{A}}$ on $\mathcal{A}$:** That the extension of μ_0 to $\mu_{\mathcal{A}}$ on the algebra $\mathcal{A}$ is well-defined and finitely additive on $\mathcal{A}$ is proved in steps 1 and 2 of the proof of the Carathéodory Extension Theorem 2 of Proposition 6.13. While that proposition assumed that μ_0 was a **pre-measure** on $\mathcal{A}'$ and thus countable additive, only finite additivity on $\mathcal{A}'$ was used in these steps.* ∎

Remark 7.16 (Proposition 7.15 on $\mathcal{A}'(\mathcal{A}_i)$ and $\mathcal{A}'(\mathcal{A}_i')$) *Continuing Remark 7.14, this proof applies with μ_0 defined on $\mathcal{A}'(\mathcal{A}_i)$, the semi-algebra defined as in (7.2) but with component sets $A_i \in \mathcal{A}_i$ with $\mathcal{A}_i$ an algebra on X_i. Of course, while this proof can be duplicated, it often need not be. Finite additivity of μ_0 on $\mathcal{A}'(\mathcal{A}_i)$ is automatically implied by the above result if $\mathcal{A}_i \subset \sigma(X_i)$ since then $\mathcal{A}'(\mathcal{A}_i) \subset \mathcal{A}'$.*

On the other hand, the above proof does not apply directly to $\mathcal{A}'(\mathcal{A}_i')$, defined with component sets $A_i \in \mathcal{A}_i'$ with $\mathcal{A}_i'$ is a semi-algebra on X_i. As noted in Remark 7.14, we cannot then be assured that $\{A_{ji}'\}_{j=1}^{M_i'} \subset \mathcal{A}'(\mathcal{A}_i')$. But as for $\mathcal{A}'(\mathcal{A}_i)$, finite additivity of μ_0 on $\mathcal{A}'(\mathcal{A}_i')$ is implied by the above result when $\mathcal{A}_i' \subset \sigma(X_i)$ since then $\mathcal{A}'(\mathcal{A}_i') \subset \mathcal{A}'$.

*An alternative proof of these results will be provided in Book III using a powerful result from integration theory known as **Lebesgue's monotone convergence theorem**, named for **Henri Léon Lebesgue** (1875–1941).*

7.3.2 Countable Additivity on the Algebra $\mathcal{A}$ for σ-Finite Spaces

Using the approach of the previous section, it is not possible to prove that μ_0 is countably additive on the semi-algebra $\mathcal{A}'$ and thus a measure. This is because given a collection $\{A_j\}_{j=1}^{\infty}$ with $A_j = \bigcup_{k=1}^{\infty} A_{kj}$, the collection of disjoint sets derived by Proposition 7.13 for each j, $\{A_{\alpha j}'\}$, can in theory be uncountable.

Specifically, since now $M = \infty$, each $A_{\alpha j}'$ set is defined:

$$A_j = \bigcup_{\alpha}' A_{\alpha j}', \qquad A_{\alpha j}' = \bigcap_{k=1}^{\infty} D_{kj},$$

where in every $A_{\alpha j}'$-set at least one D_{kj} equals A_{kj}, the remainder equal $\widetilde{A}_{kj}$. These sets can in theory be put into one-to-one correspondence with real numbers in the interval $(0,1]$ by representing each as a **binary number**: $\alpha = 0.a_1 a_2...$, and then identifying $a_k = 0$ with $D_{kj} = \widetilde{A}_{kj}$ and $a_k = 1$ with $D_{kj} = A_{kj}$. Since at least one D_{kj} equals A_{kj}, this identification omits 0 but otherwise is one-to-one between $\{A_{\alpha j}'\}$ and $(0,1]$.

But this argument "in theory" leaves open the possibility that in reality, the collection $\{A_{\alpha j}'\}$ may be countable if "most" such intersection sets are empty. The following example proves that while an uncountable collection is not necessary, it is possible.

Example 7.17 (Countably and uncountably many $A_{\alpha j}'$-sets) *Let $\{r_j\}_{j=0}^{\infty}$ be an arbitrary enumeration of the rationals $\mathbb{Q}$, and define:*

$$A_{j1} = (r_j, \infty), \qquad B_{j1} = (-\infty, r_j], \qquad I_{j2} = [-(j+1), -j) \bigcup [j, j+1).$$

Define:

$$A_j = A_{j1} \times I_{j2}, \qquad B_j = B_{j1} \times I_{j2},$$

and consider the countable collection of measurable rectangles $\{A_j, B_j\}_{j=0}^{\infty}$. Then all such rectangles are disjoint, and collectively union to a measurable rectangle:

$$\bigcup_{j=0}^{\infty} A_j \bigcup \bigcup_{j=0}^{\infty} B_j = \mathbb{R}^2.$$

Now $\bigcup_{j=0}^{\infty} I_{j2} = \mathbb{R}$, and it is an exercise to check that with each D_{j2} equal to I_{j2} or $\tilde{I}_{j2}$, the collection $\{\bigcap_{j=1}^{\infty} D_{j2}\}$ is countable and indeed equals the original collection $\{I_{j2}\}_{j=0}^{\infty}$. Hint: Such an intersection set is empty unless exactly one $D_{j2} = I_{j2}$ and the rest equal $\tilde{I}_{k2}$ for $k \neq j$.

Also $\bigcup_j A_{j1} \bigcup \bigcup_j B_{j1} = \mathbb{R}$, but the collection of intersection sets is uncountable and indeed in one-to-one correspondence with the real numbers.

To see this, consider $\{\bigcap_{k=1}^{\infty} D_{k1}\}$ where each D_{k1} equals one of A_{j1} or $\tilde{A}_{j1}$ or B_{j1} or $\tilde{B}_{j1}$. As before each intersection set includes at least one A_{j1} or one B_{j1}. Since $A_{j1} \cap B_{j1} = \emptyset$ for all j, it is clear that unless the D_{k1}-sets are chosen carefully that $\bigcap_{k=1}^{\infty} D_k = \emptyset$. Specifically, if $D_{k1} = A_{j1}$ for some k and j, then to avoid an empty intersection it is necessary that $D_{l1} = \tilde{B}_{j1}$ for some l. Analogously, if $D_{k1} = B_{j1}$ for some k and j, then it is necessary that some $D_{l1} = \tilde{A}_{j1}$.

*Thus for this intersection set to be nonempty, the $\{D_{k1}\}$-collection must contain **all** paired sets, meaning for every j this collection contains either the pair $\{A_{j1}, \tilde{B}_{j1}\}$ or the pair $\{B_{j1}, \tilde{A}_{j1}\}$. Since $A_{j1} = \tilde{B}_{j1}$ and $B_{j1} = \tilde{A}_{j1}$, the nonempty intersection sets $\bigcap_{k=1}^{\infty} D_{k1}$ can be recharacterized as for each k either $D_{k1} = A_{k1}$ or $D_{k1} = B_{k1}$. Thus:*

$$\bigcap_{k=1}^{\infty} D_{k1} = \bigcap_N A_{k1} \cap \bigcap_{\tilde{N}} B_{k1}$$
$$= \bigcap_R (r_k, \infty) \cap \bigcap_{\tilde{R}} (-\infty, r_k],$$

where $N \subset \{0, 1, 2, ...\}$ and the associated collection of rationals $R = \{r_k | k \in N\}$.

It then follows that:

$$\bigcap_{k=1}^{\infty} D_{k1} = (-\infty, \inf_{\tilde{R}} r_k] \bigcap \{\sup_R r_k, \infty),$$

where for the second set the use of "{" indicates that this interval can be open or closed depending on the set of rationals R.

Now, if R is a finite collection, the above intersection set is empty since then $\inf_{\tilde{R}} r_k = -\infty$. The same conclusion is obtained if $\tilde{R}$ is finite. More generally, for this intersection set to be nonempty, it is necessary that both R and $\tilde{R}$ be infinite, with R bounded from above, and $\tilde{R}$ bounded from below.

Since these are complementary sets of rationals, one way to achieve a nonempty intersection is to define $R = (-\infty, x] \cap \mathbb{Q}$ for given $x \in \mathbb{R}$, where this interval can be open or closed. Then defined as above:

$$\bigcap_{k=1}^{\infty} D_k = x,$$

since $\sup_R r_k = \inf_{\tilde{R}} r_k = x$.

Indeed this is the only way to achieve a nonempty intersection set since if even a single rational $r_k > x$ is added to this definition of R, $\sup_R r_k > \inf_{\tilde{R}} r_k$ and the intersection is empty.

In summary, the only nonempty intersection sets are those that intersect to given $x \in \mathbb{R}$, and thus the collection of such intersection sets is uncountable.

The above example demonstrates that a proof of countable additivity for μ_0 will require a different approach from that used for finite additivity. In this section we prove a somewhat more restrictive result but one equally useful in practice. Specifically we prove that under the assumption of σ-finiteness of the component measure spaces, recalling Definition 5.34, μ_0 is countably additive on the algebra $\mathcal{A}$ generated by $\mathcal{A}'$ and thus a measure.

It will soon be evident that sigma finiteness is not a material restriction. In the next section, it will in any event be necessary to require such σ-finiteness anyway, in order to conclude that the final extension to a measure on a sigma algebra is unique.

The proof below uses the **continuity from above** approach noted in Proposition 6.19, which there applied to finite measures. However, we will see that by assuming that the component space measures μ_i are σ-finite, this proposition can be applied in this more general setting.

Proposition 7.18 (Countable additivity; $\mu_{\mathcal{A}}$ is a measure on $\mathcal{A}$) *Let $\mathcal{A}'$ denote the collection of measurable rectangles in $X = \prod_{i=1}^{n} X_i$ defined in (7.2), and μ_0 the product set function defined in (7.3) with each μ_i assumed to be σ-finite.*

Then the extension $\mu_{\mathcal{A}}$ to the algebra $\mathcal{A}$ generated by $\mathcal{A}'$ is countably additive. Hence $\mu_{\mathcal{A}}$ is a measure on $\mathcal{A}$.

Proof. 1. Reduction to rectangles of finite measure: *Since finite additivity of $\mu_{\mathcal{A}}$ on $\mathcal{A}$ was proved in Proposition 7.15, that $\mu_{\mathcal{A}}$ is a measure will follow from countable additivity by definition.*

Let $\{B_k\}_{k=1}^{\infty} \subset \mathcal{A}$ denote a collection of finite disjoint unions of measurable rectangles,

$$B_k = \bigcup_{j=1}^{m_k} \prod_{i=1}^{n} A_{ji}^{(k)},$$

where $\{A_{ji}^{(k)}\}_{jk} \subset \sigma(X_i)$ for each i. To apply Proposition 6.19, assume that this collection is nested, $B_{k+1} \subset B_k$, and that $\bigcap_k B_k = \emptyset$. Our goal is to demonstrate that $\lim_k \mu_{\mathcal{A}}(B_k) = 0$ where $\mu_{\mathcal{A}}$ is the extension of μ_0 to the algebra $\mathcal{A}$ of Proposition 7.15. Because the measure space is σ-finite and not necessarily finite as Proposition 6.19 requires, the B_k-sets must first be decomposed into subsets of finite measure.

The extension $\mu_{\mathcal{A}}$ is defined to be finitely additive on disjoint unions of measurable rectangles in $\mathcal{A}$:

$$\mu_{\mathcal{A}}\left(\bigcup_{j=1}^{m} \prod_{i=1}^{n} A_{ji}\right) \equiv \sum_{j=1}^{m} \mu_0\left(\prod_{i=1}^{n} A_{ji}\right),$$

and this extension is welldefined by Proposition 7.15. As each μ_i is σ-finite, each X_i can be decomposed into disjoint sets of finite measure, say, $X_i = \bigcup_l Y_{li}$ with $\mu_i(Y_{li}) < \infty$ all l. Hence $X \equiv \prod_{i=1}^{n} X_i$ can be expressed as a countable disjoint union of measurable rectangles:

$$X = \prod_{i=1}^{n}\left(\bigcup_l Y_{li}\right) = \bigcup_L \left(\prod_{i=1}^{n} Y_{l(i),i}\right), \tag{1}$$

with all $\mu_0\left(\prod_{i=1}^{n} Y_{l(i),i}\right) < \infty$ by (7.3). Here $L = \{(l(1), l(2), ..., l(n))\}$ is defined as the countable collection of all n-tuples of component set indexes.

For a given rectangle $A = \prod_{i=1}^{n} A_i \in \mathcal{A}'$, then $\mu_0(A) = \prod_{i=1}^{n} \mu_i(A_i)$ by (7.3). Given disjoint decompositions as above, $X_i = \bigcup_l Y_{li}$, it follows that $\mu_i(A_i) = \sum_l \mu_i(A_i \cap Y_{li})$ since μ_i is countably additive. Further, the value of this countable sum is independent of the decompositions chosen for $\{X_i\}$. Therefore, with the set L as above:

$$\mu_0(A) = \prod_{i=1}^{n} \left[\sum_l \mu_i \left(A_i \cap Y_{li} \right) \right]$$

$$= \sum_L \mu_0 \left(\prod_{i=1}^{n} \left[A_i \cap Y_{l(i),i} \right] \right)$$

$$= \sum_L \mu_0 \left(A \cap \prod_{i=1}^{n} Y_{l(i),i} \right).$$

Because $\mu_{\mathcal{A}}$ is finitely additive on $\mathcal{A}$, this same identity holds for $\mu_{\mathcal{A}}$ defined on a disjoint union $B_k = \bigcup_{j=1}^{m_k} \prod_{i=1}^{n} A_{ji}^{(k)} \in \mathcal{A}$:

$$\mu_{\mathcal{A}}(B_k) \equiv \sum_{j=1}^{m_k} \mu_0 \left(\prod_{i=1}^{n} A_{ji}^{(k)} \right)$$

$$= \sum_{j=1}^{m_k} \sum_L \mu_0 \left(\prod_{i=1}^{n} A_{ji}^{(k)} \cap \prod_{i=1}^{n} Y_{l(i),i} \right)$$

$$= \sum_L \mu_{\mathcal{A}} \left(B_k \cap \prod_{i=1}^{n} Y_{l(i),i} \right). \tag{2}$$

Note that (2) implies that $\mu_{\mathcal{A}}$ is countably additive on the partition of any set in A over the rectangle collection in (1).

2. *Countable additivity from Proposition 6.19: The goal is now to prove that given $\{B_k\} \subset \mathcal{A}$ with $B_{k+1} \subset B_k$ and $\bigcap_k B_k = \emptyset$, that:*

$$\lim_{k \to \infty} \mu_{\mathcal{A}} \left(B_k \cap \prod_{i=1}^{n} Y_{l(i),i} \right) = 0, \tag{3}$$

for each n-tuple $(l(1), l(2), ..., l(n)) \in L$. Then by Proposition 6.19 this obtains that $\mu_{\mathcal{A}}$ is countably additive on $\mathcal{A}$ over every rectangle in the partition in (1). By the remark following (2), this obtains countable additivity on $\mathcal{A}$.

To simplify notation for the proof of (3), since $\{B_k \cap \prod_{i=1}^{n} Y_{l(i),i}\}$ is again a finite disjoint union of measurable rectangles, there is no loss of generality assuming that $\mu_{\mathcal{A}}(B_k) < \infty$ for all k and suppressing the set $\prod_{i=1}^{n} Y_{l(i),i}$.

Rewriting $B_k = \bigcup_{j=1}^{m_k} \left(A_{j1}^{(k)} \times \prod_{i=2}^{n} A_{ji}^{(k)} \right)$, we can by Proposition 2.20 assume that $\{A_{j1}^{(k)}\}_{j=1}^{m_k}$ is a disjoint collection for each k. Define a sequence of functions on X_1 by:

$$f_k(x_1) = \begin{cases} \prod_{i=2}^{n} \mu_i \left(A_{ji}^{(k)} \right), & \text{if } x_1 \in A_{j1}^{(k)}, \\ 0, & \text{if } x_1 \notin \bigcup_{j=1}^{m_k} A_{j1}^{(k)}. \end{cases}$$

Note that for $x_1 \in A_{j1}^{(k)}$ that $f_k(x_1)$ is defined by (7.3) as the product pre-measure of $\prod_{i=2}^{n} A_{ji}^{(k)}$ interpreted as a measurable rectangle on $\prod_{i=2}^{n} X_i$.

Now $B_{k+1} \subset B_k$ with $B_{k+1} = \bigcup_{l=1}^{m_{k+1}} \left(A_{l1}^{(k+1)} \times \prod_{i=2}^{n} A_{li}^{(k+1)} \right)$, and this is a disjoint union of rectangles. This implies that for any j and l, either $\prod_{i=2}^{n} A_{ji}^{(k+1)} \subset \prod_{i=2}^{n} A_{li}^{(k)}$ or these sets are disjoint. Hence if $x_1 \in A_{j1}^{(k)} \cap A_{l1}^{(k+1)}$, then $\prod_{i=2}^{n} A_{ji}^{(k+1)} \subset \prod_{i=2}^{n} A_{li}^{(k)}$, and by monotonicity of the μ_i-measures this obtains that $f_{k+1}(x_1) \leq f_k(x_1)$. In other words, $\{f_k(x_1)\}$ is a nonnegative, decreasing (i.e., nonincreasing) sequence for each $x_1 \in X_1$.

We claim that $\bigcap_k B_k = \emptyset$ assures that $f_k(x_1) \to 0$ for each $x_1 \in X_1$. To prove this, let x_1 be given, and let $\{\prod_{i=1}^{n} A_{jki}^{(k)}\}_{k=1}^{\infty}$ be a collection of rectangles with $\prod_{i=1}^{n} A_{jki}^{(k)} \subset B_k$ and $x_1 \in A_{jk1}^{(k)}$ for all k. Since $f_k(x_1)$ is nonnegative and decreasing as noted above, either $f_k(x_1) \to 0$, or there exists $\epsilon > 0$ with $f_k(x_1) \geq \epsilon$ for all k. In the first case we are done, while in the second we conclude that $\prod_{i=2}^{n} \mu_i \left(A_{jki}^{(k)} \right) \geq \epsilon$ for all k, and so by nesting $\bigcap_k \prod_{i=2}^{n} A_{jki}^{(k)} \neq \emptyset$. But $\bigcap_k B_k = \emptyset$ implies that $\bigcap_k \prod_{i=1}^{n} A_{jki}^{(k)} = \emptyset$, and so $\bigcap_k \prod_{i=2}^{n} A_{jki}^{(k)} \neq \emptyset$ and $x_1 \in \bigcap_k A_{jk1}^{(k)}$ is a contradiction. Thus $f_k(x_1) \to 0$ for all $x_1 \in X_1$. Recalling the notational convention at the beginning of this step, we have proved that $f_k(x_1) \to 0$ for all $x_1 \in Y_{l(1),1}$ where $\mu_1 \left(Y_{l(1),1} \right) < \infty$.

Next we claim that this pointwise convergence implies that for all $\epsilon > 0$:

$$\mu_1 \left(\left\{ x_1 \in \bigcup_{j=1}^{m_k} A_{j1}^{(k)} \, \middle| \, f_k(x_1) > \epsilon \right\} \right) \to 0 \text{ as } k \to \infty. \qquad (4)$$

The collection of sets $\{\bigcup_{j=1}^{m_k} A_{j1}^{(k)}\}_k$ is nested and all contained in a set $Y_{l(1),1}$ of finite measure. Further, $f_k(x_1) \to 0$ for all $x_1 \in Y_{l(1),1}$ assures that:

$$\bigcap_{k=1}^{\infty} \left(\left\{ x_1 \in \bigcup_{j=1}^{m_k} A_{j1}^{(k)} \, \middle| \, f_k(x_1) > \epsilon \right\} \right) = \emptyset.$$

So (4) follows from continuity from above of μ_1. It then follows that:

$$\mu_1 \left(\left\{ x_1 \in \bigcup_{j=1}^{m_k} A_{j1}^{(k)} \, \middle| \, f_k(x_1) > \epsilon \right\} \right) < \epsilon \text{ for } k \geq N(\epsilon).$$

Putting the pieces together, finite additivity and monotonicity yield:

$$\mu_{\mathcal{A}}(B_k) = \sum_{j=1}^{m_k} \mu_1(A_{j1}^{(k)}) \prod_{i=2}^{n} \mu_i \left(A_{ji}^{(k)} \right)$$
$$\leq \mu_1 \left(\bigcup_{j=1}^{m_k} A_{j1}^{(k)} \right) \sum_{j=1}^{m_k} \prod_{i=2}^{n} \mu_i \left(A_{ji}^{(k)} \right).$$

Splitting:

$$\bigcup_{j=1}^{m_k} A_{j1}^{(k)} = \left\{ x_1 \in \bigcup_{j=1}^{m_k} A_{j1}^{(k)} \, | \, f_k(x_1) > \epsilon \right\} \bigcup \left\{ x_1 \in \bigcup_{j=1}^{m_k} A_{j1}^{(k)} \, | \, f_k(x_1) \leq \epsilon \right\},$$

and this obtains for $k \geq N(\epsilon)$:

$$\mu_{\mathcal{A}}(B_k) \leq \mu_1\left(\left\{x_1 \in \bigcup_{j=1}^{m_k} A_{j1}^{(k)} | f_k(x_1) > \epsilon\right\}\right) \sum_{j=1}^{m_k} \prod_{i=2}^{n} \mu_i\left(A_{ji}^{(k)}\right) + \epsilon \sum_{j=1}^{m_k} \mu_1\left(A_{j1}^{(k)}\right)$$

$$\leq \mu_1\left(\left\{x_1 \in \bigcup_{j=1}^{m_k} A_{j1}^{(k)} | f_k(x_1) > \epsilon\right\}\right) \sum_{j=1}^{m_1} \prod_{i=2}^{n} \mu_i\left(A_{ji}^{(1)}\right) + \epsilon \sum_{j=1}^{m_1} \mu_1\left(A_{j1}^{(1)}\right)$$

$$\leq \epsilon \sum_{j=1}^{m_1} \prod_{i=2}^{n} \mu_i\left(A_{ji}^{(1)}\right) + \epsilon \sum_{j=1}^{m_1} \mu_1\left(A_{j1}^{(1)}\right).$$

Since ϵ is arbitrary, the result in (3) follows, completing the proof. ∎

Remark 7.19 (Proposition 7.18 on $\mathcal{A}'(\mathcal{A}_i)$ and $\mathcal{A}'(\mathcal{A}_i')$) *As noted above for finite additivity, the extension of the product set function μ_0 to $\mu_{\mathcal{A}}$ obtains that $\mu_{\mathcal{A}}$ is countably additive and thus a measure on the algebra $\mathcal{A}(\mathcal{A}_i)$, defined with component sets $A_i \in \mathcal{A}_i$ where $\mathcal{A}_i$ is an algebra on X_i. It also applies on $\mathcal{A}(\mathcal{A}_i')$ with $A_i \in \mathcal{A}_i'$, a semi-algebra on X_i with $\mathcal{A}_i' \subset \sigma(X_i)$, since then $\mathcal{A}(\mathcal{A}_i') \subset \mathcal{A}$. In the case where $\mathcal{A}_i$ is the algebra generated by $\mathcal{A}_i'$, then in fact $\mathcal{A}(\mathcal{A}_i') = \mathcal{A}(\mathcal{A}_i)$.*

7.4 Extension to a Measure on the Product Space

With the result in Proposition 7.18 providing the "required step" identified in Section 6.3, Summary of Construction Process, all that is left is to collect the "free steps."

Proposition 7.20 ($\mu_{\mathcal{A}}$ extends to a measure μ_X on $\sigma(X)$) *Given σ-finite measure spaces $\{(X_i, \sigma(X_i), \mu_i) | i = 1, ..., n\}$, let $\mathcal{A}'$ denote the semi-algebra of measurable rectangles on $X \equiv \prod_{i=1}^{n} X_i$ defined in (7.2), and μ_0 the set function defined on $\mathcal{A}'$ by (7.3).*

Then μ_0 can be extended to a measure μ_X on a complete sigma algebra $\sigma(X)$ with $\mathcal{A}' \subset \sigma(X)$. Further, the extension μ_X is unique on the sigma algebra $\sigma(X)$.

Proof. *By Proposition 7.18, μ_0 extends to a measure $\mu_{\mathcal{A}}$ on the algebra $\mathcal{A}$ generated by $\mathcal{A}'$. The measure $\mu_{\mathcal{A}}$ and algebra $\mathcal{A}$ can then be used to define an outer measure μ_A^* on the power sigma algebra on X, $\sigma(P(X))$, as in (6.3). This is then a true outer measure as in the proof of the Hahn-Kolmogorov Extension theorem of Proposition 6.4.*

The collection of Carathéodory measurable sets $\sigma(X)$ defined in (6.1) is then a complete sigma algebra by Proposition 6.2, and μ_A^* restricted to $\sigma(X)$ is a measure. With μ_X denoting the restriction of μ_A^* to $\sigma(X)$, it follows that $(X, \sigma(X), \mu_X)$ is a complete measure space.

By Proposition 6.14, the extension of $\mu_{\mathcal{A}}$ to μ_X is unique on $\sigma(\mathcal{A}) \subset \sigma(X)$, with $\sigma(\mathcal{A})$ the sigma algebra generated by the algebra $\mathcal{A}$ of finite disjoint unions of sets in $\mathcal{A}'$. Then by Proposition 6.24, μ_X is also the unique extension of $\mu_{\mathcal{A}}$ to $\sigma(X)$ since $\sigma(X) = \sigma^C(\mathcal{A})$, the completion of $\sigma(\mathcal{A})$ provided by Proposition 6.20. ∎

Notation 7.21 (Product space/measure) *The complete measure space $(X, \sigma(X), \mu_X)$ is called a **product measure space**, and μ_X is called the **product measure**. The product measure μ_X is often denoted $\mu_X = \prod_{i=1}^{n} \mu_i$, and more simply when n is small, say, $n = 2$, by $\mu_X = \mu_1 \times \mu_2$.*

Remark 7.22 (On sigma-finiteness) *The assumption of σ-finiteness in Proposition 7.20 was needed because of the approach taken in the proof of countable additivity in Proposition 7.18. However, the extension result of Proposition 7.20 remains true without the σ-finiteness restriction as will be demonstrated in Book V. However, in this more general case we can no longer be assured that the extension to μ_X is unique on $\sigma(\mathcal{A})$, recalling Proposition 6.14.*

One of the advantages of developing a product measure space by the above Carathéodory approach is that we immediately obtain the approximations of Proposition 6.5, which we state here for completeness.

Corollary 7.23 (Approximating Measurable Sets in Product Spaces) *Let $\mathcal{A}$ denote the σ-finite algebra of finite disjoint unions of measurable rectangles, and $(X, \sigma(X), \mu_X)$ the associated complete product measure space given in Proposition 7.20. For $B \in \sigma(X)$ and $\epsilon > 0$:*

1. *There is a set $A \in \mathcal{A}_\sigma$, the collection of countable unions of sets in the algebra $\mathcal{A}$, so that $B \subset A$ with:*

$$\mu_X(A) \leq \mu_X(B) + \epsilon, \text{ and } \mu_X(A - B) < \epsilon. \tag{7.10}$$

2. *There is a set $C \in \mathcal{A}_\delta$, the collection of countable intersections of sets in the algebra $\mathcal{A}$, so that $C \subset B$ with:*

$$\mu_X(B) \leq \mu_X(C) + \epsilon, \text{ and } \mu_X(B - C) < \epsilon. \tag{7.11}$$

3. *There is a set $A' \in \mathcal{A}_{\sigma\delta}$, the collection of countable intersections of sets in $\mathcal{A}_\sigma$, and $C' \in \mathcal{A}_{\delta\sigma}$, the collection of countable unions of sets in $\mathcal{A}_\delta$, so that $C' \subset B \subset A'$ and:*

$$\mu_X(A' - B) = \mu_X(B - C') = 0. \tag{7.12}$$

Proof. *Proposition 6.5.* ∎

7.5 Well-Definedness of Product Measure Spaces

In this section we address whether the σ-finite product space constructed in the prior sections can be constructed sequentially rather than in one step. For example, assume given three σ-finite measure spaces $\{(X_i, \sigma(X_i), \mu_i)|i = 1, 2, 3\}$. If $(X_1 \times X_2, \sigma(X_1 \times X_2), \mu_{X_1 \times X_2})$ is constructed first using Proposition 7.20, then $(X_1 \times X_2) \times X_3$ constructed as a second step, would this have resulted in the same complete measure space as if Proposition 7.20 was applied to obtain $X_1 \times X_2 \times X_3$ in one step? In other words, is:

$$(X_1 \times X_2 \times X_3, \sigma(X_1 \times X_2 \times X_3), \mu_{X_1 \times X_2 \times X_3}) \tag{0}$$
$$= ((X_1 \times X_2) \times X_3, \sigma(\sigma(X_1 \times X_2) \times \sigma(X_3)), \mu_{(X_1 \times X_2) \times X_3})?$$

Here $\sigma(X_1 \times X_2 \times X_3)$ is the complete sigma algebra of Carathéodory measurable sets of Proposition 7.20 using the algebra $\mathcal{A}_{123}$ generated by the semi-algebra $\mathcal{A}'_{123}$ of measurable rectangles in $X_1 \times X_2 \times X_3$ defined with $\sigma(X_1)-, \sigma(X_2)-$ and $\sigma(X_3)$-measurable component sets. With $\sigma(X_1 \times X_2)$ defined analogously with $\mathcal{A}_{12}$ and $\mathcal{A}'_{12}$, the sigma algebra $\sigma(\sigma(X_1 \times X_2) \times \sigma(X_3))$ is the complete sigma algebra of Carathéodory measurable sets of Proposition 7.20 using the algebra generated by the semi-algebra of measurable rectangles defined with $\sigma(X_1 \times X_2)-$ and $\sigma(X_3)-$ measurable component sets.

The answer is in the affirmative. To demonstrate this, we first show that the spaces that result are equivalent with respect to the smallest product sigma algebras generated at each step. We then pursue the smallest completions of these spaces as given by the completion theorem of Proposition 6.20, and apply Proposition 6.24.

For clarity, in this section we introduce a slight variation from the earlier notational conventions. Let:

$$(X_1 \times X_2 \times X_3, \sigma_0(X_1 \times X_2 \times X_3), \mu_{X_1 \times X_2 \times X_3}),$$

denote the measure space constructed by Proposition 7.20 directly from the three given measure spaces, but restricted to $\sigma_0(X_1 \times X_2 \times X_3) \subset \sigma(X_1 \times X_2 \times X_3)$, defined as the smallest sigma algebra that contains the semi-algebra $\mathcal{A}'_{123}$ defined above.

Similarly, let:

$$((X_1 \times X_2) \times X_3, \sigma_0(\sigma_0(X_1 \times X_2) \times \sigma(X_3)), \mu_{(X_1 \times X_2) \times X_3}),$$

denote the measure space produced with two applications of this proposition. The first application produces:

$$(X_1 \times X_2, \sigma_0(X_1 \times X_2), \mu_{X_1 \times X_2}),$$

with $\sigma_0(X_1 \times X_2)$ similarly defined in terms of semi-algebra $\mathcal{A}'_{12}$ of measurable rectangles in $X_1 \times X_2$ defined with $\sigma(X_1)-$ and $\sigma(X_2)$-measurable sets. Then define $\sigma_0(\sigma_0(X_1 \times X_2) \times \sigma(X_3))$ as the smallest sigma algebra containing the semi-algebra, $\mathcal{A}'_{12 \times 3}$, formed with measurable rectangles defined with $\sigma_0(X_1 \times X_2)-$ and $\sigma(X_3)$-measurable sets.

Finally, we denote by $(X_1 \times X_2 \times X_3, \sigma_0^c(X_1 \times X_2 \times X_3), \mu_{X_1 \times X_2 \times X_3}^c)$ and $((X_1 \times X_2) \times X_3, \sigma_0^c(\sigma_0(X_1 \times X_2) \times \sigma(X_3)), \mu_{(X_1 \times X_2) \times X_3}^c)$ the completion of these measure spaces by Proposition 6.20.

To prove equivalence of these complete measure spaces, we first prove that:

$$(X_1 \times X_2 \times X_3, \sigma_0^c(X_1 \times X_2 \times X_3), \mu_{X_1 \times X_2 \times X_3}^c)$$
$$= ((X_1 \times X_2) \times X_3, \sigma_0^c(\sigma_0(X_1 \times X_2) \times \sigma(X_3)), \mu_{(X_1 \times X_2) \times X_3}^c),$$

proceeding in several steps, and then tie up the loose ends. Since it is apparent that $X_1 \times X_2 \times X_3 = (X_1 \times X_2) \times X_3$ as spaces of points, the focus of the proof is on sigma algebras and measures.

1. σ_0 **Sigma Algebras**: We first show with sigma algebras defined above that:

$$\sigma_0(X_1 \times X_2 \times X_3) = \sigma_0(\sigma_0 (X_1 \times X_2) \times \sigma (X_3)). \tag{1}$$

First:

$$\sigma_0(X_1 \times X_2 \times X_3) \subset \sigma_0(\sigma_0 (X_1 \times X_2) \times \sigma (X_3)),$$

because the σ-algebra on the left is generated by the semi-algebra of measurable rectangles $A_1 \times A_2 \times A_3$ with $A_i \in \sigma(X_i)$, while the σ-algebra on the right is generated by the semi-algebra of sets $A_{12} \times A_3$ with $A_{12} \in \sigma_0(X_1 \times X_2)$. The inclusion then follows since $A_1 \times A_2 \in \sigma_0(X_1 \times X_2)$ for $A_i \in \sigma(X_i)$.

For the reverse inclusion, let $A_3 \in \sigma(X_3)$ and define $\sigma = \{A_{12} \in \sigma_0(X_1 \times X_2)|A_{12} \times A_3 \in \sigma_0(X_1 \times X_2 \times X_3)\}$. It is an exercise to check that σ is a sigma algebra, and σ contains the measurable rectangles $\{A_1 \times A_2|A_i \in \sigma(X_i)\}$ as just noted. Thus as a sigma algebra it follows that $\sigma_0(X_1 \times X_2) \subset \sigma$. Hence:

$$\sigma_0(\sigma_0 (X_1 \times X_2) \times \sigma (X_3)) \subset \sigma_0(X_1 \times X_2 \times X_3),$$

and (1) is proved.

2. **Measures on σ_0 Sigma Algebras**: For equality of the measures defined on these sigma algebras, we claim that for any measurable set A:

$$A \in \sigma_0(X_1 \times X_2 \times X_3) = \sigma_0(\sigma_0 (X_1 \times X_2) \times \sigma (X_3)),$$

that:

$$\mu_{X_1 \times X_2 \times X_3}(A) = \mu_{(X_1 \times X_2) \times X_3}(A). \tag{2}$$

To see this, note that for any measurable rectangle $A \in \mathcal{A}'_{123}$, $A \equiv A_1 \times A_2 \times A_3$ with $A_i \in \sigma(X_i)$, that by definition:

$$\mu_{X_1 \times X_2 \times X_3}(A) \equiv \prod_{i=1}^3 \mu_i(A_i).$$

But similarly:

$$\mu_{(X_1 \times X_2) \times X_3}(A) \equiv \mu_{X_1 \times X_2} (A_1 \times A_2)\, \mu_{X_3} (A_3) = \prod_{i=1}^3 \mu_i(A_i).$$

By the assumption of σ-finiteness and proposition 6.14, these measures agree on the smallest sigma algebra containing $\mathcal{A}'_{123}$, which is $\sigma_0(X_1 \times X_2 \times X_3)$, and the result follows by (1).

3. **Measures on Completed σ_0^c Sigma Algebras**: In steps 1 and 2 we proved that as measure spaces:

$$(X_1 \times X_2 \times X_3, \sigma_0(X_1 \times X_2 \times X_3), \mu_{X_1 \times X_2 \times X_3})$$
$$= ((X_1 \times X_2) \times X_3, \sigma_0(\sigma_0(X_1 \times X_2) \times \sigma(X_3)), \mu_{(X_1 \times X_2) \times X_3}).$$

Applying the completion theorem of Proposition 6.20, each version can be completed. Further, if:

$$A \in \sigma_0(X_1 \times X_2 \times X_3) = \sigma_0(\sigma_0(X_1 \times X_2) \times \sigma(X_3)),$$

then by property 2 of Proposition 6.20, and (2) above:

$$\mu^c_{X_1 \times X_2 \times X_3}(A) = \mu_{X_1 \times X_2 \times X_3}(A) = \mu_{(X_1 \times X_2) \times X_3}(A) = \mu^c_{(X_1 \times X_2) \times X_3}(A).$$

Further by the completeness theorem:

$$\sigma^c_0(\sigma_0(X_1 \times X_2) \times \sigma(X_3)) = \{A | B \subset A \subset C\},$$

where $B, C \in \sigma_0(\sigma_0(X_1 \times X_2) \times \sigma(X_3))$ and $\mu_{(X_1 \times X_2) \times X_3}(C - B) = 0$. Similarly, $\sigma^c_0(X_1 \times X_2 \times X_3)$ is defined only where $B, C \in \sigma_0(X_1 \times X_2 \times X_3)$ and $\mu_{X_1 \times X_2 \times X_3}(C - B) = 0$. But then by (1) and (2), these completed sigma algebras agree. Specifically, we have $B, C \in \sigma_0(\sigma_0(X_1 \times X_2) \times X_3)$ and $\mu_{(X_1 \times X_2) \times X_3}(C - B) = 0$ if and only if $B, C \in \sigma_0(X_1 \times X_2 \times X_3)$ and $\mu_{X_1 \times X_2 \times X_3}(C - B) = 0$.

Thus, $A \in \sigma^c_0(\sigma_0(X_1 \times X_2) \times \sigma(X_3))$ if and only if $A \in \sigma^c_0(X_1 \times X_2 \times X_3)$. In summary:

$$(X_1 \times X_2 \times X_3, \sigma^c_0(X_1 \times X_2 \times X_3), \mu^c_{X_1 \times X_2 \times X_3})$$
$$= ((X_1 \times X_2) \times X_3, \sigma^c_0(\sigma_0(X_1 \times X_2) \times \sigma(X_3)), \mu^c_{(X_1 \times X_2) \times X_3}). \tag{3}$$

4. **Measures on Carathéodory Sigma Algebras**: By Proposition 6.24, $\sigma^c_0(X_1 \times X_2 \times X_3) = \sigma(X_1 \times X_2 \times X_3)$, the sigma algebra of Carathéodory measurable sets, and $\mu^c_{X_1 \times X_2 \times X_3} = \mu_{X_1 \times X_2 \times X_3}$, the measure obtained by restricting outer measure to this sigma algebra. Thus:

$$(X_1 \times X_2 \times X_3, \sigma^c_0(X_1 \times X_2 \times X_3), \mu^c_{X_1 \times X_2 \times X_3})$$
$$= (X_1 \times X_2 \times X_3, \sigma(X_1 \times X_2 \times X_3), \mu_{X_1 \times X_2 \times X_3}).$$

Similarly,

$$((X_1 \times X_2) \times X_3, \sigma^c_0(\sigma_0(X_1 \times X_2) \times \sigma(X_3)), \mu^c_{(X_1 \times X_2) \times X_3})$$
$$= ((X_1 \times X_2) \times X_3, \sigma(\sigma_0(X_1 \times X_2) \times \sigma(X_3)), \mu_{(X_1 \times X_2) \times X_3}).$$

Hence by (3):

$$(X_1 \times X_2 \times X_3, \sigma(X_1 \times X_2 \times X_3), \mu_{X_1 \times X_2 \times X_3})$$
$$= ((X_1 \times X_2) \times X_3, \sigma(\sigma_0(X_1 \times X_2) \times \sigma(X_3)), \mu_{(X_1 \times X_2) \times X_3}). \tag{4}$$

5. **Completion of Sigma Algebras in Steps**: The result in (4) is almost the conclusion sought in (0) but falls short. In the assumed sequential construction, instead of using $\sigma_0(X_1 \times X_2)$ and $\sigma(X_3)$ to define the algebra $\mathcal{A}'_{12 \times 3}$, we would have instead used $\sigma_0^c(X_1 \times X_2) = \sigma(X_1 \times X_2)$ and $\sigma(X_3)$. Thus, the final step is to show that with the apparent notation:

$$\sigma(\sigma_0(X_1 \times X_2) \times \sigma(X_3)) = \sigma(\sigma(X_1 \times X_2) \times \sigma(X_3)), \tag{5}$$

and to do this requires the application of (7.12).

First, $\sigma_0(\sigma_0(X_1 \times X_2) \times \sigma(X_3)) \subset \sigma_0(\sigma(X_1 \times X_2) \times \sigma(X_3))$ by definition, and so it follows that:

$$\sigma(\sigma_0(X_1 \times X_2) \times \sigma(X_3)) \subset \sigma(\sigma(X_1 \times X_2) \times \sigma(X_3))$$

by the construction of the completeness theorem.

For the reverse inclusion, assume that $B \in \sigma(\sigma(X_1 \times X_2) \times \sigma(X_3))$. By Proposition 6.5, let $A, C \in \sigma_0(\sigma(X_1 \times X_2) \times \sigma(X_3))$ with $C \subset B \subset A$ and $\mu_{(X_1 \times X_2) \times X_3}(A - C) = 0$. Specifically, $A \in (\overline{\mathcal{A}}_{12 \times 3})_{\sigma\delta}$, the collection of countable intersections of countable unions of sets in $\overline{\mathcal{A}}_{12 \times 3}$, and $C \subset (\overline{\mathcal{A}}_{12 \times 3})_{\delta\sigma}$, the collection of countable unions of countable intersections of sets in $\overline{\mathcal{A}}_{12 \times 3}$, where the algebra $\overline{\mathcal{A}}_{12 \times 3}$ is defined as the collection of finite unions from the semi-algebra of measurable rectangles:

$$\overline{\mathcal{A}}'_{12 \times 3} = \{E_1 \times F_1 | E_1 \in \sigma(X_1 \times X_2), F_1 \in \sigma(X_3)\}.$$

To prove that $B \in \sigma(\sigma_0(X_1 \times X_2) \times \sigma(X_3)) = \sigma_0^c(\sigma_0(X_1 \times X_2) \times \sigma(X_3))$, we show that there exists $A', C' \in \sigma_0(\sigma_0(X_1 \times X_2) \times \sigma(X_3))$ with $C' \subset C \subset B \subset A \subset A'$ and $\mu_{(X_1 \times X_2) \times X_3}(A' - C') = 0$. In fact we will show that $A' \in (\mathcal{A}_{12 \times 3})_{\sigma\delta}$ and $C' \subset (\mathcal{A}_{12 \times 3})_{\delta\sigma}$ with algebra $\mathcal{A}_{12 \times 3}$ analogously defined with semi-algebra:

$$\mathcal{A}'_{12 \times 3} = \{E_2 \times F_2 | E_2 \in \sigma_0(X_1 \times X_2), F_2 \in \sigma(X_3)\},$$

introduced before the first step.

To this end, first note that every rectangle in $\overline{\mathcal{A}}'_{12 \times 3}$ can be approximated within measure 0 by sub- and super-sets in $\mathcal{A}'_{12 \times 3}$. In detail, if $E_1 \times F_1 \in \overline{\mathcal{A}}'_{12 \times 3}$ then by (7.12) there are sets $E'_1, E''_1 \in \sigma_0(X_1 \times X_2)$ with $E'_1 \subset E_1 \subset E''_1$ and $\mu_{X_1 \times X_2}(E''_1 - E'_1) = 0$. Thus:

$$E'_1 \times F_1 \subset E_1 \times F_1 \subset E''_1 \times F_1,$$

with $E'_1 \times F_1 \in \mathcal{A}'_{12 \times 3}, E''_1 \times F_1 \in \mathcal{A}'_{12 \times 3}$ and:

$$\mu_{(X_1 \times X_2) \times X_3}(E''_1 \times F_1 - E'_1 \times F_1) = 0,$$

by 7.3.

This now implies that the same result holds for sets in the algebra $\overline{\mathcal{A}}_{12\times3}$, which are finite unions of $\overline{\mathcal{A}}'_{12\times3}$-sets, in that they can again be approximated by sub- and super-sets in $\mathcal{A}_{12\times3}$ within measure 0. This then holds for sets in $\left(\overline{\mathcal{A}}_{12\times3}\right)_\sigma$ and $\left(\overline{\mathcal{A}}_{12\times3}\right)_\delta$ using sets in $\left(\mathcal{A}_{12\times3}\right)_\sigma$ and $\left(\mathcal{A}_{12\times3}\right)_\delta$, and finally to sets in $\left(\overline{\mathcal{A}}_{12\times3}\right)_{\sigma\delta}$ and $\left(\overline{\mathcal{A}}_{12\times3}\right)_{\delta\sigma}$ using sets in $\left(\mathcal{A}_{12\times3}\right)_{\sigma\delta}$ and $\left(\mathcal{A}_{12\times3}\right)_{\delta\sigma}$. Thus A, C above can be approximated by A', C' as asserted, and so $B \in \sigma(\sigma_0(X_1 \times X_2) \times \sigma(X_3))$ and (5) is proved.

By induction, the same conclusion follows for any collection of σ-finite measure spaces, $\{(X_i, \sigma(X_i), \mu_i) | i = 1, ..., n\}$, but we avoid the combinatorial challenge of stating the most general result.

Proposition 7.24 *Given three σ-finite measure spaces, $\{(X_i, \sigma(X_i), \mu_i) | i = 1, 2, 3\}$, let $(X_1 \times X_2 \times X_3, \sigma(X_1 \times X_2 \times X_3), \mu_{X_1 \times X_2 \times X_3})$ denote the product measure space constructed in Proposition 7.20, and $((X_1 \times X_2) \times X_3, \sigma(\sigma(X_1 \times X_2) \times \sigma(X_3)), \mu_{(X_1 \times X_2) \times X_3})$ denote the product measure space constructed in two applications of Proposition 7.20. Then:*

$$(X_1 \times X_2 \times X_3, \sigma(X_1 \times X_2 \times X_3), \mu_{X_1 \times X_2 \times X_3}) \tag{7.13}$$
$$= ((X_1 \times X_2) \times X_3, \sigma(\sigma(X_1 \times X_2) \times \sigma(X_3)), \mu_{(X_1 \times X_2) \times X_3}).$$

7.6 Lebesgue and Borel Product Spaces

In this section we develop three important applications of the product measure space results of the prior sections. We note that all of the component space measures below are σ-**finite measure spaces**, and thus the earlier results apply. Chapter 8 develops a more general approach to Borel measures on $\mathbb{R}^n$ which includes as special cases the Borel and Lebesgue product measures below.

Chapter 8 will also address properties of general Borel measures and these product measures in Section 8.3. We thus defer a discussion of properties of Lebesgue and Borel product measures to this more general context.

1. **Product of Lebesgue Spaces**

Starting with the complete Lebesgue measure space $(\mathbb{R}, \mathcal{M}_L(\mathbb{R}), m)$, it is possible to construct the n-**dimensional Lebesgue measure space** $(\mathbb{R}^n, \mathcal{M}_L(\mathbb{R}^n), m^n)$, where $\mathbb{R}^n$ denotes the usual space of all real n-tuples:

$$\mathbb{R}^n = \{(x_1, x_2, ..., x_n) | x_j \in \mathbb{R}\},$$

$\mathcal{M}_L(\mathbb{R}^n)$ denotes the complete sigma algebra of Lebesgue measurable sets in $\mathbb{R}^n$ given in Proposition 7.20, and m^n denotes n-dimensional Lebesgue measure.

The development begins with defining a set function m_0^n on the measurable rectangles in $\mathbb{R}^n$. A **Lebesgue measurable rectangle** in $\mathbb{R}^n$ is a set A so that as in (7.2),

$$A = \prod_{i=1}^{n} A_i, \text{ with } A_i \in \mathcal{M}_L(\mathbb{R}).$$

The collection of Lebesgue measurable rectangles in $\mathbb{R}^n$, denoted $\mathcal{A}'$, is a semi-algebra. The **Lebesgue product set function** m_0^n is defined as in (7.3) by:

$$m_0^n(A) = \prod_{i=1}^{n} m(A_i).$$

This set function is then finitely additive on $\mathcal{A}'$, and by Proposition 7.18 extends to a measure $m_{\mathcal{A}}^n$ on the associated algebra $\mathcal{A}$.

Then by Proposition 7.20, $m_{\mathcal{A}}^n$ extends to a measure m^n on a complete sigma algebra $\mathcal{M}_L(\mathbb{R}^n)$, and since $m_{\mathcal{A}}^n$ is σ-finite, this extension is unique on $\sigma(\mathcal{A}) \subset \mathcal{M}_L(\mathbb{R}^n)$, the smallest sigma algebra which contains $\mathcal{A}$. Further, the complete sigma algebra $\mathcal{M}_L(\mathbb{R}^n)$ is the completion of $\sigma(\mathcal{A})$ in this product measure, and hence m^n is the unique extension of m_0^n to $\mathcal{M}_L(\mathbb{R}^n)$.

As $\mathcal{A}'$ contains the semi-algebra $\mathcal{A}'(\mathcal{A}_i')$ of **n-dimensional right semi-closed rectangles**,

$$\mathcal{A}'(\mathcal{A}_i') = \left\{ \prod_{i=1}^{n} (a_i, b_i] \right\},$$

for which

$$m_0^n(A) = \prod_{i=1}^{n} (b_i - a_i),$$

$\sigma(\mathcal{A})$ also contains all n-dimensional open, closed and left semi-closed rectangles, analogously defined.

The smallest sigma algebra that contains the n-dimensional open rectangles and thus the open sets is the **Borel sigma algebra** of Definition 2.13, $\mathcal{B}(\mathbb{R}^n)$, and hence:

$$\mathcal{B}(\mathbb{R}^n) \subset \sigma(\mathcal{A}) \subset \mathcal{M}_L(\mathbb{R}^n).$$

See also Proposition 8.1. This construction therefore obtains the complete Lebesgue measure space $(\mathbb{R}^n, \mathcal{M}_L(\mathbb{R}^n), m^n)$, and the Borel space $(\mathbb{R}^n, \mathcal{B}(\mathbb{R}^n), m^n)$. The notation m^n is consistent with Notation 7.21, $\mu_X = \prod_{i=1}^{n} \mu_i$, since here $\mu_i = m$ for all i.

2. Product of Borel Spaces

Starting with a complete Borel measure space $(\mathbb{R}, \mathcal{M}_F(\mathbb{R}), \mu_F)$, we can now similarly derive the **n-dimensional Borel measure space** $(\mathbb{R}^n, \mathcal{M}_F(\mathbb{R}^n), \mu_F^n)$, where $\mathcal{M}_F(\mathbb{R}^n)$ denotes the complete sigma algebra in $\mathbb{R}^n$ of measurable sets as in Proposition 7.20, and μ_F^n denotes n-dimensional Borel product measure.

The development begins with defining a set function μ_F^n on the measurable rectangles in $\mathbb{R}^n$. A **measurable rectangle** in $\mathbb{R}^n$ is a set A so that as in (7.2):

$$A = \prod_{i=1}^{n} A_i, \text{ with } A_i \in \mathcal{M}_F(\mathbb{R}).$$

Then $\mathcal{A}'$, the collection of measurable rectangles in $\mathbb{R}^n$, is a semi-algebra. The **Borel product set function** μ_0^n is defined as in (7.3) by:

$$\mu_0^n(A) = \prod_{i=1}^{n} \mu_F(A_i).$$

This set function is finitely additive on $\mathcal{A}'$, and by Proposition 7.18 extends to a measure $\mu_{\mathcal{A}}^n$ on the algebra $\mathcal{A}$ generated by $\mathcal{A}'$.

Then by Proposition 7.20, $\mu_{\mathcal{A}}^n$ extends to a measure μ_F^n on a complete sigma algebra $\mathcal{M}_F(\mathbb{R}^n)$. Since μ_F^n is σ-finite by item 3 of Definition 5.1, this extension is unique on $\sigma(\mathcal{A}) \subset \mathcal{M}_F(\mathbb{R}^n)$, the smallest sigma algebra which contains $\mathcal{A}$. Further, the complete sigma algebra $\mathcal{M}_F(\mathbb{R}^n)$ is the completion of $\sigma(\mathcal{A})$ in this product measure, and hence μ_F^n is the unique extension of μ_0^n to $\mathcal{M}_F(\mathbb{R}^n)$.

As $\mathcal{A}'$ contains the collection of n-**dimensional right semi-closed rectangles**, $\mathcal{A}'$ $(\mathcal{A}_i')$ defined above, and for which:

$$\mu_0^n(A) = \prod_{i=1}^{n} (F(b_i) - F(a_i)),$$

$\sigma(\mathcal{A})$ also contains all n-dimensional open, closed and left semi-closed intervals, analogously defined.

As noted above (see also Proposition 8.1), the smallest sigma algebra that contains the n-dimensional open rectangles and thus the open sets is the **Borel sigma algebra**, $\mathcal{B}(\mathbb{R}^n)$, and hence:

$$\mathcal{B}(\mathbb{R}^n) \subset \sigma(\mathcal{A}) \subset \mathcal{M}_F(\mathbb{R}^n).$$

This construction therefore obtains the completion of a Borel measure space $(\mathbb{R}^n, \mathcal{M}_F(\mathbb{R}^n), \mu_F^n)$, and the Borel space $(\mathbb{R}^n, \mathcal{B}(\mathbb{R}^n), \mu_F^n)$. The notation μ_F^n is consistent with Notation 7.21, $\mu_X = \prod_{i=1}^{n} \mu_i$, since here $\mu_i = \mu_F$ for all i.

Note that here, the measure μ_F^n is a Borel measure on $\mathbb{R}^n$ by Definition 8.4 since if A is compact, $A \subset \prod_{i=1}^{n} (a_i, b_i]$ for a bounded rectangle. Thus $\mu_F^n(A) < \infty$ by monotonicity and the above formula for $\mu_F^n\left(\prod_{i=1}^{n} (a_i, b_i]\right)$. The same is true in item 3.

3. General Products of Borel Spaces

In item 2, all of the Borel measure spaces were notationally represented as $(\mathbb{R}, \mathcal{M}_{\mu_F}(\mathbb{R}), \mu_F)$, implying that each was generated from the same right continuous, increasing function, $F(x)$. However, the same construction works if we instead begin with $\{(\mathbb{R}, \mathcal{M}_{F_i}(\mathbb{R}), \mu_{F_i})\}$, a finite collection of Borel measure spaces defined with different such functions $F_i(x)$. The notational options for the resulting measure are then cumbersome. Consistent with Notation 7.21, we could denote this product measure $\mu_F \equiv \prod_{i=1}^{n} \mu_{F_i}$, or $\mu_F \equiv \mu_{\prod F_i}$. This latter convention is more consistent with the Chapter 8 investigation of Borel measures on $\mathbb{R}^n$, where the multivariate function $\prod_{i=1}^{n} F_i(x_i)$ will be generalized to appropriate $F(x_1, ..., x_n)$.

8

Borel Measures on $\mathbb{R}^n$

The examples of the prior chapter presented an approach to constructing a Borel measure μ_F^n on $\mathbb{R}^n$ starting with a 1-dimensional Borel measure space, $(\mathbb{R}, \mathcal{M}_F(\mathbb{R}), \mu_F)$, or a finite collection of such spaces $\{(\mathbb{R}, \mathcal{M}_{F_i}(\mathbb{R}), \mu_{F_i})\}$, and then applying the results that lead to Proposition 7.20. For example, a right continuous increasing function $F(x)$ defined on $\mathbb{R}$ obtains a set function on right semi-closed intervals defined in (5.4) by $|(a, b]|_F = F(b) - F(a)$. Thus induces a set function on a right semi-closed rectangle in $\mathbb{R}^n$ defined by:

$$\mu_0 \left(\prod_{i=1}^{n} (a_i, b_i] \right) = \prod_{i=1}^{n} (F(b_i) - F(a_i)).$$

By Propositions 5.20 and 5.23, $\mu_F((a_i, b_i]) = F(b_i) - F(a_i)$, and so this is consistent with (7.3). In the more general case of a finite collection of such spaces, the result is:

$$\mu_0 \left(\prod_{i=1}^{n} (a_i, b_i] \right) = \prod_{i=1}^{n} (F_i(b_i) - F_i(a_i)).$$

There is in fact a more general way to define the measure of such rectangles which involves a multivariate function $F(x_1, ..., x_n)$, suitably restricted in terms of its **continuity** and **monotonicity** properties. Such a function generalizes the earlier applications in which $F(x_1, ..., x_n) = \prod_{i=1}^{n} F(x_i)$ or $F(x_1, ..., x_n) = \prod_{i=1}^{n} F_i(x_i)$. Thus, the results of this chapter will apply to Borel product measures.

8.1 Rectangle Collections that Generate $\mathcal{B}(\mathbb{R}^n)$

As noted in Section 7.6 and defined in Definition 2.13, the n-dimensional Borel sigma algebra $\mathcal{B}(\mathbb{R}^n)$ is the smallest σ-algebra that contains all the open sets in $\mathbb{R}^n$. By complementarity, it is also the smallest σ-algebra that contains the closed sets in $\mathbb{R}^n$. To parallel the development in $\mathbb{R}$ we can also characterize this σ-algebra in terms of the semi-algebra of **right semi-closed rectangles**, defined by:

$$\mathcal{A}' \equiv \left\{ A \subset \mathbb{R}^n | A = \prod_{i=1}^{n} (a_i, b_i], \text{ with } -\infty \leq a_i \leq b_i \leq \infty \right\}.$$

When $b = \infty$, we interpret $(a, \infty] = (a, \infty)$. Then $\mathcal{A}'$ is a semi-algebra as was proved in Corollary 7.3.

Though not a semi-algebra, the class of **bounded right semi-closed rectangles**,

$$\mathcal{A}'_B \equiv \left\{ A \subset \mathbb{R}^n | A = \prod_{i=1}^{n} (a_i, b_i], \text{ with } -\infty < a_i \leq b_i < \infty \right\},$$

DOI: 10.1201/9781003257745-8

generates $\mathcal{A}'$ by disjoint countable unions. Thus if $\sigma(\mathcal{A}'_B)$ denotes the **smallest σ-algebra** that contains $\mathcal{A}'$, then $\sigma(\mathcal{A}'_B) = \sigma(\mathcal{A}')$.

Moreover:

Proposition 8.1 ($\mathcal{B}(\mathbb{R}^n)$-generating rectangle collections) *Given the above notation:*

$$\sigma(\mathcal{A}') = \sigma(\mathcal{A}'_B) = \mathcal{B}(\mathbb{R}^n). \tag{8.1}$$

Proof. *As noted above, $\mathcal{A}'_B$ generates $\mathcal{A}'$ by disjoint countable unions. For example, if all $-\infty < a_i < \infty$:*

$$\prod_{i=1}^n (a_i, \infty) = \prod_{i=1}^n \left[\bigcup_{j=1}^\infty (a_i + j - 1, a_i + j] \right]$$
$$= \bigcup_J \prod_{i=1}^n (a_i + j_k - 1, a_i + j_k],$$

where $J = \{(j_1, ..., j_n) | 1 \le j_k \text{ all } k\}$. Thus $\sigma(\mathcal{A}') = \sigma(\mathcal{A}'_B)$.

Also, $\sigma(\mathcal{A}'_B) \subset \mathcal{B}(\mathbb{R}^n)$ since $\prod_{i=1}^n (a_i, b_i]$ is a countable intersection of bounded open sets. Taking rational $\epsilon > 0$:

$$\prod_{i=1}^n (a_i, b_i] = \bigcap_{\epsilon > 0} \prod_{i=1}^n (a_i, b_i + \epsilon).$$

Conversely, every such bounded open rectangle is generated by $\mathcal{A}'_B$-rectangles. With $b \equiv \min\{b_i\}$, we have with rational ϵ:

$$\prod_{i=1}^n (a_i, b_i) = \bigcup_{b > \epsilon > 0} \prod_{i=1}^n (a_i, b_i - \epsilon].$$

If $G \subset \mathbb{R}^n$ is open, then for any $q \in G$ with all rational coordinates there is an $r_q > 0$ so that $B_{r_q}(q) \subset G$, where $B_{r_q}(q)$ is the open ball of of Definition 2.10. Further, as in the proof of Proposition 2.12, we can take r_q to equal the supremum of all rational r_k with $B_{r_k}(q) \subset G$. Then G is a union of such balls as proved there, and we now show each such ball is a union of open rectangles.

Letting $\epsilon_q = \sqrt{r_q/n}$ obtains that:

$$\prod_{i=1}^n (q_i - \epsilon_q, q_i + \epsilon_q) \subset B_{r_q}(q).$$

If $p \in B_{r_q}(q) - \prod_{i=1}^n (q_i - \epsilon_q, q_i + \epsilon_q)$ with all rational coordinates, then since $B_{r_q}(q)$ is an open set there exists $B_{r_p}(p) \subset B_{r_q}(q)$, with r_p defined as above. Letting $\epsilon_p = \sqrt{r_p/n}$, then $B_{r_p}(p)$ contains the associated open rectangle. We now claim that with rectangles so constructed, and all p with rational coordinates:

$$B_{r_q}(q) = \bigcup_{p \in B_{r_q}(q)} \prod_{i=1}^n (p_i - \epsilon_{p_i}, p_i + \epsilon_{p_i}).$$

The proof is by contradiction. If there exists arbitrary $x \in B_{r_q}(q)$ that is not in this union, then there exists $B_{r_x}(x) \subset B_{r_q}(q)$ and thus $p \in B_{r_x}(x)$ with all rational coordinates. By density we can *assume that* $|x - p| < r_x/4$, *and then with rational* $r_p < r_x/4$, *it follows that* $x \in B_{r_p}(p) \subset B_{r_x}(x)$, *a contradiction.* ∎

Remark 8.2 (Countable $\mathcal{A}'_{\mathbb{Q}}$**)** *Define* $\mathcal{A}'_{\mathbb{Q}} \subset \mathcal{A}'_B$ *as the class of bounded right semi-closed rectangles with rational* a_i, b_i. *Then also* $\sigma(\mathcal{A}'_{\mathbb{Q}}) = \mathcal{B}(\mathbb{R}^n)$ *since* $B_{r_q}(q)$ *in the proof above is expressed as a countable union of* $\mathcal{A}'_{\mathbb{Q}}$*-sets. Hence, the Borel sigma algebra can also be generated by a countable collection of bounded rectangles.*

Exercise 8.3 *Prove that* $\mathcal{A}'_B$ *and* $\mathcal{A}'_{\mathbb{Q}}$ *are closed under finite unions and intersections. In other words, if* $\{A_j\}_{j=1}^n \subset \mathcal{A}'_Z$ *then* $\bigcap_j A_j \in \mathcal{A}'_Z$ *and* $\bigcup_j A_j \in \mathcal{A}'_Z$ *where* $\mathcal{A}'_Z$ *denotes* $\mathcal{A}'_B$ *or* $\mathcal{A}'_{\mathbb{Q}}$. *Are any of these four statements true for countable collections?*

8.2 Borel Measures and Induced Functions

In Chapter 5 it was shown that there is effectively a one-to-one correspondence between Borel measures μ on $\mathbb{R}$, and the collection of right continuous, increasing functions F. For a right semi-closed interval this correspondence is defined by:

$$\mu\left[(a, b]\right] = F(b) - F(a).$$

Specifically:

1. Given Borel μ defined on $\mathcal{B}(\mathbb{R})$, F_μ is defined as in (5.1) in general, or as in (5.3) when $\mu\left[\mathbb{R}\right] < \infty$. It was then shown in Proposition 5.7 that F_μ is increasing and right continuous, and in Section 5.3 that this correspondence is effectively one-to-one. That is, if F also generates μ, then $F = F_\mu + c$ for some $c \in \mathbb{R}$.

2. Given F which is increasing and right continuous and defining $\mu_F[(a, b]]$ as above on the semi-algebra of right semi-closed intervals, this definition can be extended by Proposition 5.23 to a measure on $\mathcal{B}(\mathbb{R})$. Again, F and $F + c$ generate the same μ_F, since they agree on semi-closed intervals.

The goal of this Chapter 8 is to reproduce this development for $\mathcal{B}(\mathbb{R}^n)$. We begin with a study of **finite Borel measures** for result 1 above, for which the primary application is to probability measures, and then address the general case. For result 2, we address the general case.

The definition of Borel measure on $\mathbb{R}^n$ is stated here for completeness. is identical with that given in Definition 5.1 but applied to $\mathcal{B}(\mathbb{R}^n)$ rather than $\mathcal{B}(\mathbb{R})$. Recall that by **compact** it is meant that every open cover of A has a finite subcover. By a generalization of the Heine-Borel theorem of Proposition 2.27, a set $A \subset \mathbb{R}^n$ is compact if and only if it is closed and bounded. See **Reitano** (2010).

Definition 8.4 (Borel measure on $\mathbb{R}^n$**)** *A Borel measure on* $\mathbb{R}^n$ *is a nonnegative set function* μ *defined on the Borel sigma algebra* $\mathcal{B}(\mathbb{R}^n)$, *taking values in the nonnegative extended real numbers,* $\overline{\mathbb{R}}^+ \equiv \mathbb{R}^+ \bigcup \{\infty\}$, *and which satisfies the following properties:*

1. $\mu(\emptyset) = 0$.

2. **Countable Additivity**: *If $\{A_j\}$ is a countable collection of pairwise disjoint sets in the sigma algebra $\mathcal{B}(\mathbb{R}^n)$, then:*

$$\mu\left(\bigcup_j A_j\right) = \sum_j \mu(A_j).$$

3. *For any compact set A, $\mu(A) < \infty$.*

*Then $(\mathbb{R}^n, \mathcal{B}(\mathbb{R}^n), \mu)$ is called a **Borel measure space**.*

As noted in Remark 5.2, a Borel measure by Definition 8.4 is not simply a measure on the Borel sigma algebra $\mathcal{B}(\mathbb{R}^n)$ due to the added restriction in item 3. For example, if $\{r_j\}_{j=1}^{\infty}$ is an enumeration of the points in $\mathbb{R}^n$ with all rational components, and we define $\mu(r_j) = 1$ and in general $\mu(A) = \sum_{r_j \in A} \mu(r_j)$, then μ is a measure on the power sigma algebra $\sigma(P(\mathbb{R}^n))$ of Example 2.7, and thus also a measure on $\mathcal{B}(\mathbb{R}^n)$. But μ does not satisfy item 3 of Definition 5.1 since the only compact sets with finite measure are the sets that contain finitely many r_j. So μ is not a Borel measure.

The addition of the restriction in item 3 is not universally included by all authors in their definitions of Borel measure. But this restriction eliminates measures on $\mathcal{B}(\mathbb{R}^n)$ with behaviors far outside the applications of interest in these books. In particular, all probability measures satisfy this requirement since then $\mu(\mathbb{R}^n) = 1$.

We begin with a study of a finite Borel measure on $\mathbb{R}^n$ and derive the implied properties on an associated function $F \equiv F(x_1, ..., x_n)$ that generalize "increasing and right continuous" in the 1-dimensional case. The next section will then generalize this development to general Borel measures. In the final section we show that any function with these properties induces a Borel measure.

8.2.1 Functions Induced by Finite Borel Measures

Given a **finite Borel measure** μ defined on $\mathcal{B}(\mathbb{R}^n)$, where finite means that $\mu(\mathbb{R}^n) < \infty$, define a **function F_μ induced by** μ on $\mathbb{R}^n$ by:

$$F_\mu(x) = \mu[A_x] \equiv \mu\left[\prod_{i=1}^{n}(-\infty, x_i]\right], \tag{8.2}$$

where $x = (x_1, ..., x_n)$. In other words, $F_\mu(x)$ is defined as the μ-measure of the right-semi-closed rectangle:

$$A_x \equiv \prod_{i=1}^{n}(-\infty, x_i],$$

and generalizes what was denoted $\bar{F}_\mu(x)$ in (5.3). The function F_μ is also called **a distribution function associated with** μ.

By Proposition 2.45, every measure μ is continuous from above and this imposes a continuity condition on every F_μ as defined in (8.2) as follows. Let $x^{(m)} \in \mathbb{R}^n$ with

$x_i^{(m)} \geq x_i$ for all i and $x^{(m)} \rightarrow x$ as $m \rightarrow \infty$. Then $A_x = \bigcap_m A_{x^{(m)}}$, and since μ is a finite measure, continuity from above applies to obtain:

$$\mu[A_x] = \lim_{m \rightarrow \infty} \mu\left[A_{x^{(m)}}\right].$$

This translates to a property of the function F_μ induced by μ in (8.2) whereby this function is said to be **continuous from above**.

In some texts, this property of a function F is also called **right continuous** or **generalized right continuous**. However, since this property is a direct corollary of continuity from above of the associated measure, this terminology seems fitting in the current context.

Definition 8.5 (Continuous from above) *A function F is said to be **continuous from above** at $x = (x_1, ..., x_n)$ if given a sequence $x^{(m)} = (x_1^{(m)}, ..., x_n^{(m)})$ with $x_i^{(m)} \geq x_i$ for all i and m, and $x^{(m)} \rightarrow x$ as $m \rightarrow \infty$, then:*

$$F(x) = \lim_{m \rightarrow \infty} F(x^{(m)}). \tag{8.3}$$

Also, F is continuous from above if the above property is true for all x.

If μ is a **product measure** defined with finite Borel measures, then by Proposition 7.20:

$$\begin{aligned}
F_\mu(x) &= \mu\left[\prod_{i=1}^{n}(-\infty, x_i]\right] \\
&= \prod_{i=1}^{n} \mu_i(-\infty, x_i] \\
&= \prod_{i=1}^{n} F_i(x_i).
\end{aligned}$$

Thus $F_\mu(x)$ is continuous from above since all F_i are right continuous by Proposition 5.7.

More generally as noted above, the distribution function associated with every finite Borel measure as defined in (8.2) is continuous from above.

We next investigate an appropriate notion of **increasing** for F_μ. By definition, $\mu[A] \geq 0$ for $A \in \mathcal{B}(\mathbb{R}^n)$, and this is then also true for bounded right semi-closed rectangles, $A = \prod_{i=1}^{n}(a_i, b_i]$. In order to translate the property that $\mu\left[\prod_{i=1}^{n}(a_i, b_i]\right] \geq 0$ to a property of F_μ requires an identity between $\mu\left[\prod_{i=1}^{n}(a_i, b_i]\right]$ and values of $\mu\left[\prod_{i=1}^{n}(-\infty, x_i]\right]$ for various x. We motivate the final result with two intuitive special cases.

Example 8.6

*1. **Borel product measure**: Let $G(x)$ be a right continuous, increasing and bounded function on $\mathbb{R}$, and μ_G the induced finite Borel measure of Proposition 5.23. For notational simplicity assume that $G(-\infty) = 0$. Let μ be defined as the associated Borel product measure on $\mathbb{R}^n$ of Proposition 7.20 using (5.3) and (7.3).*

$$\mu[A_x] \equiv \prod_{i=1}^{n} \mu_G(-\infty, x_i] = \prod_{i=1}^{n} G(x_i).$$

In other words, in this special case:

$$F_\mu(x) = \prod_{i=1}^{n} G(x_i).$$

Then by (5.2) and (7.3):

$$\mu\left[\prod_{i=1}^{n}(a_i, b_i]\right] \equiv \prod_{i=1}^{n}[G(b_i) - G(a_i)].$$

The product on the right-hand side produces a summation of 2^n terms:

$$\prod_{i=1}^{n}[G(b_i) - G(a_i)] = \sum_x sgn(x)\prod_{i=1}^{n} G(x_i).$$

Each $x = (x_1, ..., x_n)$ in the summation is one of the 2^n vertices of $\prod_{i=1}^{n}(a_i, b_i]$, so $x_i = a_i$ or $x_i = b_i$. In addition, $sgn(x)$ is defined to equal -1 if the number of a_i-components of x is odd, and equals $+1$ otherwise.

Thus:

$$\mu\left[\prod_{i=1}^{n}(a_i, b_i]\right] = \sum_x sgn(x)\mu[A_x],$$

or in terms of the defining function F:

$$\mu\left[\prod_{i=1}^{n}(a_i, b_i]\right] = \sum_x sgn(x)F_\mu(x). \tag{8.4}$$

2. Probability theory: *To avoid using more advanced concepts from a general integration theory to be pursued in Books III and V, this example is formally limited to be an application of Riemann integration theory, with which the reader may be more familiar. Let $f(x)$ be a nonnegative continuous function of $x = (x_1, ..., x_n) \in \mathbb{R}^n$, and assume that defined as a Riemann integral:*

$$\int_{-\infty}^{\infty} \cdots \int_{-\infty}^{\infty} f(y_1, ..., y_n)dy_1...dy_n = 1,$$

To avoid (important) technicalities, this integral is defined as an iterated integral, meaning we integrate one variable at a time.

Define the function:

$$F(x) = \int_{-\infty}^{x_n} \cdots \int_{-\infty}^{x_1} f(y_1, ..., y_n)dy_1...dy_n.$$

*In the terminology of probability theory, f is a **joint density function** and F is the associated **joint distribution function**.*

*With n sequential applications of one version of the **fundamental theorem of calculus** on the derivative of an indefinite integral (proposition 10.40, **Reitano** (2010)):*

$$f(x) = \frac{\partial^n F}{\partial x_1 \dots \partial x_n}.$$

Then n applications of a second version of the fundamental theorem of calculus on the definite integral of a derivative (proposition 10.33, **Reitano** (2010)) yield with the above notation:

$$\int_{a_n}^{b_n} \cdots \int_{a_1}^{b_1} f(y_1, \dots, y_n)\,dy_1 \dots dy_n = \sum_x sgn(x)F(x).$$

The integral on the left can be interpreted as the probability that the under-lying random variable resides in $\prod_{i=1}^n (a_i, b_i]$, or equivalently $\prod_{i=1}^n [a_i, b_i]$ for continuous f.

In these special cases, the Borel measure of a bounded right semi-closed rectangle, or the probability of an outcome in such, can be expressed as a **signed summation of the measures** of unbounded right semi-closed rectangles A_x, where x varies over the vertices of $\prod_{i=1}^n (a_i, b_i]$. Hence, the nonnegativity of the Borel measure (or probability) of this set imposes a condition on the associated function F, called the n-**increasing condition** (sometimes called the n-**nondecreasing condition**).

Definition 8.7 (n-**increasing condition**) *A function F is said to be n-**increasing**, or to satisfy the n-**increasing condition**, if given any bounded right semi-closed rectangle* $A = \prod_{i=1}^n (a_i, b_i]$:

$$\sum_x sgn(x)Fx) \geq 0. \tag{8.5}$$

Each $x = (x_1, \dots, x_n)$ in the summation is one of the 2^n vertices of A, so $x_i = a_i$ or $x_i = b_i$, and $sgn(x)$ equals -1 if the number of a_i-components of x is odd, and equals $+1$ otherwise.

If μ is a **product measure** defined with finite Borel measures, then $F_\mu(x) = \prod_{i=1}^n F_i(x_i)$ is n-increasing by item 1 of Example 8.6 since all F_i are increasing by Proposition 5.7.

In one dimension, so $n = 1$, (8.4) reflects (5.2) applied to $A = (a, b]$, where:

$$\mu((a, b]) = F_\mu(b) - F_\mu(a).$$

Thus the 1-increasing condition reduces to the condition that $F_\mu(b) \geq F_\mu(a)$. For $n = 2$, (8.4) becomes:

$$\mu\left[\prod_{i=1}^2 (a_i, b_i]\right] = F_\mu(b_1, b_2) - F_\mu(a_1, b_2) - F_\mu(b_1, a_2) + F_\mu(a_1, a_2).$$

It is an exercise to justify this formula geometrically by sketching the region in $\mathbb{R}^2$ defined by each unbounded cell. For $n > 2$, the measure formula in (8.4) and corollary n-increasing condition on F are harder to intuit.

We show below that the result in (8.4) is true for all finite Borel measures, where F is defined in (8.2). The proof relies on the **inclusion-exclusion formula** attributed to **Abraham de Moivre** (1667–1754). This formula provides the measure of a general finite union of sets, and unsurprisingly reduces to finite additivity when these sets are disjoint.

Proposition 8.8 (Inclusion-Exclusion formula) *Given measurable sets* $\{A_j\}_{j=1}^n$:

$$\mu\left[\bigcup_{j=1}^n A_j\right] = \sum_{j=1}^n \mu[A_j] - \sum_{i<j} \mu\left[A_i \bigcap A_j\right]$$

$$+ \sum_{i<j<k} \mu\left[A_i \bigcap A_j \bigcap A_k\right] \tag{8.6}$$

$$- \cdots + (-1)^{n+1} \mu\left[\bigcap_{j=1}^n A_j\right].$$

Proof. *Using induction, first note that:*

$$\mu\left(A\bigcup B\right) = \mu(A) + \mu(B) - \mu\left(A\bigcap B\right).$$

This follows by finite additivity of μ, *since let* $A' \equiv A - (A \bigcap B)$ *and* $B' \equiv B - (A \bigcap B)$. *Then:*

$$A\bigcup B = A'\bigcup B'\bigcup\left(A\bigcap B\right),$$

is a disjoint union and thus $\mu(A\bigcup B)$ *is a sum of measures of these sets. Substituting* $\mu(A') = \mu(A) - \mu(A\bigcap B)$, *since* $A = A'\bigcup(A\bigcap B)$ *is a disjoint union, and similarly for* $\mu(B')$, *yields the result for* $n = 2$.

For the induction step, write $\bigcup_{j=1}^{n+1} A_j = \left(\bigcup_{j=1}^n A_j\right)\bigcup A_{n+1}$ *and recall that* $\left(\bigcup_{j=1}^n A_j\right)\bigcap A_{n+1} = \bigcup_{j=1}^n \left(A_j \bigcap A_{n+1}\right)$ *by De Moivre's formulas. Applying the result for* $n = 2$:

$$\mu\left[\bigcup_{j=1}^{n+1} A_j\right] = \mu\left[\bigcup_{j=1}^n A_j\right] + \mu[A_{n+1}] - \mu\left[\bigcup_{j=1}^n \left(A_j\bigcap A_{n+1}\right)\right].$$

Now (8.6) *can be applied to both* n-*unions. The result follows by properly pairing results.*
   First:

$$\mu\left[\bigcup_{j=1}^n A_j\right] + \mu[A_{n+1}] = \sum_{j=1}^{n+1} \mu[A_j] - \sum_{i<j\leq n} \mu\left[A_i \bigcap A_j\right]$$

$$+ \sum_{i<j<k\leq n} \mu\left[A_i \bigcap A_j \bigcap A_k\right]$$

$$- \cdots + (-1)^{n+1} \mu\left[\bigcap_{j=1}^n A_j\right].$$

Also:

$$-\mu\left[\bigcup_{j=1}^n \left(A_j\bigcap A_{n+1}\right)\right] = -\sum_{j=1}^n \mu\left[A_j\bigcap A_{n+1}\right]$$

$$+ \sum_{i<j\leq n} \mu\left[A_i \bigcap A_j \bigcap A_{n+1}\right]$$

$$-\sum_{i<j<k\leq n}\mu\left[A_i\bigcap A_j\bigcap A_k\bigcap A_{n+1}\right]$$

$$+\ldots-(-1)^{n+1}\mu\left[\bigcap_{j=1}^{n+1}A_j\right].$$

Adding respective terms completes the induction. For example:

$$-\sum_{i<j\leq n}\mu\left[A_i\bigcap A_j\right]-\sum_{j=1}^{n}\mu\left[A_j\bigcap A_{n+1}\right]=-\sum_{i<j\leq n+1}\mu\left[A_i\bigcap A_j\right].\quad\blacksquare$$

Proposition 8.9 (Finite Borel induced $F_\mu(x)$ is n-increasing) *Let μ be a finite Borel measure on $\mathcal{B}(\mathbb{R}^n)$ and F_μ the induced function defined in (8.2).*
Then for any bounded right semi-closed rectangle $\prod_{i=1}^{n}(a_i,b_i]$:

$$\mu\left[\prod_{i=1}^{n}(a_i,b_i]\right]=\sum_x sgn(x)F_\mu(x)\geq 0,\tag{8.7}$$

where each $x=(x_1,\ldots,x_n)$ is one of the 2^n vertices of $\prod_{i=1}^{n}(a_i,b_i]$, so $x_i=a_i$ or $x_i=b_i$, and $sgn(x)$ is defined as -1 if the number of a_i-components of x is odd, and $+1$ otherwise.

Proof. *First:*

$$\prod_{i=1}^{n}(a_i,b_i]=A_{(b_1,\ldots,b_n)}-\left[\bigcup_{j=1}^{n}A_{x^{(j)}}\right],$$

where $x_i^{(j)}=b_i$ for $i\neq j$ and $x_j^{(j)}=a_j$. This follows because if $x\in A_{(b_1,\ldots,b_n)}-\prod_{i=1}^{n}(a_i,b_i]$ then $x_i\leq a_i$ for at least one i, and thus $x\in A_{x^{(j)}}$ for some j. Now since $\bigcup_{j=1}^{n}A_{x^{(j)}}\subset A_{(b_1,\ldots,b_n)}$:

$$\mu\left[\prod_{i=1}^{n}(a_i,b_i]\right]=\mu\left[A_{(b_1,\ldots,b_n)}\right]-\mu\left[\bigcup_{j=1}^{n}A_{x^{(j)}}\right]$$

$$=F_\mu(b_1,\ldots,b_n)-\mu\left[\bigcup_{j=1}^{n}A_{x^{(j)}}\right].$$

The measure of $\bigcup_{j=1}^{n}A_{x^{(j)}}$ can be evaluated using the inclusion-exclusion formula:

$$\mu\left[\bigcup_{j=1}^{n}A_{x^{(j)}}\right]=\sum_{j=1}^{n}\mu\left[A_{x^{(j)}}\right]-\sum_{i<j}\mu\left[A_{x^{(i)}}\bigcap A_{x^{(j)}}\right]$$

$$+\sum_{i<j<k}\mu\left[A_{x^{(i)}}\bigcap A_{x^{(j)}}\bigcap A_{x^{(k)}}\right]$$

$$-\cdots+(-1)^{n+1}\mu\left[\bigcap_{j=1}^{n}A_{x^{(j)}}\right].$$

For the first summation, $\mu\left[A_{x^{(j)}}\right]=F_\mu(x^{(j)})$ by definition.

Then, $A_{x^{(i)}} \bigcap A_{x^{(j)}} = A_{x^{(i,j)}}$, where $x_l^{(i,j)} = b_l$ for $l \neq i, j$ and $x_l^{(i,j)} = a_l$ for $l = i, j$, and so:

$$-\mu \left[A_{x^{(i)}} \bigcap A_{x^{(j)}} \right] = -\mu \left[A_{x^{(i,j)}} \right] \equiv -F_\mu(x^{(i,j)}).$$

Similarly $A_{x^{(i)}} \bigcap A_{x^{(j)}} \bigcap A_{x^{(k)}} = A_{x^{(i,j,k)}}$, where $x_l^{(i,j,k)} = b_l$ for $l \neq i, j, k$ and $x_l^{(i,j,k)} = a_l$ for $l = i, j, k$, and so $\mu \left[A_{x^{(i,j,k)}} \right] \equiv F_\mu(x^{(i,j,k)})$. This is then true for all such intersections.

Substituting into the above expression for $\mu \left[\bigcup_{j=1}^n A_{x^{(j)}} \right]$, and then that result into the formula for $\mu \left[\prod_{i=1}^n (a_i, b_i] \right]$ yields (8.7). ∎

The following proposition summarizes the results of this section.

Proposition 8.10 (Properties of induced functions: Finite Borel measures) *Let μ be a finite Borel measure on $\mathcal{B}(\mathbb{R}^n)$ and F_μ the induced function defined in (8.2). Then:*

1. *F_μ is **continuous from above**: If $x^{(m)} \in \mathbb{R}^n$ with $x_i^{(m)} \geq x_i$ for all i and $x^{(m)} \to x$ as $m \to \infty$, then as in (8.3):*

$$F_\mu(x) = \lim_{m \to \infty} F_\mu(x^{(m)}).$$

2. *For any bounded right semi-closed rectangle $\prod_{i=1}^n (a_i, b_i]$, then as in (8.7):*

$$\mu \left[\prod_{i=1}^n (a_i, b_i] \right] = \sum_x sgn(x) F_\mu(x),$$

*and hence F_μ is n-**increasing** as in (8.5):*

$$\sum_x sgn(x) F_\mu(x) \geq 0.$$

In these summations, each $x = (x_1, ..., x_n)$ is one of the 2^n vertices of $\prod_{i=1}^n (a_i, b_i]$, so $x_i = a_i$ or $x_i = b_i$, and $sgn(x)$ is defined as -1 if the number of a_i-components of x is odd, and $+1$ otherwise.

8.2.2 Functions Induced by General Borel Measures

In this section we develop the framework for induced functions of general Borel measures. The prior section addressed finite Borel measures, and thus the induced functions generalized the 1-dimensional approach seen in (5.3). For general Borel measures, we must instead use the approach seen in (5.1), where:

$$F_\mu(x) \equiv \begin{cases} -\mu((x, 0]), & x < 0, \\ 0, & x = 0, \\ \mu((0, x]), & x > 0. \end{cases}$$

To simplify the generalization, it is useful to introduce some notation. For x real, define the **justified right semi-closed interval** $(0, x]^+$ as follows:

$$(0, x]^+ \equiv \begin{cases} (x, 0], & x < 0, \\ \varnothing, & x = 0, \\ (0, x], & x > 0. \end{cases} \tag{8.8}$$

Then $F_\mu(x)$ above can be expressed:

$$F_\mu(x) = s(x)\mu((0, x]^+), \tag{8.9}$$

where the sign function $s(x)$ is defined:

$$s(x) \equiv \begin{cases} -1, & x < 0. \\ 0, & x = 0, \\ 1, & x > 0. \end{cases} \tag{8.10}$$

Recalling (5.2), that $\mu(a, b] = F_\mu(b) - F_\mu(a)$, the above representation for $F_\mu(x)$ obtains:

$$\mu(a, b] = s(b)\mu((0, b]^+) - s(a)\mu((0, a]^+). \tag{8.11}$$

We next define a generalization of (5.1) as follows. Given a Borel measure μ on $\mathcal{B}(\mathbb{R}^n)$, define a **function $\tilde{F}_\mu$ induced by μ on $\mathbb{R}^n$** by:

$$\tilde{F}_\mu(x) \equiv \prod_{i=1}^{n} s(x_i)\mu\left[\prod_{i=1}^{n} (0, x_i]^+\right]. \tag{8.12}$$

The function $\tilde{F}_\mu$ is also called **a distribution function associated with μ**.

Note that $\tilde{F}_\mu(x)$ is the **signed measure** of the justified right semi-closed rectangle determined by x and the origin 0. This rectangle is defined as the product of justified right semi-closed intervals determined by the x_i-components. The sign of this measure is defined as the product of the signs of the x-components.

Note also that $\tilde{F}_\mu(x)$ is well defined since every such rectangle is bounded, and thus contained in a compact set. By monotonicity of μ and property 3 of Definition 8.4, $\mu\left[\prod_{i=1}^{n} (0, x_i]^+\right]$ is finite for all x.

The sign convention in (8.12) on the measure of such rectangles generalizes that in the 1-dimensional case, in the sense that (5.1) is obtained if $n = 1$. To see how this function is reflected in the measure of a general right semi-closed rectangle $\prod_{i=1}^{n} (a_i, b_i]$, we begin with two examples.

Example 8.11

1. Borel product measure: Assume that the Borel measure μ on $\mathcal{B}(\mathbb{R}^n)$ is a Borel product measure of Proposition 7.20, derived from the Borel measure spaces $\{(\mathbb{R}, \mathcal{B}(\mathbb{R}), \mu_i)\}_{i=1}^{n}$. Then:

$$\mu\left[\prod_{i=1}^{n} (a_i, b_i]\right] = \prod_{i=1}^{n} \mu_i(a_i, b_i].$$

In this case, (8.12) and (8.9) obtain:

$$\tilde{F}_\mu(x) \equiv \prod_{i=1}^n s(x_i)\mu\left[\prod_{i=1}^n (0,x_i]^+\right]$$

$$= \prod_{i=1}^n s(x_i)\mu_i((0,x_i]^+)$$

$$= \prod_{i=1}^n F_{\mu_i}(x_i).$$

Here $F_{\mu_i}(x_i)$ is the function induced by μ_i originally defined in (5.1), and restated in (8.9).
  Applying (8.11):

$$\mu\left[\prod_{i=1}^n (a_i,b_i]\right] = \prod_{i=1}^n \left[s(b_i)(\mu(0,b_i]^+) - s(a_i)(\mu(0,a_i]^+)\right].$$

This product expands to a summation of 2^n terms, each of which reflects one of the vertices of this rectangle and has the general form:

$$\prod_{k=1}^m s(b_{j_k})\mu((0,b_{j_k}]^+) \prod_{k=m+1}^n \left(-s(a_{j_k})\right)\mu((0,a_{j_k}]^+),$$

where $0 \le m \le n$.
  Comparing with (8.12), and using the earlier notational convention:

$$\mu\left[\prod_{i=1}^n (a_i,b_i]\right] = \sum_x sgn(x)\tilde{F}_\mu(x), \qquad (8.13)$$

where each $x = (x_1,...,x_n)$ is one of the 2^n vertices of $\prod_{i=1}^n (a_i,b_i]$, so $x_i = a_i$ or $x_i = b_i$, and $sgn(x)$ is defined as -1 if the number of a_i-components of x is odd, and $+1$ otherwise.
  This is identical to the result obtained in the case of finite Borel measures of Proposition 8.9, but using the induced function in (8.12).
  2. A special case for a general Borel measure: For a general Borel measure μ, there is a special case for which the result in (8.13) is relatively transparent. Let $A = \prod_{i=1}^n (a_i,b_i]$ and assume that $a_i < 0 < b_i$ for all i. Then $(a_i,b_i] = (0,a_i]^+ \bigcup (0,b_i]^+$, and so:

$$A = \prod_{i=1}^n \left[(0,a_i]^+ \bigcup (0,b_i]^+\right]$$

$$= \bigcup_x \prod_{i=1}^n (0,x_i]^+,$$

where as always, the x-union is over all 2^n vertices of this rectangle.
  In this special case, A is seen to be a disjoint union of rectangles, and so by finite additivity:

$$\mu(A) = \sum_x \mu\left[\prod_{i=1}^n (0,x_i]^+\right].$$

This is (8.13) because here, $s(b_i) = 1$, $s(a_i) = -1$, and so for all x:

$$sgn(x)\tilde{F}_\mu(x) \equiv sgn(x) \prod_{i=1}^n s(x_i)\mu \left[\prod_{i=1}^n (0, x_i]^+ \right]$$

$$= \prod_b s(b_i) \prod_a (-s(a_i)) \, \mu \left[\prod_{i=1}^n (0, x_i]^+ \right]$$

$$= \mu \left[\prod_{i=1}^n (0, x_i]^+ \right].$$

By simple examples in $\mathbb{R}^2$, one can confirm that A is not a disjoint union when $a_i < b_i < 0$ or $0 < a_i < b_i$. So, the general proof requires a different approach.

With these examples as background, we now prove the generalization of Proposition 8.10. For the proof we use the **characteristic function of a set** A, denoted $\chi_A(y)$. This function is also called the **indicator function** of the set A, and was introduced in Definition 3.20. The key insight is that $A = B$ if and only if $\chi_A(y) = \chi_B(y)$ for all y. See also Exercise 3.21.

However, more will be needed. The author is not aware of a direct proof of this result using only the results of this book. But of course this result belongs here. The given proof is relatively straightforward if the reader will accept a result that will be seen in Book III for the Lebesgue integral, yet not fully developed until Book V for integrals with respect to general measures. Fortunately, the needed result should seem quite plausible now.

To motivate with the Riemann integral, if $f(x) = 1$ then recalling the notation from Chapter 1:

$$(\mathcal{R}) \int_a^b f(x)dx = b - a = m[a, b],$$

where m denotes Lebesgue measure. This same result is true for the Lebesgue integral, and we can rewrite this equation as:

$$(\mathcal{L}) \int \chi_{[a,b]}(x)dx = m[a, b].$$

It is natural to wonder, is this identity also true if $[a, b]$ is replaced by an arbitrary Lebesgue measurable set A?

For the Riemann approach, one can both define the above integral and prove existence for all continuous functions without too much mathematical machinery. The reason for this is that continuity is a very strong condition on a function, and an interval is a very simple type of set.

For Lebesgue and general measures, the process is more subtle. The approach is axiomatic in that one explicitly **defines** the value of the integral for certain special functions, and then proceeds to prove that this definition extends to a wide class of functions of interest. One also defines the integral to have an important property with respect to this special class of functions, again with the goal to prove that this property extends to the wider class.

Thus at the start of the integration theories of Books III and V we will take as **axioms**:

1. **Integral of characteristic functions of measurable sets**: If $(X, \sigma(X), \mu)$ is a measure space and $A \in \sigma(X)$, then:

$$\int \chi_A(x) d\mu \equiv \mu(A).$$

2. **Linearity**: If $\{A_i\}_{i=1}^n \subset \sigma(X)$ and $\{a_i\}_{i=1}^n \subset \mathbb{R}$ (or $\mathbb{C}$), then:

$$\int \sum_{i=1}^n a_i \chi_{A_i}(x) d\mu \equiv \sum_{i=1}^n a_i \int \chi_{A_i}(x) d\mu.$$

While item 1 is unambiguous in its statement, axiom 2 initially raises the question of well-definedness and challenges its label as an axiom. In detail, if $\{A_i'\}_{i=1}^m \subset \sigma(X)$ and $\{a_i'\}_{i=1}^m \subset \mathbb{R}$, and for all x:

$$\sum_{i=1}^n a_i \chi_{A_i}(x) = \sum_{i=1}^m a_i' \chi_{A_i'}(x),$$

does this axiom obtain the same result for the associated integrals? The answer is "yes," but this must be proved.

If the reader accepts these axioms for now, the following proof will be complete, subject to confirmation of the above results in future books.

Proposition 8.12 (Properties of induced functions: General Borel measures)
Let μ be a Borel measure on $\mathcal{B}(\mathbb{R}^n)$ and $\tilde{F}_\mu$ the induced function defined in (8.12).
Then:

1. *$\tilde{F}_\mu$ is **continuous from above**: If $x^{(m)} \in \mathbb{R}^n$ with $x_i^{(m)} \geq x_i$ for all i and $x^{(m)} \to x$ as $m \to \infty$, then as in (8.3):*

$$\tilde{F}_\mu(x) = \lim_{m \to \infty} \tilde{F}_\mu(x^{(m)}).$$

2. *For any bounded right semi-closed rectangle $\prod_{i=1}^n (a_i, b_i]$, then as in (8.13):*

$$\mu \left[\prod_{i=1}^n (a_i, b_i] \right] = \sum_x sgn(x) \tilde{F}_\mu(x),$$

*and hence $\tilde{F}_\mu$ is n-**increasing** as in (8.5):*

$$\sum_x sgn(x) \tilde{F}_\mu(x) \geq 0.$$

In these summations, each $x = (x_1, ..., x_n)$ is one of the 2^n vertices of $\prod_{i=1}^n (a_i, b_i]$, so $x_i = a_i$ or $x_i = b_i$, and $sgn(x)$ is defined as -1 if the number of a_i-components of x is odd, and $+1$ otherwise.

Proof. 1. *Continuity*: *For continuity, let x be given and so by (8.12):*

$$\tilde{F}_\mu(x) \equiv \prod_{i=1}^n s(x_i)\mu\left[\prod_{i=1}^n (0,x_i]^+\right].$$

This product can be split between negative and nonnegative terms:

$$\tilde{F}_\mu(x) \equiv \prod_{i=1}^n s(x_i)\mu\left[\prod_{i=1}^k (0,x_{j_i}]\prod_{i=k+1}^n (x_{j_i},0]\right].$$

Here $(j_1,...,j_n)$ is a permutation of $(1,...,n)$, $x_{j_i} \geq 0$ for $1 \leq i \leq k$ and $x_{j_i} < 0$ for $k+1 \leq i \leq n$.

Now let $x^{(m)} \in \mathbb{R}^n$ with $x_i^{(m)} \geq x_i$. Writing $x_i^{(m)} = x_i + \epsilon_i^{(m)}$, since we are interested in the limit it can be assumed that $x_{j_i} + \epsilon_{j_i}^{(m)} < 0$ for $k+1 \leq i \leq n$. Then $\tilde{F}_\mu(x)$ and $\tilde{F}_\mu(x^{(m)})$ have the same $\prod_{i=1}^n s(x_i)$-factors, and so we focus on the measures of the respective rectangles.

Denote:

$$A = \prod_{i=1}^k (0,x_{j_i}]\prod_{i=k+1}^n (x_{j_i},0],$$
$$A_m = \prod_{i=1}^k (0,x_{j_i} + \epsilon_{j_i}^{(m)}]\prod_{i=k+1}^n (x_{j_i} + \epsilon_{j_i}^{(m)},0].$$

Our goal is to prove that $\mu(A_m) \to \mu(A)$.

If $k = n$ then $\mu(A_m) \to \mu(A)$ by continuity from above since then $A \subset A_m$ and $A = \bigcap_m A_m$. This result applies since all A_m are bounded and thus have finite measure by item 3 of Definition 8.4. Similarly, if $k = 0$ then $\mu(A_m) \to \mu(A)$ by continuity from below since then $A_m \subset A$ and $A = \bigcup_m A_m$.

Now assume that $1 \leq k \leq n-1$. It follows from De Morgan's laws that:

$$A\bigcup(A_m - A) = A_m\bigcup(A - A_m),$$

and thus as disjoint unions:

$$\mu(A) + \mu(A_m - A) = \mu(A_m) + \mu(A - A_m). \tag{1}$$

To complete the proof that $\mu(A_m) \to \mu(A)$, it is enough given (1) to show that $\mu(A_m - A) \to 0$ and $\mu(A - A_m) \to 0$.

To this end:

$$A_m - A = \prod_{i=1}^k (x_{j_i}, x_{j_i} + \epsilon_{j_i}^{(m)}]\prod_{i=k+1}^n (x_{j_i} + \epsilon_{j_i}^{(m)},0].$$

By monotonicity:

$$\mu(A_m - A) \leq \mu\left[\prod_{i=1}^k (x_{j_i}, x_{j_i} + \epsilon_{j_i}^{(m)}]\prod_{i=k+1}^n (x_{j_i},0]\right],$$

and since $k \geq 1$, $\mu(A_m - A) \to 0$ by continuity from above.

Similarly:

$$A - A_m = \prod_{i=1}^{k} (0, x_{j_i}] \prod_{i=k+1}^{n} (x_{j_i}, x_{j_i} + \epsilon_{j_i}^{(m)}],$$

and since $k \leq n - 1$, $\mu(A - A_m) \to 0$ by continuity from above.

2. Identity (8.13): *Let $A = \prod_{i=1}^{n} (a_i, b_i]$, a bounded right semi-closed rectangle. Considering the three relative positions of a right semi-closed interval $(a, b]$ relative to 0, it is an exercise to show that:*

$$\chi_{(a,b]} = s(b)\chi_{(0,b]^+} - s(a)\chi_{(0,a]^+}.$$

Then with $y = (y_1, ..., y_n)$:

$$\chi_A(y) = \prod_{i=1}^{n} \chi_{(a_i,b_i]}(y_i)$$
$$= \prod_{i=1}^{n} \left(s(b_i)\chi_{(0,b_i]^+}(y_i) - s(a_i)\chi_{(0,a_i]^+}(y_i) \right).$$

As in Example 8.11, this product expands to a summation of 2^n terms, each of which reflects one of the vertices of this rectangle and has the general form:

$$\prod_{k=1}^{m} s(b_{j_k})\chi_{(0,b_{j_k}]^+}(y_{j_k}) \prod_{k=m+1}^{n} \left(-s(a_{j_k}) \right) \chi_{(0,a_{j_k}]^+}(y_{j_k}).$$

Combining obtains:

$$\chi_A(y) = \sum_x \prod_b s(x_i)\chi_{(0,x_i]^+}(y_i) \prod_a (-s(x_i)) \chi_{(0,x_i]^+}(y_i).$$

As above, this is a summation over the 2^n vertices, where for each vertex x, the associated products reflect the a_i or b_i-components. Rewriting:

$$\chi_A(y) = \sum_x sgn(x) \prod_{i=1}^{n} s(x_i)\chi_{\prod_{i=1}^{n} (0,x_i]^+}(y), \tag{2}$$

where $sgn(x)$ is defined as -1 if the number of a_i-components of x is odd, and $+1$ otherwise.

We can integrate (2) with respect to μ and utilize the axioms above. Then:

$$\int \chi_A(y)d\mu = \mu(A), \quad \int \chi_{\prod_{i=1}^{n} (0,x_i]^+}(y)d\mu = \mu \left(\prod_{i=1}^{n} (0, x_i]^+ \right),$$

and (8.13) now follows from (2) and linearity of the integral. ∎

8.2.3 Borel Measures Induced by Functions

In this section we prove that the necessary conditions on F as seen in Propositions 8.10 and 8.12 are also sufficient to assure that a Borel measure μ_F exists which satisfies (8.7). Specifically, we show that if F is **continuous from above** as in (8.3) of Definition 8.5, and satisfies the *n*-**increasing condition** in (8.5) of Definition 8.7, then *F* induces a Borel measure μ_F on $\mathcal{B}(\mathbb{R}^n)$ which satisfies (8.13).

This derivation will require several propositions. The added complexity in this result is caused by the fact that the *n*-increasing condition in (8.5) only provides an insight into the definition of the induced measure μ_F on $\mathcal{A}'_B$, the class of **bounded right semi-closed rectangles** studied in Proposition 8.1:

$$\mathcal{A}'_B \equiv \left\{ A \in \mathcal{B}(\mathbb{R}^n) | A = \prod_{i=1}^{n} (a_i, b_i], \text{ with } -\infty < a_i \le b_i < \infty \right\}.$$

Indeed, given this *n*-increasing condition it is compelling to explicitly define the associated set function μ_0 on $A \in \mathcal{A}'_B$ by (8.7) (or equivalently (8.13)), and this will be the approach taken.

This approach will generalize the 1-dimensional Borel construction initiated in (5.4) where we defined the set function $\mu_0(a, b] = |(a, b]|_F$ on $\mathcal{A}'$ by:

$$|(a, b]|_F = F(b) - F(a).$$

In this 1-dimensional case, unbounded intervals were manageable in the sense that there were only two types, and such an intervals either had finite measure or not. On $\mathbb{R}^n$, unbounded sets are more complicated, and so we have chosen to work with $\mathcal{A}'_B$ to avoid the associated complexities.

However, while $\mathcal{A}'_B$ generates the Borel sigma algebra as proved in Proposition 8.1, it is not a semi-algebra. Consequently, even once this definition of μ_0 on $\mathcal{A}'_B$ is proved to possess the desired additivity properties, the extension results of Chapter 6 cannot be immediately applied to complete the derivation. Instead, it will take extra steps to show that these earlier extension results apply, given this more limited conclusion.

The first result shows that μ_0 so defined is finitely additive and countably subadditive on $\mathcal{A}'_B$. One consequence of the observation that $\mathcal{A}'_B$ is not a semi-algebra is that Proposition 6.18 does not then apply. That is we cannot simply conclude from this result that μ_0 has a unique extension to a measure μ_F on the algebra generated by this collection.

Proposition 8.13 (μ_0 **is finitely additive, countably subadditive on** $\mathcal{A}'_B$) *Given a real-valued function* $F : \mathbb{R}^n \to \mathbb{R}$ *that is continuous from above as in (8.3) of Definition 8.5, and satisfies the n-increasing condition of (8.5) of Definition 8.7, define a set function* μ_0 *on the class of bounded right semi-closed rectangles* $A = \prod_{i=1}^{n} (a_i, b_i] \in \mathcal{A}'_B$ *by:*

$$\mu_0[A] = \sum_x sgn(x)F(x). \tag{8.14}$$

Each $x = (x_1, ..., x_n)$ *in the summation is one of the* 2^n *vertices of* $\prod_{i=1}^{n} (a_i, b_i]$*, so* $x_i = a_i$ *or* $x_i = b_i$*, and* $sgn(x)$ *is defined as* -1 *if the number of* a_i*-components of x is odd, and* $+1$ *otherwise.*

Then μ_0 *is finitely additive and countably subadditive on* $\mathcal{A}'_B$.

Proof. *By (8.5) it follows that* $\mu_0[A] \ge 0$ *for* $A \in \mathcal{A}'_B$. *To prove that* $\mu_0[\emptyset] = 0$, *let* $a_i = b_i$ *for some i, so* $A = \emptyset$. *The summation in (8.14) can be split into two summations of* 2^{n-1} *terms each, the first summation containing all terms with* $x_i = a_i$, *and the second all terms with* $x_i = b_i$.

Since $a_i = b_i$ for A, the terms in these summations can be paired, they have opposite signs by definition, and thus $\mu_0[\emptyset] = 0$.

1. ***Finite Additivity of μ_0 on Partitions of Rectangles**: With $A = \prod_{i=1}^{n}(a_i, b_i]$, define a partition of each interval by:*

$$a_i = c_{i,0} < c_{i,1} < \cdots < c_{i,m_i} = b_i,$$

so $(a_i, b_i] = \bigcup_{j=0}^{m_i-1} (c_{i,j}, c_{i,j+1}]$. Then:

$$A = \bigcup_N C_J,$$

where the union contains $\prod_{i=1}^{n} m_i$ disjoint rectangles. These rectangles are defined over all n-tuples of the index set $N = \{(J(1), J(2), \ldots, J(n))|\ 0 \le J(i) \le m_i - 1\}$, and given $J \in N$, $C_J \equiv \prod_{i=1}^{n} (c_{i,J(i)}, c_{i,J(i)+1}]$.

To show finite additivity is to prove that:

$$\mu_0[A] = \sum_N \mu_0[C_J]. \tag{1}$$

Let $sgn_J(x)$ be defined relative to the 2^n vertices of the rectangle C_J, meaning that if x is a vertex of $C_J \equiv \prod_{i=1}^{n} (c_{i,J(i)}, c_{i,J(i)+1}]$ then $sgn_J(x) = -1$ if the number of $c_{i,J(i)}$-components of x is odd, and $+1$ otherwise. An application of (8.14) yields:

$$\sum_N \mu_0[C_J] = \sum_N \sum_{x \in C_J} sgn_J(x)F(x)$$
$$= \sum_x F(x) \sum_N sgn_J(x).$$

For the last expression, $\sum_x$ denotes the sum over all x that are vertices of one or more C_J, and for each such x, $\sum_N sgn_J(x)$ denotes the sum over all $J \in N$ with x a vertex of C_J.

To evaluate this inner summation, denote by $a(x)$ the number of components of x equal to $a_i = c_{i,0}$, by $b(x)$ the number of components of x equal to $b_i = c_{i,m_i}$, and by $c(x)$ the number of such components equal to $c_{i,j}$ for $1 \le j \le m_i - 1$. Then for given x, the number of subrectangles of A with vertex x is $2^{c(x)}$. This follows because given x, each subrectangle with vertex x is a product of n intervals, with $a(x)$ intervals fixed and equal to $(a_i, c_{i,1}]$, $b(x)$ intervals fixed and equal to $(c_{i,m_i-1}, b_i]$, and $c(x)$ intervals equal to either $(c_{i,j}, c_{i,j+1}]$ or $(c_{i,j-1}, c_{i,j}]$. Thus the number of such subrectangles is determined by $c(x)$.

There are now two cases to consider:

(a) *If x is not a vertex of A, then $\sum_N sgn_J(x) = 0$:*
    In this case $c(x) \ge 1$ and hence there are an even number of sub-rectangles with vertex x. Arbitrarily choose the first component of x that equals $c_{i,j}$ for $1 \le j \le m_i - 1$ and divide this collection of subrectangles into two groups. The first group all have interval $(c_{i,j}, c_{i,j+1}]$ in this component, the second group all have interval $(c_{i,j-1}, c_{i,j}]$. By construction, if C_J and $C_{J'}$ are any such pair of subrectangles which differ only by this component interval, $sgn_J(x) = -sgn_{J'}(x)$, and thus $\sum_N sgn_J(x) = 0$.

(b) *If x is a vertex of A, then $\sum_N sgn_J(x) = sgn(x)$, where $sgn(x)$ is defined relative to A: In this case x has $a(x)$ components equal to a_i, and $b(x) = n - a(x)$ components equal to b_i. As $c(x) = 0$, there is only one subrectangle C_J that contains x, and of necessity it is the subrectangle formed with $a(x)$ intervals of type $(a_i, c_{i,1}]$, and $b(x) = n - a(x)$ intervals of type $(c_{i,m_i-1}, b_i]$. Thus considered as a vertex of A or C_J, both $sgn(x)$ and $sgn_J(x)$ are defined relative to the parity of the number of components of x equal to a_i, so $sgn_J(x) = sgn(x)$.*

From this analysis it follows that:

$$\sum_N \mu_0 [C_J] = \sum_x F(x) sgn(x)$$
$$= \mu_0[A],$$

and (1) is proved.

2. **Finite Additivity of** μ_0 **on** $\mathcal{A}'_B$: *Recall that by Definition 6.6, we need only prove finite additivity when the union of disjoint $\mathcal{A}'_B$-sets is in $\mathcal{A}'_B$. Given disjoint:*

$$\{A_k\}_{k=1}^m = \left\{ \prod_{i=1}^n (a_i^{(k)}, b_i^{(k)}] \right\}_{k=1}^m \subset \mathcal{A}'_B,$$

with $\bigcup_{k=1}^m A_k = A = \prod_{i=1}^n (a_i, b_i]$, it follows from Proposition 7.6 that $(a_i, b_i] = \bigcup_{k=1}^m (a_i^{(k)}, b_i^{(k)}]$ for every i, though these unions are in general not disjoint by Proposition 7.9. However, the collection of endpoints $\left\{ a_i^{(k)}, b_i^{(k)} \right\}_{k=1}^m$ forms a partition of $(a_i, b_i]$ for each i, and this partition can be used to create a disjoint rectangle partition of A. Thus from part 1, the measure of A equals the sum of the measures of the rectangles formed by this partition.

On the other hand, this partition also induces a partition of each A_k. For example, $A_1 = \prod_{i=1}^n (a_i^{(1)}, b_i^{(1)}]$, and as each interval $(a_i^{(1)}, b_i^{(1)}]$ is partitioned by the collection $\left\{ a_i^{(k)}, b_i^{(k)} \right\}_{k=1}^m$, A_1 is in turn partitioned by a subset of the original partition. Another application of (1) then yields that the measure of each A_k is the sum of the measures of the corresponding subset of the original partition.

The proof is complete by noting that the disjointness of the A_k sets assures that no rectangle in the partition is used more than once, while the assumption that the union of such A_k is the rectangle A assures that every rectangle in the partition is used exactly once.

3. **Monotonicity and Finite Subadditivity of** μ_0 **on** $\mathcal{A}'_B$: *Part 4 will require that μ_0 be finitely subadditive, which is a corollary of the following result on monotonicity. If $A, \{A_k\}_{k=1}^m \subset \mathcal{A}'_B$ and $A \subset \bigcup_{k=1}^m A_k$, where this union need not be disjoint, then we claim that:*

$$\mu_0 [A] \leq \sum_{k=1}^m \mu_0 [A_k]. \tag{2}$$

Continuing with the notation of part 2, the collection of endpoints $\left\{ a_i, b_i, \left\{ a_i^{(k)}, b_i^{(k)} \right\}_{k=1}^m \right\}$ forms a partition of $(a(i), b(i)]$ for each i, where $a(i)$ is defined as the minimum of the

a_i-points in this set, and $b(i)$ is the maximum of the b_i-points. Thus $(a(i), b(i)] = \bigcup_{j=0}^{m_i-1}$ $(c_{i,j}, c_{i,j+1}]$, where $\{c_{i,j}\}_{j=0}^{m_i}$ represents a reordering of these endpoints, the number of which may differ with i depending on repetitions.

Thus:

$$A \subset \bigcup_{k=1}^{m} A_k \subset \prod_{i=1}^{n} (a(i), b(i)] = \bigcup_{N} C_J,$$

where the C_J-rectangles form a disjoint union defined by the interval partitions. Now by construction, for each J either $C_J \subset A$ or $C_J \bigcap A = \emptyset$, and for each k and J either $C_J \subset A_k$ or $C_J \bigcap A_k = \emptyset$. In addition, since $A \subset \bigcup_{k=1}^{m} A_k$, it follows that if $C_J \subset A$, then $C_J \subset A_k$ for some k.

Combining results, A equals a disjoint union of a subset of C_J-rectangles, so by item 2 the measure of A equals the sum of the measures of these C_J-rectangles. The same is true of the measure of each A_k. Thus since μ_0 is nonnegative on the C_J-rectangles that are contained in some A_k but not contained in A, the monotonicity result of (2) follows.

Finite subadditivity of μ_F is now a corollary. Given arbitrary $\{A_k\}_{k=1}^{m} \subset \mathcal{A}'_B$ with $\bigcup_{k=1}^{m} A_k \in \mathcal{A}'_B$, then by monotonicity with $A = \bigcup_{k=1}^{m} A_k$:

$$\mu_0 \left[\bigcup_{k=1}^{m} A_k \right] \leq \sum_{k=1}^{m} \mu_0 [A_k]. \tag{3}$$

4. **Countable Subadditivity of μ_0 on $\mathcal{A}'_B$:** As noted in part 2, we need only prove countable subadditivity when the countable union of disjoint $\mathcal{A}'_B$-sets is in $\mathcal{A}'_B$. Given disjoint $\{A_k\}_{k=1}^{\infty} = \left\{ \prod_{i=1}^{n} (a_i^{(k)}, b_i^{(k)}] \right\}_{k=1}^{\infty} \subset \mathcal{A}'_B$, assume that $\bigcup_{k=1}^{\infty} A_k = A \in \mathcal{A}'_B$ with $A = \prod_{i=1}^{n} (a_i, b_i]$. For any $\epsilon > 0$, there are $\delta > 0$ and $\delta^{(k)} > 0$ so that:

$$\mu_0 \left[\prod_{i=1}^{n} (a_i + \delta, b_i] \right] \geq \mu_0 [A] - \epsilon, \tag{4}$$

$$\mu_0 \left[\prod_{i=1}^{n} (a_i^{(k)}, b_i^{(k)} + \delta^{(k)}] \right] \leq \mu_0 [A_k] + \epsilon/2^k.$$

The first statement in (4) follows from the n-increasing and continuity from above conditions on F. First:

$$\prod_{i=1}^{n} (a_i, b_i] = \prod_{i=1}^{n} \left[(a_i, a_i + \delta] \bigcup (a_i + \delta, b_i] \right]$$

$$= \prod_{i=1}^{n} (a_i + \delta, b_i] + \bigcup_{k} \prod_{i=1}^{n} (c_i^{(k)}, d_i^{(k)}].$$

This k-union is over $2^n - 1$ disjoint rectangles, each of which has at least one factor equal to $(a_i, a_i + \delta]$. By finite additivity, the μ_0-measure of this union is the sum of the measures. If $(c_j^{(k)}, d_j^{(k)}] = (a_j, a_j + \delta]$, then by continuity from above the measure of this rectangle can be made arbitrarily small for δ small. In particular, the sum of the measures of all such rectangles can be made smaller than ϵ. The second statement in (4) is derived similarly.

By construction:

$$\prod_{i=1}^{n} [a_i + \delta, b_i] \subset \bigcup_{k=1}^{\infty} \left[\prod_{i=1}^{n} (a_i^{(k)}, b_i^{(k)} + \delta^{(k)}) \right],$$

and hence by the Heine-Borel theorem the compact set $\prod_{i=1}^{n} [a_i + \delta, b_i]$ *has a finite open cover. Re-indexing with the same notation:*

$$\prod_{i=1}^{n} [a_i + \delta, b_i] \subset \bigcup_{k=1}^{N} \left[\prod_{i=1}^{n} (a_i^{(k)}, b_i^{(k)} + \delta^{(k)}) \right],$$

and so:

$$\prod_{i=1}^{n} (a_i + \delta, b_i] \subset \bigcup_{k=1}^{N} \left[\prod_{i=1}^{n} (a_i^{(k)}, b_i^{(k)} + \delta^{(k)}] \right]. \tag{5}$$

Using (4), then (5), (3) and (4):

$$\mu_0 [A] - \epsilon \leq \mu_0 \left[\prod_{i=1}^{n} (a_i + \delta, b_i] \right]$$

$$\leq \mu_0 \left(\bigcup_{k=1}^{N} \left[\prod_{i=1}^{n} (a_i^{(k)}, b_i^{(k)} + \delta^{(k)}] \right] \right)$$

$$\leq \sum_{k=1}^{N} \mu_0 \left[\prod_{i=1}^{n} (a_i^{(k)}, b_i^{(k)} + \delta^{(k)}] \right]$$

$$\leq \sum_{k=1}^{\infty} \mu_0 [A_k] + \epsilon.$$

Since ϵ *is arbitrary, this obtains countable subadditivity:*

$$\mu_0 \left[\bigcup_{k=1}^{\infty} A_k \right] \leq \sum_{k=1}^{\infty} \mu_0 [A_k]. \qquad \blacksquare$$

We next show that μ_0 can be extended from $\mathcal{A}'_B$ to a measure on the Borel sigma algebra $\mathcal{B}(\mathbb{R}^n)$. The approach will be to define an outer measure μ_F^* on the power sigma algebra $\sigma(P(\mathbb{R}^n))$ using μ_0 and sets in $\mathcal{A}'_B$. For this we recall Definitions 5.16 and (5.18), but here modifying the class of sets used from an algebra $\mathcal{A}$ to the class $\mathcal{A}'_B$.

As $\mathcal{A}'_B$ is not a semi-algebra, we must prove that μ_F^* so defined is an outer measure by Definition 6.1, to then be able to apply Proposition 6.2, the Carathéodory Extension Theorem I.

Proposition 8.14 (Defining an outer measure μ_F^***)** *Given a real-valued function* $F : \mathbb{R}^n \rightarrow \mathbb{R}$ *that is continuous from above by Definition 8.5, and satisfies the n-increasing condition of Definition 8.7, let* μ_0 *denote the associated set function on the class of bounded right semi-closed rectangles* $\mathcal{A}'_B$ *as given in (8.14). For any set* $A \subset \mathbb{R}^n$, *define the set function* $\mu_F^*(A)$ *by:*

$$\mu_F^*(A) = \inf \left\{ \sum_n \mu_0(A_n) \mid A \subset \bigcup A_n, \ A_n \in \mathcal{A}'_B \right\}. \tag{8.15}$$

*Then μ_F^* is an outer measure in the sense of Definition 6.1 and called the μ_F^*-**outer measure induced by** μ_0 and $\mathcal{A}_B'$.*

Proof. *As noted in the proof of Proposition 8.13, $\emptyset \in \mathcal{A}_B'$ and $\mu_0(\emptyset) = 0$, so $\mu_F^*(\emptyset) = 0$. Also, μ_F^* is monotonic since if $A \subset B$ and $B \subset \bigcup_n B_n$ then $A \subset \bigcup_n B_n$ and so $\mu_F^*(A) \leq \mu_F^*(B)$.*

For countable subadditivity, if $A = \bigcup_n A_n$ and $\mu_F^*(A_n) = \infty$ for some n, then $\mu_F^*(A) \leq \sum_n \mu_F^*(A_n)$, so assume $\mu_F^*(A_n) < \infty$ for all n. Given $\epsilon > 0$, for each n choose $\{A_{n,k}\}_k \subset \mathcal{A}_B'$ so that $A_n \subset \bigcup_k A_{n,k}$ and:

$$\sum_k \mu_F(A_{n,k}) \leq \mu_F^*(A_n) + \epsilon/2^n.$$

This is possible by the definition of infimum since $\mu_F^*(A_n) < \infty$. Then $A = \bigcup_n A_n \subset \bigcup_{n,k} A_{n,k}$, and by the definition of μ_F^*:

$$\mu_F^*(A) \leq \sum_{n,k} \mu_F(A_{n,k}) \leq \sum_n \mu_F^*(A_n) + \epsilon.$$

This inequality applies whether the summation on the right is convergent or not. Subadditivity now follows since ϵ is arbitrary.   ∎

The final step is to collect the steps offered by the extension theory. In the statement of this result, that μ_F is a Borel measure means subject to Definition 8.4, which requires more than that μ_F be defined on $\mathcal{B}(\mathbb{R}^n)$.

Proposition 8.15 (μ_0 extends to a Borel measure μ_F) *Given a real-valued function $F : \mathbb{R}^n \to \mathbb{R}$ that is continuous from above as in (8.3) of Definition 8.5, and satisfies the n-increasing condition of (8.5) of Definition 8.7, let μ_0 be the associated set function defined on the class of bounded right semi-closed rectangles $\mathcal{A}_B'$ by (8.14).*

Then μ_0 can be extended to a Borel measure μ_F on the Borel sigma algebra $\mathcal{B}(\mathbb{R}^n)$, and to a measure on the complete sigma algebra of Carathéodory measurable sets $\mathcal{M}_F(\mathbb{R}^n)$ with $\mathcal{B}(\mathbb{R}^n) \subset \mathcal{M}_F(\mathbb{R}^n)$.

Proof. *Recall Proposition 6.2, the Carathéodory Extension Theorem I, and denote by $\mathcal{M}_F(\mathbb{R}^n)$ the collection of all subsets of $\sigma(P(\mathbb{R}^n))$ that are Carathéodory measurable with respect to the outer measure μ_F^* by (6.1). Thus $A \in \mathcal{M}_F(\mathbb{R}^n)$ if for all $E \in \sigma(P(\mathbb{R}^n))$:*

$$\mu_F^*(E) = \mu_F^*(A \bigcap E) + \mu_F^*(\widetilde{A} \bigcap E). \tag{1}$$

Then $\mathcal{M}_F(\mathbb{R}^n)$ is a complete sigma-algebra by that theorem, and if μ_F denotes the restriction of μ_F^* to $\mathcal{M}_F(\mathbb{R}^n)$, then μ_F is a measure and hence $(\mathbb{R}^n, \mathcal{M}_F(\mathbb{R}^n), \mu_F)$ is a complete measure space.

We now prove that as defined on $M_F(\mathbb{R}^n)$, that μ_F extends the set function μ_0 defined on $\mathcal{A}_B'$. For this we show that $\mathcal{A}_B' \subset M_F(\mathbb{R}^n)$ and $\mu_F^*(A) = \mu_0(A)$ for all $A \in \mathcal{A}_B'$.

1. *$\mathcal{A}_B' \subset \mathcal{M}_F(\mathbb{R}^n)$: Let $A \in \mathcal{A}_B'$. If $\mu_F^*(E) = \infty$, then (1) is automatically satisfied by subadditivity of μ_F^*, so assume $\mu_F^*(E) < \infty$. As subadditivity also yields that $\mu_F^*(E) \leq \mu_F^*(A \bigcap E) + \mu_F^*(\widetilde{A} \bigcap E)$, only the reverse inequality must be addressed.*

Given $\epsilon > 0$, let $\{A_n\} \subset \mathcal{A}'_B$ be chosen so that $E \subset \bigcup_n A_n$ and $\sum_n \mu_0(A_n) \le \mu_F^*(E) + \epsilon$. This choice is possible by the definition of infimum since $\mu_F^*(E) < \infty$. Now $A \bigcap A_n \in \mathcal{A}'_B$ and since $A \bigcap A_n \subset A_n$, we recall the construction in part 3 of the proof of proposition 8.13. Specifically, there exists disjoint sets $\{B_{n,j}\}_{j=1}^{N_n} \subset \mathcal{A}'_B$ that are disjoint from $A \bigcap A_n$ so that:

$$A_n = \left(A \bigcap A_n\right) \bigcup \left(\bigcup_{j=1}^{N_n} B_{n,j}\right). \tag{2}$$

Now $E \subset \bigcup_n A_n$ and so $A \bigcap E \subset \bigcup_n (A \bigcap A_n)$. As $\tilde{A} \bigcap A_n = A_n - (A \bigcap A_n)$, the splitting of A_n in (2) obtains $\tilde{A} \bigcap A_n \subset \bigcup_{j=1}^{N_n} B_{n,j}$, and then from $E \subset \bigcup_n A_n$ it follows that $\tilde{A} \bigcap E \subset \bigcup_n \bigcup_{j=1}^{N_n} B_{n,j}$. Using the definition and countable subadditivity of μ_F^*, the finite additivity of μ_0 on $\mathcal{A}'_B$, and (2):

$$\mu_F^*\left(A \bigcap E\right) + \mu_F^*\left(\tilde{A} \bigcap E\right) \le \sum_n \mu_0\left(A \bigcap A_n\right) + \sum_n \sum_{j=1}^{N_n} \mu_0\left(B_{n,j}\right)$$

$$= \sum_n \mu_0\left[\left(A \bigcap A_n\right) \bigcup \left(\bigcup_{j=1}^{N_n} B_{n,j}\right)\right]$$

$$= \sum_n \mu_0\left(A_n\right)$$

$$\le \mu_F^*(E) + \epsilon.$$

The result follows since $\epsilon > 0$ is arbitrary.

2. *$\mu_F^*(A) = \mu_0(A)$ for all $A \in \mathcal{A}'_B$: If $\{A_n\}_n \subset \mathcal{A}'_B$ and $A \subset \bigcup_n A_n$ then $A = \bigcup_n (A \bigcap A_n)$, so using countable subadditivity then monotonicity:*

$$\mu_0(A) \le \sum_n \mu_0\left(A \bigcap A_n\right) \le \sum_n \mu_0(A_n),$$

and thus $\mu_0(A) \le \mu_F^*(A)$. But then $\mu_F^*(A) \le \mu_0(A)$ by (8.15), choosing only one set $A_n = A$.

In summary, $(\mathbb{R}^n, \mathcal{M}_F(\mathbb{R}^n), \mu_F)$ is a complete measure space, $\mathcal{A}'_B \subset \mathcal{M}_F(\mathbb{R}^n)$, and $\mu_F^*(A) = \mu_0(A)$ for all $A \in \mathcal{A}'_B$. As $\mathcal{A}'_B$ generates $\mathcal{B}(\mathbb{R}^n)$ by Proposition 8.1, it follows that $\mathcal{B}(\mathbb{R}^n) \subset \mathcal{M}_F(\mathbb{R}^n)$ and $(\mathbb{R}^n, \mathcal{B}(\mathbb{R}^n), \mu_F)$ is a measure space.

To prove that μ_F is a Borel measure by Definition 8.4 requires only a consideration of item 3 of that definition. If $K \subset \mathcal{B}(\mathbb{R}^n)$ is compact, then it is closed and bounded by the Heine-Borel theorem of Proposition 2.27, and so $K \subset A \in \mathcal{A}'_B$ for some bounded rectangle. Then by monotonicity of μ_F, and that $\mu_F = \mu_0$ on $\mathcal{A}'_B$, and (8.14):

$$\mu_F(K) \le \mu_0[A] \le \sum_x |F(x)| < \infty.$$

Finiteness follows as a finite sum of values of a real-valued function. ∎

8.3 Properties of Borel Measures on $\mathbb{R}^n$

In this section we investigate properties of general Borel measures on $\mathbb{R}^n$. The reader is reminded of the remark following Definition 8.4 that we require Borel measures to be **finite on compact sets**, a restriction not universally required by all authors but one that eliminates measures that are far outside the applications in these books. Examples of the behavior of measures that do not satisfy this requirement were given in Remark 5.2, and following Definition 8.4.

It should be noted that the results of this section automatically apply to all **Lebesgue and Borel product measures** introduced in Section 7.6. If μ is a Borel product measure of Chapter 7, derived from the Borel measure spaces $\{(\mathbb{R}, \mathcal{B}(\mathbb{R}), \mu_i)\}_{i=1}^n$, then by Proposition 7.20:

$$\mu\left[\prod_{i=1}^n (a_i, b_i]\right] = \prod_{i=1}^n \mu_i(a_i, b_i].$$

And as noted in Example 8.11, the induced function $\tilde{F}_\mu(x)$ of (8.12) satisfies:

$$\tilde{F}_\mu(x) = \prod_{i=1}^n F_{\mu_i}(x_i), \tag{1}$$

where $F_{\mu_i}(x_i)$ is the function induced by μ_i in (5.1). In the case of Lebesgue measures, $F_{\mu_i}(x_i) = x_i$ for all i.

Conversely, if we begin with $F(x)$ defined as in (1) in terms of $F(x_i)$, as a product of right continuous, increasing functions, then $F(x)$ is continuous from above as noted following Definition 8.5. Further, the Borel measure induced by $F(x)$ is a product measure of the measures induced by the $F(x_i)$. This follows because if $\prod_{i=1}^n (a_i, b_i] \in \mathcal{A}_B'$, then by (8.14):

$$\mu_F[A] \equiv \sum_x sgn(x)F(x)$$
$$= \prod_{i=1}^n (F(b_i) - F(a_i))$$
$$= \prod_{i=1}^n \mu_{F_i}(a_i, b_i].$$

This also proves that such $F(x)$ is n-increasing.

8.3.1 Uniqueness and Consistency

The development leading to Proposition 8.15 circumvented the algebra to sigma algebra approach of the Hahn-Kolmogorov extension theorem of Proposition 6.4. And this therefore calls into question the applicability of the uniqueness theorem of Proposition 6.14. In this section we prove that the extension of μ_0 to μ_F is unique, and then address the consistency of Borel constructions, extending the discussion of Section 5.3.

The approach taken in the following proof is to show that Proposition 6.14 does indeed apply because the above construction is equivalent to a construction via the associated algebra.

Proposition 8.16 (Uniqueness of μ_F**)** *Given a real-valued function* $F\colon\mathbb{R}^n \to \mathbb{R}$ *that is continuous from above as in Definition 8.5 and satisfies the n-increasing condition of Definition 8.7, let* μ_0 *be the associated set function defined on* $\mathcal{A}'_B$ *by* (8.14).

Then the Proposition 8.15 extension of μ_0 *to the Borel measure* μ_F *on* $\mathcal{B}(\mathbb{R}^n)$ *or* $\mathcal{M}_F(\mathbb{R}^n)$ *is unique. That is, if* μ *is a measure on* $\mathcal{B}(\mathbb{R}^n)$ *or* $\mathcal{M}_F(\mathbb{R}^n)$ *, respectively, with* $\mu = \mu_0$ *on* $\mathcal{A}'_B$ *, then* $\mu = \mu_F$ *on* $\mathcal{B}(\mathbb{R}^n)$ *or* $\mathcal{M}_F(\mathbb{R}^n)$ *, respectively.*

Proof. *Let* $\mathcal{A}'$ *denote the semi-algebra of right semi-closed rectangles:*

$$\mathcal{A}' \equiv \left\{ A \subset \mathbb{R}^n | A = \prod_{i=1}^{n} (a_i, b_i], \text{ with } -\infty \leq a_i \leq b_i \leq \infty \right\},$$

and $\mathcal{A}$ *the associated algebra of finite disjoint unions of* $\mathcal{A}'$*-sets.*

Define a set function $\bar{\mu}_0$ *on* $\mathcal{A}$ *by* $\bar{\mu}_0(A) = \mu_F(A)$*. Then since* $\mathcal{A} \subset \mathcal{B}(\mathbb{R}^n) \subset \mathcal{M}_F(\mathbb{R}^n)$*, it follows that* $\bar{\mu}_0$ *is a measure on* $\mathcal{A}$*, and since* $\mathcal{A}'_B \subset \mathcal{A}$ *we have* $\bar{\mu}_0(A) = \mu_0(A)$ *for all* $A \in \mathcal{A}'_B$*. Now if* $A \in \mathcal{A}$*, then* $A = \bigcup_{i=1}^{m} A_i$ *with disjoint* $A_i \in \mathcal{A}'$*. Further since* $\mathcal{A}'_B$ *generates* $\mathcal{A}'$ *by countable disjoint unions by Proposition 8.1, each such* $A_i = \bigcup_j A_i^{(j)}$ *with disjoint* $A_i^{(j)} \in \mathcal{A}'_B$*. Thus by countable additivity, for all* $A \in \mathcal{A}$*:*

$$\bar{\mu}_0(A) = \sum_{i=1}^{m} \sum_j \mu_0\left(A_i^{(j)}\right). \tag{1}$$

By Proposition 6.4, the measure $\bar{\mu}_0$ *on* $\mathcal{A}$ *gives rise to an outer measure* $\bar{\mu}_0^*$ *defined on* $B \in \sigma\left(P\left(\mathbb{R}^n\right)\right)$ *by* (6.3):

$$\bar{\mu}_0^*(B) = \inf\left\{ \sum_k \bar{\mu}_0(A_k) \mid B \subset \bigcup_k A_k \right\},$$

where $A_k \in \mathcal{A}$*. Then by* (1), $\bar{\mu}_0^*(B)$ *can be equivalently expressed:*

$$\bar{\mu}_0^*(B) = \inf\left\{ \sum_k \sum_{i=1}^{m} \sum_j \mu_0\left(A_{k,i}^{(j)}\right) \mid B \subset \bigcup_k \bigcup_{i=1}^{m} \bigcup_j A_{k,i}^{(j)} \right\}, \tag{2}$$

where $A_{k,i}^{(j)} \in \mathcal{A}'_B$*.*

As the union of $\mathcal{A}'_B$*-sets in* (2) *is countable, it follows that for all* $B \in \sigma\left(P\left(\mathbb{R}^n\right)\right)$*:*

$$\bar{\mu}_0^*(B) = \mu_F^*(B),$$

where $\mu_F^*(B)$ *is given in* (8.15). *Thus a set* $B \in \sigma\left(P\left(\mathbb{R}^n\right)\right)$ *is Carathéodory measurable with respect to the outer measure* $\bar{\mu}_0^*$ *if and only if it is Carathéodory measurable with respect to the outer measure* μ_F^**. With* $X = \mathbb{R}^n$*, this proves that the complete sigma algebra* $\mathcal{C}(\mathbb{R}^n)$ *of Proposition 6.4 satisfies* $\mathcal{C}(\mathbb{R}^n) = \mathcal{M}_F(\mathbb{R}^n)$*, and further that the measure* $\bar{\mu} \equiv \bar{\mu}_0^*$ *on* $\mathcal{C}(\mathbb{R}^n)$ *of this result satisfies* $\bar{\mu} = \mu_F$*.*

If μ *is a measure on* $\mathcal{B}(\mathbb{R}^n)$ *with* $\mu = \mu_0 = \bar{\mu}_0$ *on* $\mathcal{A}'_B$*, then by the same argument,* $\mu = \mu_0 = \bar{\mu}_0$ *on* $\mathcal{A}$*. Now* $\bar{\mu}_0$ *is a* σ*-finite measure on* $\mathcal{A}$ *since* $\mathbb{R}^n \in \mathcal{A}$*, and hence equals a countable union of disjoint* $\mathcal{A}'_B$*-sets by the above argument. But then all such sets have finite measure by the last paragraph of the proof of Proposition 8.15. Thus Proposition 6.14 applies to* *derive that $\mu = \mu_F$ on $\sigma(\mathcal{A})$, the smallest sigma algebra that contains $\mathcal{A}$, or equivalently by Proposition 8.1, $\mu = \mu_F$ on $\mathcal{B}(\mathbb{R}^n)$.*

Finally, assume that μ is a measure on $\mathcal{M}_F(\mathbb{R}^n)$ with $\mu = \mu_0$ on $\mathcal{A}$. Then as just proved, $\mu = \mu_F$ on $\sigma(\mathcal{A}) = \mathcal{B}(\mathbb{R}^n)$. Then by Proposition 6.24, $\mu^C = \mu_F$ on $\sigma^C(\mathcal{A}) = \mathcal{M}_F(\mathbb{R}^n)$, where C denotes completion with respect to μ. But since μ is a measure on $\mathcal{M}_F(\mathbb{R}^n)$ and $\mu = \mu^C$ on $\mathcal{B}(\mathbb{R}^n)$, it follows from the construction of Proposition 6.14 that $\mu = \mu^C$ on $\mathcal{M}_F(\mathbb{R}^n)$. ∎

The following corollary generalizes the discussion of Section 5.3.

Corollary 8.17 (Consistency of Borel measure constructions) *If μ is a Borel measure on $\mathcal{B}(\mathbb{R}^n)$, $\tilde{F}_\mu(x)$ the induced function defined in (8.12), and $\mu_{\tilde{F}_\mu}$ the associated Borel measure of Proposition 8.15, then:*

$$\mu = \mu_{\tilde{F}_\mu} \text{ on } \mathcal{B}(\mathbb{R}^n).$$

Further, if μ^C denotes the completion of μ of Proposition 6.20, and $\mathcal{B}^C(\mathbb{R}^n)$ the associated completion of $\mathcal{B}^C(\mathbb{R}^n)$, then $\mathcal{B}^C(\mathbb{R}^n) = \mathcal{M}_{\tilde{F}_\mu}(\mathbb{R}^n)$ and:

$$\mu^C = \mu_{\tilde{F}_\mu} \text{ on } \mathcal{M}_{\tilde{F}_\mu}(\mathbb{R}^n).$$

Conversely, given real-valued $F : \mathbb{R}^n \to \mathbb{R}$ that is continuous from above as in Definition 8.5 and satisfies the n-increasing condition of Definition 8.7, let μ_F be the associated Borel measure of Proposition 8.15, and $\tilde{F}_{\mu_F}(x)$ the induced function defined in (8.12). Then:

$$\tilde{F}_{\mu_F}(x) = F(x) + c \text{ for } c \in \mathbb{R}.$$

Proof. *By item 2 of Proposition 8.12 and (8.14) we obtain that $\mu = \mu_{\tilde{F}_\mu}$ on $\mathcal{A}'_B$, and so $\mu = \mu_{\tilde{F}_\mu}$ on $\mathcal{B}(\mathbb{R}^n)$ by proposition 8.16. Now μ^C is a measure on $\mathcal{B}^C(\mathbb{R}^n)$ by Proposition 6.20, and Proposition 6.24 obtains that $\mathcal{B}^C(\mathbb{R}^n) = \mathcal{M}_{\tilde{F}_\mu}(\mathbb{R}^n)$. Thus $\mu^C = \mu_{\tilde{F}_\mu}$ on $\mathcal{M}_{\tilde{F}_\mu} \mathbb{R}^n$.*

Given F as above, $\mu_F = \mu_{\tilde{F}_{\mu_F}}$ on $\mathcal{B}(\mathbb{R}^n)$ as was just proved, and thus in particular $\mu_F = \mu_{\tilde{F}_{\mu_F}}$ on $\mathcal{A}'_B$. Let $G(x) \equiv \tilde{F}_{\mu_F}(x) - F(x)$, and note that $G(x)$ is real-valued and continuous from above, and satisfies the n-increasing condition since given $A = \prod_{i=1}^n (a_i, b_i] \in \mathcal{A}'_B$:

$$\sum_x sgn(x)G(x) = \sum_x sgn(x)\tilde{F}_{\mu_F}(x) - \sum_x sgn(x)F(x) = 0.$$

This summation is over all 2^n vertices of this rectangle, with $sgn(x)$ defined as -1 if the number of a_i-components of x is odd, and otherwise is defined as $+1$. This equality holds if $b_i = a_i$ for one or more i, and also holds for $A = \prod_{i=1}^n (a_i, b_i]^+$ recalling (8.8), and this allows $b_i < a_i$ for one or more i.

Now define the operator $\Delta_{\Delta a_i}$ on G by:

$$\Delta_{\Delta a_i}G(a) = G(a_1, ..., a_i + \Delta a_i, a_{i+1}, ..., a_n),$$

and consider the operator $\Delta \equiv \prod_{i=1}^{n} \left(\Delta_{\Delta a_i} - 1 \right)$. Note that:

$$\prod_{i=1}^{n} \left(\Delta_{\Delta a_i} - 1 \right) G(a) = \sum_x sgn(x) G(x),$$

where the summation on the right is defined relative to $A = \prod_{i=1}^{n} (a_i, a_i + \Delta a_i]^+$.
Thus $\prod_{i=1}^{n} \left(\Delta_{\Delta a_i} - 1 \right) G(a) = 0$ for all a and all $\{\Delta a_i\}$, and for nonzero $\{\Delta a_i\}$:

$$\prod_{i=1}^{n} \left(\frac{\Delta_{\Delta a_i} - 1}{\Delta a_i} \right) G(a) = 0.$$

We can then take limits of $\Delta a_i \to 0$ in any order and obtain that for all x:

$$\frac{\partial^n G}{\partial x_1 \dots \partial x_n} = 0.$$

Thus $G(x) = c$ a constant, and the proof is complete. ∎

It should be noted that the above proof does not imply that if $F : \mathbb{R}^n \to \mathbb{R}$ satisfies the n-increasing condition and is continuous from above, that F is differentiable in the sense of the existence of $\frac{\partial^n F}{\partial x_1 \dots \partial x_n}$. It simply states that given any two such functions derived from the same measure by (8.12), that the difference is so differentiable.

8.3.2 Approximating Borel Measurable Sets

With the derivation in the proof of Proposition 8.16, we are now able to quote the approximation result of Proposition 6.5 in the current context.

Proposition 8.18 (Approximations with Borel sub/supersets) *Let $\mathcal{A}$ denote the algebra of finite disjoint unions of right semi-closed rectangles. If $B \in \mathcal{M}_F(\mathbb{R}^n)$, then given $\epsilon > 0$:*

1. *There is a set $A \in \mathcal{A}_\sigma$, the collection of countable unions of sets in the algebra $\mathcal{A}$, so that $B \subset A$ and:*

$$\mu_F(A) \le \mu_F(B) + \epsilon, \qquad \mu_F(A - B) \le \epsilon. \tag{8.16}$$

2. *There is a set $C \in \mathcal{A}_\delta$, the collection of countable intersections of sets in the algebra $\mathcal{A}$, so that $C \subset B$ and:*

$$\mu_F(B) \le \mu_F(C) + \epsilon, \qquad \mu_F(B - C) \le \epsilon. \tag{8.17}$$

3. *There is a set $A' \in \mathcal{A}_{\sigma\delta}$, the collection of countable intersections of sets in $\mathcal{A}_\sigma$, and $C' \in \mathcal{A}_{\delta\sigma}$, the collection of countable unions of sets in $\mathcal{A}_\delta$, so that $C' \subset B \subset A'$ and:*

$$\mu_F(A' - B) = \mu_F(B - C') = 0. \tag{8.18}$$

Proof. *Recalling the first few paragraphs of the proof of Proposition 8.16, we obtained the conclusion that with $X = \mathbb{R}^n$, the complete sigma algebra $\mathcal{C}(\mathbb{R}^n)$ of Proposition 6.4 satisfies $\mathcal{C}(\mathbb{R}^n) = \mathcal{M}_F(\mathbb{R}^n)$, and the measure $\bar{\mu} \equiv \bar{\mu}_0^*$ on $\mathcal{C}(\mathbb{R}^n)$ of that result satisfies $\bar{\mu} = \mu_F$. In other words, the Proposition 8.15 construction of the complete measure space $(\mathbb{R}^n, \mathcal{M}_F(\mathbb{R}^n), \mu_F)$ is consistent with the general construction of Proposition 6.4. Thus Proposition 6.5 applies as restated above.* ∎

8.3.3 Continuity and Regularity

Borel measures on $\mathbb{R}^n$, like all measures, are **continuous from above** and **continuous from below** by Proposition 2.45, so we only reference this property here for completeness.

Regularity properties of Lebesgue and Borel measures on $\mathbb{R}$ were addressed in Sections 2.7 and 5.5, respectively. The goal of this section is to extend these results and show that Borel measures on $\mathbb{R}^n$ are also **regular**. As noted in the introduction, this result and those above then also apply to Borel and Lebesgue product measures.

To simplify the somewhat long proof below, we independently state and prove a technical result that will be needed. It provides a refinement of the results in Proposition 2.20.

Proposition 8.19 (Rectangle unions to disjoint rectangle unions) *Given a countable collection of bounded right semi-closed rectangles $\{C_j\}_{j=1}^\infty \subset \mathcal{A}_B'$, there exists a disjoint collection of bounded right semi-closed rectangles $\{B_j\}_{j=1}^\infty \subset \mathcal{A}_B'$ so that:*

$$\bigcup_{j=1}^\infty C_j = \bigcup_{j=1}^\infty B_j.$$

Further, for each m there exists $\{j_k^{(m)}\}_{k=1}^{N(m)}$ with $N(m) \le \infty$ so that:

$$\bigcup_{j=1}^m C_j = \bigcup_{k=1}^{N(m)} B_{j_k^{(m)}}.$$

Proof. *Let $C_j = \prod_{i=1}^n (a_i^{(j)}, b_i^{(j)}]$ with $-\infty < a_i^{(j)} \le b_i^{(j)} < \infty$. For each i, order the distinct points in the set $\{a_i^{(j)}, b_i^{(j)}\}_{j=1}^\infty$ and relabel them $\{c_i^{(j)}\}_{j=-\infty}^\infty$ where $c_i^{(j)} \le 0$ for $j \le 0$ and $c_i^{(j)} > 0$ for $j \ge 1$. In other words, $c_i^{(j)} < c_k^{(j)}$ if $i < k$. This $c_i^{(j)}$-set can be finite or countable, so the notation reflects the general case.*

Then $\{(c_k^{(j)}, c_{k+1}^{(j)}]\}_k$ is a finite or countable disjoint collection. By construction, each interval $(c_k^{(j)}, c_{k+1}^{(j)}]$ is either contained in a given $(a_i^{(j)}, b_i^{(j)}]$ or is disjoint from this interval. Further, each $(a_i^{(j)}, b_i^{(j)}]$ is a finite or countable union of such intervals:

$$(a_i^{(j)}, b_i^{(j)}] = \bigcup_{k=m_j(i)}^{n_j(i)} (c_k^{(j)}, c_{k+1}^{(j)}],$$

and so $-\infty \le m_j(i) < n_j(i) \le \infty$.

Hence:

$$C_j = \prod_{i=1}^{n} (a_i^{(j)}, b_i^{(j)}]$$
$$= \prod_{i=1}^{n} \bigcup_{k=m_j(i)}^{n_j(i)} (c_k^{(j)}, c_{k+1}^{(j)}]$$
$$= \bigcup_{K_j} \prod_{i=1}^{n} (c_{k(i)}^{(j)}, c_{k(i)+1}^{(j)}],$$

where $K_j = \{(k_j(1), ..., k_j(n) | m_j(i) \le k_j(i) \le n_j(i)\}$. This K_j-union is disjoint by construction, and so each C_j is a finite or countable disjoint union of bounded right semi-closed rectangles.
   Thus:

$$\bigcup_{j=1}^{\infty} C_j = \bigcup_{j=1}^{\infty} \bigcup_{K_j} \prod_{i=1}^{n} (c_{k(i)}^{(j)}, c_{k(i)+1}^{(j)}], \tag{1}$$

and we note that the double union in (1) contains at most countably many bounded rectangles.
   We can now define the $\{B_j\}_{j=1}^{\infty} \subset \mathcal{A}'_B$ iteratively, the goal of which is to include every distinct rectangle in (1). First, we include the finite (or countable) collection $\{\prod_{i=1}^{n} (c_{k(i)}^{(j)}, c_{k(i)+1}^{(j)}]\}_{K_1}$. Then add every unique set from the collection $\{\prod_{i=1}^{n} (c_{k(i)}^{(j)}, c_{k(i)+1}^{(j)}]\}_{K_2}$, then every unique set in $\{\prod_{i=1}^{n} (c_{k(i)}^{(j)}, c_{k(i)+1}^{(j)}]\}_{K_3}$, and so forth. By construction this is a disjoint countable collection of $\mathcal{A}'_B$-sets for which every C_j is a union, and the proof is complete. ∎

We are now ready for the main result.

Proposition 8.20 (Regularity of Borel Measures) *Borel measure μ_F is **regular** on $\mathcal{M}_F(\mathbb{R}^n)$. Specifically, if $A \in \mathcal{M}_F(\mathbb{R}^n)$:*

   1. *μ_F is **outer regular:***

$$\mu_F(A) = \inf_{A \subset G} \mu_F(G), \quad G \text{ open.} \tag{8.19}$$

   2. *μ_F is **inner regular:***

$$\mu_F(A) = \sup_{F \subset A} \mu_F(F), \quad F \text{ compact.} \tag{8.20}$$

Proof. 1. *Outer Regularity:* *If $\mu_F(A) = \infty$, then (8.19) follows by monotonicity of μ_F, since then $\mu_F(G) = \infty$ for any G with $A \subset G$.*
   So assume $A \in \mathcal{M}_F(\mathbb{R}^n)$ and $\mu_F(A) < \infty$. If $A = \prod_{i=1}^{n} (a_i, b_i] \in \mathcal{A}'_B$, a bounded right semi-closed rectangle, let $G_m = \prod_{i=1}^{n} (a_i, b_i + 1/m]$. It then follows that G_m is open, $A \subset G_m$, and for any $\epsilon' > 0$ there is an $m(\epsilon')$ so that:

$$\mu_F(A) \le \mu_F(G_{m(\epsilon')}) \le \mu_F(A) + \epsilon', \quad A \in \mathcal{A}'_B. \tag{1}$$

The first inequality follows from monotonicity of μ_F, and the second from continuity from above since $\bigcap_m G_m = A$. Continuity from above of Proposition 2.45 applies here since $G_1 \subset \prod_{i=1}^n [a_i, b_i + 1]$, a compact set with finite measure by Definition 8.4. Hence, (8.19) follows for $A \in \mathcal{A}'_B$.

For general measurable A with $\mu_F(A) < \infty$, Proposition 8.18 applies so that for any $\epsilon > 0$ there is a set $B \in \mathcal{A}_\sigma$, the collection of countable unions of sets in the algebra $\mathcal{A}$, so that $A \subset B$ and:

$$\mu_F(A) \leq \mu_F(B) \leq \mu_F(A) + \epsilon. \tag{2}$$

Recall that $\mathcal{A}$ is the algebra of finite disjoint unions of right semi-closed rectangles, and each such right semi-closed interval is at most a countable disjoint union of $\mathcal{A}'_B$-sets. Thus $B \in \mathcal{A}_\sigma$ is a countable union of bounded right semi-closed rectangles:

$$B = \bigcup_{j=1}^\infty B_j, \quad B_j \in \mathcal{A}'_B.$$

We can assume the B_j-sets are disjoint by Proposition 8.19.

Applying (1) to each B_j using $\epsilon' = \epsilon/2^j$ obtains open G_j with $B_j \subset G_j$ and:

$$\mu_F(B_j) \leq \mu_F(G_j) \leq \mu_F(B_j) + \epsilon/2^j. \tag{3}$$

With $G_\epsilon \equiv \bigcup_{j=1}^\infty G_j$, then $A \subset B \subset G_\epsilon$ obtains $\mu_F(A) \leq \mu_F(B) \leq \mu_F(G_\epsilon)$. Further, $\{B_j\}$ are disjoint so $\mu_F(B) = \sum_{j=1}^\infty \mu_F(B_j)$, while by countable subadditivity and (3):

$$\mu_F(G_\epsilon) \leq \sum_{j=1}^\infty \mu_F(G_j) \leq \mu_F(B) + \epsilon.$$

Combining these results with (2):

$$\mu_F(A) \leq \mu_F(G_\epsilon) \leq \mu_F(A) + 2\epsilon. \tag{4}$$

This proves (8.19) for $\mu_F(A) < \infty$ with an infimum over such G_ϵ. But $\mu_F(A) \leq \mu_F(G)$ for any G with $A \subset G$, and so (8.19) follows with an infimum over all open G.

Since $G_\epsilon = A \bigcup (G_\epsilon - A)$, a disjoint union, the final inequality in (4) also shows that:

$$\mu_F(G_\epsilon - A) \leq 2\epsilon, \tag{5}$$

a result that will be used below.

   2. Inner Regularity: *To prove (8.20), apply the result in (4) to $\widetilde{A}$ to obtain an open G_ϵ so that $\widetilde{A} \subset G_\epsilon$ and:*

$$\mu_F(\widetilde{A}) \leq \mu_F(G_\epsilon) \leq \mu_F(\widetilde{A}) + 2\epsilon.$$

Define closed $F_\epsilon = \widetilde{G_\epsilon}$. Then $F_\epsilon \subset A$ and:

$$A - F_\epsilon = A \bigcap G_\epsilon = G_\epsilon - \widetilde{A}.$$

Since $G_\epsilon = \tilde{A} \bigcup \left(G_\epsilon - \tilde{A}\right)$, *(5) yields* $\mu_F(A - F_\epsilon) \leq 2\epsilon$, *and then* $A = F_\epsilon \bigcup (A - F_\epsilon)$ *obtains:*

$$\mu_F(A) - 2\epsilon \leq \mu_F(F_\epsilon) \leq \mu_F(A). \tag{6}$$

This proves (8.20) with closed F_ϵ, *and this extends to all closed subsets of A by monotonicity.*

Now define $F_\epsilon^{(m)} = F_\epsilon \bigcap I_m$ *with* $I_m \equiv \{x \in \mathbb{R}^n | m - 1 \leq \sum_{j=1}^n |x_j| < m\}$. *Then* $F_\epsilon^{(m)} \subset F_\epsilon$ *for all m, and* $F_\epsilon^{(m)}$ *is bounded. Also* $F_\epsilon = \bigcup_m F_\epsilon^{(m)}$, *a disjoint union, so countable additivity obtains:*

$$\mu_F(F_\epsilon) = \sum_{m=1}^\infty \mu_F\left(F_\epsilon^{(m)}\right). \tag{7}$$

If $\mu_F(A) < \infty$, *then* $\mu_F(F_\epsilon) < \infty$ *by (6), and so by (7) there is an N with* $\sum_{m=N+1}^\infty \mu_F\left(F_{\delta(\epsilon)}^{(m)}\right) < \epsilon$. *Then by countable additivity and monotonicity:*

$$
\begin{aligned}
\epsilon &> \sum_{m=N+1}^\infty \mu_F\left(F_\epsilon^{(m)}\right) \\
&= \mu_F\left(F_\epsilon \bigcap \left\{N \leq \sum_{j=1}^n |x_j|\right\}\right) \\
&\geq \mu_F\left(F_\epsilon \bigcap \left\{N + 1 < \sum_{j=1}^n |x_j|\right\}\right).
\end{aligned} \tag{8}
$$

Define $\bar{F}_\epsilon^N = F_\epsilon \bigcap \left\{\sum_{j=1}^n |x_j| \leq N + 1\right\}$. *Then* $\bar{F}_\epsilon^N \subset F_\epsilon$ *is closed and bounded and hence compact, and* $\mu_F(F_\epsilon) - \mu_F(\bar{F}_\epsilon^N) < \epsilon$ *by (8). Combining with (6) obtains:*

$$\mu_F(A) - 3\epsilon \leq \mu_F(\bar{F}_\epsilon^N) \leq \mu_F(A).$$

This proves (8.20) with compact $\bar{F}_\epsilon^N$, *and then for all compact* $F \subset A$ *by monotonicity.*

If $\mu_F(A) = \infty$ *then* $\mu_F(F_\epsilon) = \infty$ *by (6), and thus* $\sum_{m=1}^N \mu_F\left(F_\epsilon^{(m)}\right)$ *is unbounded in N by (7). But with* $\bar{F}_\epsilon^N$ *as above:*

$$
\begin{aligned}
\bigcup_{m=1}^N F_\epsilon^{(m)} &= F_\epsilon \bigcap \left\{\sum_{j=1}^n |x_j| < N\right\} \\
&\subset \bar{F}_\epsilon^N.
\end{aligned}
$$

Thus by monotonicity, $\mu_F(\bar{F}_\epsilon^N)$ *is unbounded with compact* $\bar{F}_\epsilon^N$ *and (8.20) is proved in this case.* ∎

9

Infinite Product Spaces

In this chapter, infinite dimensional product measures spaces are studied, where infinite meants countably infinite. Thus the finite products of measure spaces of Chapter 7 are generalized to **countable products**. In the same way that points in a finite product space can be identified with n-tuples of variates, points in countably infinite product spaces are identified with sequences of variates.

As will be seen, this generalization is far from straightforward. Having infinitely many dimensions raises definitional issues almost immediately for the definition of measurable rectangles. There are also technical difficulties in the proof of countable additivity on an algebra (see Section 9.5.1 for a summary), and for this we require a more general algebra formulation than that provided by the measurable rectangles.

Though we begin with a more general investigation, the final steps of this development will be limited to **products of probability spaces on** $\mathbb{R}$. The restriction to finite measure spaces or probability spaces is near universal. Our restriction to probability spaces on $\mathbb{R}$ can be relaxed somewhat to more general probability spaces. For example, if the component spaces are probability spaces on $\mathbb{R}^n$, the development below can largely be implemented with additional notational complexity, but otherwise no new ideas.

For more generality, these probability spaces must be assumed to be complete metric spaces, and the associated probability measures must then be assumed to be inner regular. As $\mathbb{R}$ (and $\mathbb{R}^n$) are complete metric spaces, and all probability measures on these spaces are Borel measures and thus regular, we obtain the necessary assumptions without explicit mention.

Uncountable products of probability spaces, the realm of stochastic processes, are studied beginning in Book VII.

9.1 Naive Attempt at a First Step

We begin by informally addressing the generalization of Chapter 7 to the case of $n = \infty$. Given measure spaces $\{(X_i, \sigma(X_i), \mu_i)\}_{i=1}^{\infty}$, define the **product space** $X = \prod_{i=1}^{\infty} X_i$ by:

$$X = \{(x_1, x_2, \ldots) | x_i \in X_i\}.$$

Next, **temporarily define a measurable rectangle** in X as a set A, where:

$$A = \prod_{i=1}^{\infty} A_i, \text{ with } A_i \in \sigma(X_i),$$

DOI: 10.1201/9781003257745-9

and denote by $\mathcal{A}'$ **the collection of measurable rectangles** in X. Finally, given a measurable rectangle $\prod_{i=1}^{\infty} A_i$, define a **product set function** μ_0 by:

$$\mu_0(A) = \prod_{i=1}^{\infty} \mu_i(A_i).$$

With this set-up it is natural to think that the major challenge of implementing the development of Chapter 7 will be working with the infinite products needed in the definition of the product set function. But it is not difficult to characterize when such products make sense.

Exercise 9.1 (On infinite products) *Given $\{p_n\}_{n=1}^{\infty}$ with $p_n > 0$ for all n, consider $\prod_{n=1}^{\infty} p_n$. Show that:*

1. *$\prod_{n=1}^{\infty} p_n$ has a well defined value in $(0, \infty)$ if and only if $\sum_{n=1}^{\infty} \ln p_n$ converges.*
2. *Generalize part 1 to characterize when $\prod_{n=1}^{\infty} p_n = 0$ or $\prod_{n=1}^{\infty} p_n = \infty$.*
3. *Given an example to show how $\prod_{n=1}^{\infty} p_n$ need not converge.*

While it is one thing to understand when the above definition of μ_0 makes sense, it is yet another to develop a sigma algebra on which μ_0 can be extended to be a measure. In fact, the major challenge that arises with this framework is the status of $\mathcal{A}'$ as a collection of sets, and specifically whether it is a semi-algebra that can support the extension of μ_0 from $\mathcal{A}'$ to a sigma algebra that contains it.

We now show that $\mathcal{A}'$ **is not a semi-algebra.**

1. $\mathcal{A}'$ is closed under finite intersections: As was the case in the finite dimensional product spaces, if $A, B \in \mathcal{A}'$, then $A \bigcap B \in \mathcal{A}'$ since with the above notation:

$$A \bigcap B = \prod_{i=1}^{\infty} \left(A_i \bigcap B_i \right).$$

2. If $A \in \mathcal{A}'$, then $\tilde{A}$ is not in general a finite disjoint union of $\mathcal{A}'$-sets: More specifically, to express $\tilde{A}$ as a disjoint union can in general require **uncountably** many $\mathcal{A}'$-sets. To see this, if $A = \prod_{i=1}^{\infty} A_i$, then since $X = \prod_{i=1}^{\infty} X_i$ and $X_i = A_i \bigcup \tilde{A}_i$:

$$X = \prod_{i=1}^{\infty} \left(A_i \bigcup \tilde{A}_i \right)$$
$$= \bigcup_I \prod_{i=1}^{\infty} D_i,$$

where the union is over the set of all possible **indicator sequences,** I. Each sequence identifies the associated component sets of the product in that for each i, D_i equals A_i or $\tilde{A}_i$. Hence formally:

$$\tilde{A} = X - A$$
$$= \bigcup_I' \prod_{i=1}^{\infty} D_i,$$

where this union is now over all indicator sets for which at least one $D_i = \tilde{A}_i$. Of course, if $D_i \neq D'_i$ for any i:

$$\prod_{i=1}^{\infty} D_i \bigcap \prod_{i=1}^{\infty} D'_i = \emptyset,$$

so $\tilde{A}$ is here expressed as a disjoint union.

Thus in general, $\tilde{A}$ is an uncountable union of disjoint sets since I can be put in one-to-one correspondence with all real numbers in $[0, 1)$. To see this, let $x \in [0, 1)$, and express x in base 2 by:

$$x_{(2)} = 0.x(1)x(2).......,$$

where each $x(i)$ equals 0 or 1. To make this assignment uniquely defined, we choose the finite expansion over the infinite expansion when there is the choice. For example, $0.10000... = 0.01111....$ and we chose the first representation.

There is then a one-to-one correspondence between all such $x_{(2)} \in [0, 1)$ and the collection of all indicator sequences in the above union by identifying $x(i) = 0$ with $D_i = \tilde{A}_i$ and $x(i) = 1$ with $D_i = A_i$. Hence, this union representation for $\tilde{A}$ is in general an uncountable union. For obvious reasons this is not an auspicious start for the development of a measure space in which only countably many operations are contemplated.

Remark 9.2 (A solution within the problem) *While $\tilde{A}$ is in general an uncountable union, it will be a finite union in certain cases. Specifically, if $A_i = X_i$ for $i > n$, then $\tilde{A}$ is a finite union of $\mathcal{A}'$-sets as proved in Proposition 9.6. This insight will alter the development below, where we will begin with a less ambitious collection of measurable rectangles that will generate the desired sigma algebra and measure space.*

Then, perhaps ironically, the last investigation of this chapter results in (9.9) of Proposition 9.24, which will confirm that our naive attempt to define $\mu\left(\prod_{i=1}^{\infty} A_i\right)$ above, is the correct answer after all.

9.2 Semi-Algebra $\mathcal{A}'$

Given the above discussion, the development of an infinite dimensional measure space must begin with a more modest collection of measurable rectangles, and one which at the minimum will ensure that only countably many set operations will be required. As noted in Remark 9.2, the idea is to define a measurable rectangle as one for which only finitely many components are restricted, and thus the remaining components are taken to be X_i. It will then be the case that the above problem with uncountable unions in the representation of $\tilde{A}$ for $A \in \mathcal{A}'$ will be avioded.

Definition 9.3 (Product space; cylinder sets) *Given measure spaces:*

$$\{(X_i, \sigma(X_i), \mu_i)\}_{i=1}^{\infty},$$

*define the **product space*** $X = \prod_{i=1}^{\infty} X_i$ *by:*

$$X = \{(x_1, x_2, \ldots) | x_i \in X_i\}. \tag{9.1}$$

*A **finite dimensional measurable rectangle** A in X, also called a **cylinder set**, is defined for any n and n-tuple of positive integers $J = (j(1), j(2), \ldots, j(n))$ by:*

$$A = \{x \in X | x_{j(i)} \in A_{j(i)}\}, \tag{9.2}$$

*where $A_{j(i)} \in \sigma(X_{j(i)})$. The cylinder set in (9.2) will be said **to be defined by** J and $\prod_{i=1}^{n} A_{j(i)}$.*

The collection of cylinder sets in X is denoted by $\mathcal{A}'$.

*The **product set function** μ_0 is defined on $\mathcal{A}'$ as follows. If $A \in \mathcal{A}'$ is defined by J and $\prod_{i=1}^{n} A_{j(i)}$, then:*

$$\mu_0(A) = \prod_{i=1}^{n} \mu_{j(i)}(A_{j(i)}). \tag{9.3}$$

Notation 9.4 (Projection maps) *There seems to be no best notational convention for displaying the fact that a cylinder set A is defined by n, a given n-tuple J, and a set $\prod_{i=1}^{n} A_{j(i)}$. Of course, this defining set is a measurable rectangle in the sense of (7.2) applied to the finite dimensional product space of $\prod_{i=1}^{n} X_{j(i)}$. One notational approach, though cumbersome, is to denote A by $A\left(\prod_{i=1}^{n} A_{j(i)}\right)$. This notation suggests that A is a function from the semi-algebras of measurable rectangles of all finite dimensional product spaces $\prod_{i=1}^{n} X_{j(i)}$, and this is an intuitive description of how such measurable rectangles in these finite dimensional product spaces give rise to cylinder sets in the infinite dimensional product space.*

*A related and intuitively better approach is to use **coordinate functions** or **projection maps**, $\{\pi_j\}_{j=1}^{\infty}$, defined on $x \in X$ by:*

$$\pi_j(x) \equiv x_j.$$

In other words π_j identifies the jth component of x, which by the above notation is an element of the space X_j.

Thus π_j is a mapping:

$$\pi_j : X \to X_j,$$

and if $A_j \subset X_j$, then $\pi_j^{-1}(A_j)$ is the cylinder set defined by $J = \{j\}$ and A_j in the sense that:

$$\pi_j^{-1}(A_j) = \{x \in X | x_j \in A_j\}.$$

More generally, given n-vector $J \equiv (j(1), \ldots, j(n))$, define the projection mapping:

$$\pi_J \equiv \prod_{i=1}^{n} \pi_{j(i)},$$

by:

$$\pi_J : X \to \prod_{i=1}^{n} X_{j(i)},$$

where:

$$\pi_J(x) = (x_{j(1)}, ..., x_{j(n)}).$$

*Then given a measurable rectangle $\prod_{i=1}^{n} A_{j(i)} \subset \prod_{i=1}^{n} X_{j(i)}$, the **cylinder set** A defined by J and $\prod_{i=1}^{n} A_{j(i)}$ can be represented:*

$$A = \pi_J^{-1} \left(\prod_{i=1}^{n} A_{j(i)} \right). \tag{9.4}$$

Then $\mathcal{A}'$ is a countable union over all such J:

$$\mathcal{A}' = \bigcup_J \pi_J^{-1} \left(\mathcal{A}'_J \right),$$

where $\mathcal{A}'_J$ is the semi-algebra of measurable rectangles in $\prod_{i=1}^{n} X_{j(i)}$ as given in Definition 7.1. It is an exercise to show that A can also be defined as as the intersection of cylinder sets:

$$A = \bigcap_{i=1}^{n} \pi_{j(i)}^{-1}(A_{j(i)}). \tag{9.5}$$

Remark 9.5 (Nonuniqueness in $\mathcal{A}'$)

1. *A potential issue that arises now and must be addressed below is that a set $A \in \mathcal{A}'$ can be defined an infinite number of ways. Indeed, given A defined as in (9.2), we also have for any index $j(n+1) \notin (j(1), j(2), ..., j(n))$ that:*

$$A = \{x \in X | x_{j(i)} \in A_{j(i)}, i \leq n; \; x_{j(n+1)} \in X_{j(n+1)}\}.$$

   The advantage of this nonuniqueness is that when needed or even if just convenient, there is no loss of generality in assuming that the index set is the fixed set:

$$J = (1, 2, ..., N),$$

   where $N \geq \max[j(i)]$.

   The key point to remember is that in (9.2), a given A_i set can be constraining, meaning $A_i \subsetneq X_i$, or nonconstraining, so $A_i = X_i$. But that said, a cylinder set can only identify and specify conditions, even if vacuous, on a finite number of indices.

2. *Given this lack of uniqueness in the potential index set identified in item 1, there is also a potential problem with the Definition of 9.3 which must be addressed below. For example, if $\mu_{j(n+1)}(X_{j(n+1)}) = \infty$, this will create a well-definedness problem for (9.3).*

3. With A as in (9.2), note that $\widetilde{A}$ is given by J and $\widetilde{\prod_{i=1}^{n} A_{j(i)}}$. As seen in Proposition 7.2 and in (1) in the proof below, $\widetilde{\prod_{i=1}^{n} A_{j(i)}}$ is formally a union of $2^n - 1$ measurable rectangles. Given point 1 above, it is worth a moment to reflect on the impact on $\widetilde{A}$ of adding $A_{j(n+1)} = X_{j(n+1)}$ to the definition of A. It then follows that $\widetilde{\prod_{i=1}^{n+1} A_{j(i)}}$ is formally a union of $2^{n+1} - 1$ measurable rectangles, each of which equals $\prod_{i=1}^{n+1} D_{j(i)}$ where each $D_{j(i)}$ equals $A_{j(i)}$ or $\widetilde{A}_{j(i)}$, and this union omits the case where $D_{j(i)} = A_{j(i)}$ for all i. Of course $\widetilde{X}_{j(n+1)} = \emptyset$ and thus any such rectangle that contains $\widetilde{X}_{j(n+1)}$ is also empty.

   Of these $2^{n+1} - 1$ measurable rectangles, there are 2^n with $D_{j(n+1)} = \widetilde{X}_{j(n+1)} = \emptyset$, since all possibilities are allowed for $1 \leq i \leq n$, while for $D_{j(n+1)} = X_{j(n+1)}$ there $2^n - 1$, since all possibilities are allowed for $1 \leq i \leq n$ except $D_{j(i)} = A_{j(i)}$ for all $i \leq n$. Thus omitting empty sets, this formal union reduces to a union of $2^{n+1} - 1 - 2^n = 2^n - 1$ measurable rectangles. These are then seen to be the measurable rectangles of $\widetilde{\prod_{i=1}^{n} A_{j(i)}}$, each with the added component of $D_{j(n+1)} = X_{j(n+1)}$.

   In summary, this definitional ambiguity has no effect on complementation.

4. Finally, note that if $A \in \mathcal{A}'$ is defined by J and $\prod_{i=1}^{n} A_{j(i)}$, then (9.3) states that the μ_0-measure of $A \in \mathcal{A}'$ equals the μ_0-measure of $\prod_{i=1}^{n} A_{j(i)} \in \mathcal{A}'_J$ as given in Definition 7.1. The nonuniqueness in the representation of A raises a question that is addressed below.

As a first step in the development of a measure space, the following result will likely not surprise.

Proposition 9.6 ($\mathcal{A}'$ is a semi-algebra) *The collection of cylinder sets $\mathcal{A}'$ is a semi-algebra.*

Proof. First, $\emptyset, X \in \mathcal{A}'$ since for any j we have $\emptyset, X_j \in \sigma(X_j)$, and so by definition $\emptyset = \pi_j^{-1}(\emptyset)$ and $X = \pi_j^{-1}(X_j)$. Now given A in (9.2) and B defined by K and $\prod_{i=1}^{m} B_{k(i)}$, consider $A \bigcap B$. Let $\{l(i)\}$ denote the union of the coordinate index sets, noting that this set has $N \leq n + m$ indexes. Define $A'_{l(i)} = A_{j(i)}$ if $l(i) = j(i)$ for some i, and $A'_{l(i)} = X_{l(i)}$ otherwise, and similarly for $B'_{l(i)}$. Then by Remark 9.5, A is given by $L \equiv (1, ..., N)$ and $\prod_{i=1}^{N} A'_{l(i)}$, B is given by L and $\prod_{i=1}^{N} B'_{l(i)}$, and:

$$A \bigcap B = \pi_L^{-1}\left[\prod_{i=1}^{N} A'_{l(i)}\right] \bigcap \pi_L^{-1}\left[\prod_{i=1}^{N} B'_{l(i)}\right]$$

$$= \pi_L^{-1}\left[\prod_{i=1}^{N}\left(A'_{l(i)} \bigcap B'_{l(i)}\right)\right].$$

Thus $A \bigcap B \in \mathcal{A}'$.

   With A as in (9.2), $\widetilde{A}$ is given by J and $\widetilde{\prod_{i=1}^{n} A_{j(i)}}$ as note in Remark 9.5. Now $X_{j(i)} = A_{j(i)} \bigcup \widetilde{A}_{j(i)}$, and we obtain the finite disjoint partition of $\prod_{i=1}^{n} X_{j(i)}$:

$$\prod_{i=1}^{n} X_{j(i)} = \prod_{i=1}^{n} A_{j(i)} \bigcup \left[\bigcup' \prod_{i=1}^{n} D_{j(i)}\right],$$

where each $D_{j(i)}$ equals $A_{j(i)}$ or $\widetilde{A}_{j(i)}$, and the union is formally over all $2^n - 1$ possibilities excluding the case where $D_{j(i)} = A_{j(i)}$ for all i. Here, "formally" means that some rectangles may be empty as noted in item 3 of Remark 9.5. This obtains

$$\widetilde{\prod_{i=1}^{n} A_{j(i)}} = \bigcup{}' \prod_{i=1}^{n} D_{j(i)}, \tag{1}$$

and so:

$$\widetilde{A} = \pi_J^{-1}\left(\bigcup{}' \prod_{i=1}^{n} D_{j(i)}\right)$$
$$= \bigcup{}' \pi_J^{-1}\left(\prod_{i=1}^{n} D_{j(i)}\right).$$

Thus $\widetilde{A}$ is a finite disjoint union of elements of $\mathcal{A}'$, completing the proof. ∎

9.3 Finite Additivity of $\mu_{\mathcal{A}}$ on $\mathcal{A}$

In order to investigate the possibility that μ_0 is a pre-measure on $\mathcal{A}'$, the ambiguity in the representation of $A \in \mathcal{A}'$ must be addressed. This ambiguity was noted in Remark 9.5 and was used to simplify the proof that $\mathcal{A}'$ is a semi-algebra. Given A defined by J and $\prod_{i=1}^{n} A_{j(i)}$, one can define $A' = A$ by adding an arbitrary but finite number of additional factors to its specification, setting $A'_{l(i)} = A_{j(k)}$ if $l(i) = j(k)$ for some k, and $A'_{l(i)} = X_{l(i)}$ otherwise. This ambiguity is then a major problem for the well-definedness of the product set function in (9.3) since $\mu_0(A)$ may then also include any finite number of $\mu_{l(i)}(X_{l(i)})$-factors.

Of course, the solution is as obvious as is the problem. By requiring $\{\mu_i\}$ to be probability measures, meaning that $\mu_i(X_i) = 1$ for all i, this redundancy is no longer an issue. We make this assumption from this point forward, and prove the well-definedness of μ_0 in the next proposition.

Proposition 9.7 (Well-definedness of μ_0 on $\mathcal{A}'$) *Given probability spaces $\{(X_i, \sigma(X_i), \mu_i)\}_{i=1}^{\infty}$, the product set function μ_0 in (9.3) is well defined in that if $A \in \mathcal{A}'$ is defined by:*

$$A = \prod_{i=1}^{n} A_{j(i)} = \prod_{i=1}^{m} A_{k(i)},$$

then:

$$\prod_{i=1}^{n} \mu_{j(i)}(A_{j(i)}) = \prod_{i=1}^{m} \mu_{k(i)}(A_{k(i)}). \tag{9.6}$$

Proof. *By the definition of cylinder set, it follows that if $j(i) = k(i')$ for some i, i', then $A_{j(i)} = A_{k(i')}$. If there exists $j(i)$ with $j(i) \neq k(i')$ for all i', then $A_{j(i)} = X_{j(i)}$. Similarly if there exists $k(i)$ with $k(i) \neq j(i')$ for all i', then $A_{k(i)} = X_{k(i)}$. Thus since $\mu_i(X_i) = 1$ for all i, the result follows.* ∎

With well-definedness assured, we now turn to finite additivity.

Proposition 9.8 (μ_0 extends to finitely additive $\mu_{\mathcal{A}}$ on $\mathcal{A}$) *Given probability spaces* $\{(X_i, \sigma(X_i), \mu_i)\}_{i=1}^{\infty}$, *the product set function μ_0 defined in (9.3) is finitely additive on $\mathcal{A}'$, and extends to a well-defined and finitely additive set function $\mu_{\mathcal{A}}$ on the algebra $\mathcal{A}$ generated by $\mathcal{A}'$.*

Proof. *To show that μ_0 is finitely additive on $\mathcal{A}'$, let $\{A_j\}_{j=1}^{n} \subset \mathcal{A}'$ be given disjoint cylinders with each A_j defined by $K_j \equiv (k_j(1), k_j(2), ..., k_j(m_j))$ and $\prod_{i=1}^{m_j} A_{k_j(i)}^{(j)}$. Assume that $A \equiv \bigcup_{j=1}^{n} A_j \in \mathcal{A}'$, with A defined by $K \equiv (k(1), k(2), ..., k(m))$ and $\prod_{i=1}^{m} B_{k(i)}$. Without loss of generality assume for notational simplicity that all index sets are expressed in increasing order.*

To express each as a cylinder set defined in terms of a common set of components $(1, 2, ..., M)$, let:

$$M = \max\{k_1(m_1), ..., k_n(m_n), k(m)\},$$

and thus M is the largest index that appears in the definitions of A or any A_j. As in the proof of Proposition 9.6, let $J = (1, 2, ..., M)$ and write each $A_j = \pi_J^{-1}\left(\prod_{l=1}^{M} A_l^{(j)}\right)$ where either $A_l^{(j)} = A_{k_j(i)}^{(j)}$ if $l = k_j(i)$ for some i, and $A_l^{(j)} = X_l$ otherwise. We similarly express $A = \pi_J^{-1}\left(\prod_{l=1}^{M} B_l\right)$.

These infinite dimensional product sets can be considered as product sets in the M-dimensional product space $\prod_{l=1}^{M} X_l$. Specifically, $\{\pi_J A_j\}_{j=1}^{n} \subset \mathcal{A}'_J$, where $\mathcal{A}'_J$ is the semi-algebra of measurable rectangles in $\prod_{i=1}^{M} X_{j(i)}$ as given in Definition 7.1. These sets are disjoint, and $\bigcup_{j=1}^{n} \pi_J A_j = \pi_J A \in \mathcal{A}'_J$. Define $\mu_0^{(M)}$ as the product set function μ_0 applied as in (9.3) but to $\prod_{l=1}^{M} X_l$. The result of Proposition 7.15 in the finite M-dimensional case can now be applied to conclude that:

$$\mu_0^{(M)}(\pi_J A) = \sum_{j=1}^{n} \mu_0^{(M)}(\pi_J A_j). \tag{1}$$

This completes the proof of finite additivity of μ_0 on $\mathcal{A}'$ since for each set Y in (1), $\mu_0^{(M)}(\pi_J Y) = \mu_0(Y)$ by construction.

To define the extension from μ_0 on $\mathcal{A}'$ to $\mu_{\mathcal{A}}$ on the algebra $\mathcal{A}$, recall (Exercise 6.10) that $\mathcal{A}$ is the collection of all finite disjoint unions of sets in $\mathcal{A}'$. Then if $B \in \mathcal{A}$, say, $B = \bigcup_{j=1}^{n} A_j$ with each $A_j \in \mathcal{A}'$, the extension of μ_0 to $\mu_{\mathcal{A}}$ on $\mathcal{A}$ is defined:

$$\mu_{\mathcal{A}}(B) \equiv \sum_{j=1}^{n} \mu_0(A_j).$$

To prove that this definition of $\mu_{\mathcal{A}}$ is well defined and independent of the representation of B as a union of $\mathcal{A}'$-sets, assume that $\{A_j\}_{j=1}^{n}$, $\{C_k\}_{k=1}^{m}$ are disjoint collections of sets in $\mathcal{A}'$ with union B. Consider the collection $\{A_j \bigcap C_k\}_{j,k}$. This is again a finite disjoint collection of $\mathcal{A}'$-sets with union B, and also both $A_j = \bigcup_k (A_j \bigcap C_k)$ and $C_k = \bigcup_j (A_j \bigcap C_k)$. Thus by finite additivity on $\mathcal{A}'$:

$$\sum_{j=1}^{n} \mu_0(A_j) = \sum_{j=1, k=1}^{n, m} \mu_0(A_j \bigcap C_k) = \sum_{k=1}^{m} \mu_0(C_k).$$

This definition then assures finite additivity on $\mathcal{A}$ because if disjoint $\{B_k\}_{k=1}^n \subset \mathcal{A}$, say $B_k = \bigcup_{j=1}^{n_k} A_j^{(k)}$, then disjointness of $\{B_k\}_{k=1}^n$ assures disjointness of $\{A_j^{(k)}\}_{j,k}$ as sets in $\mathcal{A}'$. Thus finite additivity of $\mu_{\mathcal{A}}$ on $\mathcal{A}$ follows from finite additivity of μ_0 on $\mathcal{A}'$. ∎

9.4 Free Countable Additivity on Finite Spaces

There is a special case in which the above result of finite additivity assures countable additivity, and that is when each $(X_i, \sigma(X_i), \mu_i) = (Y, \sigma(Y), p)$. Here Y is a finite collection of points, say $Y = \{y_1, y_2, ..., y_n\}$, p is a probability measure defined on Y with notational convention:

$$p_j \equiv p(y_j), \quad \sum_{j=1}^n p_j = 1,$$

and $\sigma(Y)$ is a sigma algebra which is typically defined as the power sigma algebra, $\sigma(P(Y))$, with the probability of all sets defined by additivity.

For notational simplicity, we assume that the space Y and the probability measure p is the same for each copy of this space, but this is not necessary for the result below. What is necessary more generally is that the number of elements in the spaces is bounded, so if $(X_i, \sigma(X_i), \mu_i) = (Y_j, \sigma(Y_j), p_j)$ where Y_j has n_j elements, the proof below would require that $n_j \leq N$ for some N. Other than making the notation more complex, there are no hidden challenges in proving the more general case, so we focus on the simpler and more commonly applied situation.

Notation 9.9 ($\prod_{i=1}^\infty Y_i \equiv Y^{\mathbb{N}}$) *In cases of products of identical spaces, the probability space $X = \prod_{i=1}^\infty Y_i$ is often denoted:*

$$X = Y^{\mathbb{N}},$$

where $\mathbb{N}$ is notation for the natural numbers $\{1, 2, 3...\}$. This notation generalizes the notational convention of $\mathbb{R}^n$.

Example 9.10 (Common models) *$Y = \{0, 1\}$ is a typical model for the binomial probability space, such as that for a coin flip, where for example $Y = 1$ denotes heads or H, and $Y = 0$ denotes tails or T. For a "fair" coin $p_j = 1/2$ for $j = H, T$. Also $Y = \{2, ..., 11, 12\}$ could be used to represent the probability space of outcomes from the roll of a pair of dice, where $p_j = a_j/36$ with $a_j = \min\{j - 1, 13 - j\}$.*

Similarly, $Y = \{1, 2, ..., 52\}$ could denote the probability space of outcomes from drawing one card out of a standard deck of playing cards suitably ordered. Of course the elements of Y need not be numeric and in the latter case one could assign with subscripts denoting Clubs, Diamonds, Hearts and Spades:

$$Y = \{2_C, ..., J_C, Q_C, K_C, A_C, 2_D, .., 2_H, ...A_S\}.$$

Assuming well-shuffled, $p_j = 1/52$ for all j.

In the first case, $Y^{\mathbb{N}}$ is the space of all sequences of 0s and 1s which can be understood as "realizations" of an infinite sequence of coin flips. This space can be also identified with the interval $[0, 1]$, through the representation of such numbers in binary arithmetic or base-2, and a suitable convention for ambiguities in such representations as noted in item 2 of Section 9.1.

In the second example, $Y^{\mathbb{N}}$ is the space of all sequences of digits $2, ..., 12$, and this can be identified with the results of an infinite series of paired dice rolls, or with the interval $[0, 1]$ in base-11. To have single digit representation of these numbers, one can identify:

$$(2, ..., 12) \leftrightarrow \{0, 1, 2, ..., 9, T\}.$$

For the deck of playing cards, $Y^{\mathbb{N}}$ is the collection of infinite draws from a deck of cards, where each draw is replaced in the deck and reshuffled before the next draw to maintain the original probability assignments.

Proposition 9.8 proved that the product set function μ_0 defined in (9.3) on the semi-algebra $\mathcal{A}'$ of cylinder sets defined by (9.2), is finitely additive and has a well-defined and finitely additive extension $\mu_{\mathcal{A}}$ to the algebra $\mathcal{A}$ generated by $\mathcal{A}'$. We now show that in the special case of finite probability spaces defined above, this finite additivity result assures that $\mu_{\mathcal{A}}$ is also countably additive on $\mathcal{A}$ and hence is a measure on $\mathcal{A}$. By Propositions 6.4 and 6.14, $\mu_{\mathcal{A}}$ has a unique extension to a probability measure on the smallest sigma algebra that contains $\mathcal{A}$.

The essence of the proof is that in this special case of finite probability spaces, it is impossible to have a countable disjoint union of sets in $\mathcal{A}$ equal a set in $\mathcal{A}$ unless all but a finite number of these sets is empty. In other words, the assumption that $\bigcup_{j=1}^{\infty} A_j = A \in \mathcal{A}$ for disjoint sets $\{A_j\}_{j=1}^{\infty}$ is very restrictive on this collection, and requires that at most a finite number of these sets be nonempty.

Proposition 9.11 ($\mu_{\mathcal{A}}$ is a measure on $\mathcal{A}$ for $Y^{\mathbb{N}}$) *Let $(X_i, \sigma(X_i), \mu_i) = (Y, \sigma(Y), p)$ for all i, where Y is a finite collection of points, say, $Y = \{y_1, y_2, ..., y_n\}$, and p is a probability measure defined on Y. Let $\mathcal{A}'$ denote the semi-algebra of cylinder sets on $Y^{\mathbb{N}}$ defined in (9.2) and $\mathcal{A}$ the algebra generated by $\mathcal{A}'$.*

Then the finitely additive set function $\mu_{\mathcal{A}}$ of Proposition 9.8 is a measure on $\mathcal{A}$, and has a unique extension to a measure $\mu_{Y^{\mathbb{N}}}$ on $\sigma(\mathcal{A})$, the sigma algebra generated by $\mathcal{A}$.

Proof. *We prove by contradiction that if $\{A_j\}_{j=1}^{\infty} \subset \mathcal{A}$ are disjoint sets with $\bigcup_{j=1}^{\infty} A_j \in \mathcal{A}$, then there is an integer N so that $A_j = \emptyset$ for $j \geq N$. So by finite additivity:*

$$\mu_{\mathcal{A}} \left(\bigcup_{j=1}^{\infty} A_j \right) = \sum_{j=1}^{\infty} \mu_{\mathcal{A}} (A_j),$$

since $\mu_{\mathcal{A}}(\emptyset) = 0$. This proves that $\mu_{\mathcal{A}}$ is a measure on the algebra $\mathcal{A}$, and the final conclusion follows from Propositions 6.4 and 6.14.

To this end, assume that $\{A_j\}_{j=1}^{\infty} \subset \mathcal{A}$ are such disjoint, nonempty sets with union $\bigcup_{j=1}^{\infty} A_j = A \in \mathcal{A}$. Define $\{A'_n\}_{n=1}^{\infty} \subset \mathcal{A}$ by:

$$A'_n = A - \bigcup_{j=1}^{n} A_j.$$

Then $A'_n \in \mathcal{A}$ since:

$$A'_n = A \bigcap \left(\bigcap_{j=1}^{n} \tilde{A}_j \right),$$

and an algebra is closed under complementation and finite intersections.

If $A'_n = \emptyset$ for some n, then the proof is complete with $N = n + 1$, and thus we assume each A'_n is nonempty. Now $A'_{n+1} \subset A'_n$ for all n, and hence by De Moivre's laws:

$$\begin{aligned} \emptyset &= A - A \\ &= A - \bigcup_{j=1}^{\infty} A_j \\ &= \bigcap_{n=1}^{\infty} A'_n. \end{aligned} \tag{1}$$

The desired contradiction will be obtained by showing that given nonempty sets $\{A'_n\}_{n=1}^{\infty} \subset \mathcal{A}$ with $A'_{n+1} \subset A'_n$ for all n, it must be the case that $\bigcap_{n=1}^{\infty} A'_n \neq \emptyset$.

For this, choose $\{w_n\}_{n=1}^{\infty} \subset Y^{\mathbb{N}}$ so that $w_n \in A'_n$ for each n. This is possible since each A'_n is nonempty. As elements of $Y^{\mathbb{N}}$:

$$w_n = (w_{n1}, w_{n2}, w_{n3}, \ldots),$$

where each $w_{nm} \in Y$. To simplify notation, let π_m denote the projection mapping of Notation 9.4 and defined on $Y^{\mathbb{N}}$ by:

$$\pi_m(w_n) = w_{nm}.$$

Consider the collection of first components, $\{\pi_1(w_n)\}_{n=1}^{\infty}$.

Since Y is finite, at least one element of Y, say, y'_1, is a member of this collection infinitely often. Let $\{n_{1,j}\}_{j=1}^{\infty}$ denote the unbounded subsequence of the n-sequence for which $\pi_1(w_{n_{1,j}}) = y'_1$ for all j. Next, consider the second components of these identified points, $\{\pi_2(w_{n_{1,j}})\}_{j=1}^{\infty}$. Again, there is at least one element of Y, say y'_2, that is a member of this collection infinitely often. Let $\{n_{2,j}\}_{j=1}^{\infty} \subset \{n_{1,j}\}_{j=1}^{\infty}$ denote the unbounded subsequence of the $n_{1,j}$-sequence for which $\pi_2(w_{n_{2,j}}) = y'_2$, and so $\pi_k(w_{n_{2,j}}) = y'_k$ for $k \leq 2$. Without loss of generality, we choose $n_{2,1} > n_{1,1}$, and hence $n_{2,1} \geq 2$.

Continuing in this fashion, we construct a collection of unbounded subsequences of the original n-sequence: $\{\{n_{m,j}\}_{j=1}^{\infty}\}_{m=1}^{\infty}$, with $\{n_{m+1,j}\}_{j=1}^{\infty} \subset \{n_{m,j}\}_{j=1}^{\infty}$ for all m. Also, for any m, $\pi_k(w_{n_{m,j}}) = y'_k$ for $k \leq m$ and all j. Also as above, we choose $n_{m+1,1} > n_{m,1}$ for each m, and hence $n_{m,1} \geq m$ for each m.

We now prove that $w \equiv (y'_1, y'_2, y'_3, \ldots) \in \bigcap_{n=1}^{\infty} A'_n$, and hence this intersection is not empty, contradicting (1). To prove this, let A'_n be given. Then since $A'_n \in \mathcal{A}$, it equals a finite disjoint union of measurable rectangles, say $A'_n = \bigcup_{k=1}^{N_n} B_{nk}$. Each $B_{nk} \in \mathcal{A}'$, and hence each defines restrictions on only a finite number of coordinates. Consequently, A'_n is defined by restrictions on at most a finite number of coordinates, say coordinates $1, 2, \ldots, N$. By adding Y-factors to the B_{nk}-rectangles as noted in Remark 9.5, we can without loss of generality assume that $N \geq n$.

Now recall the collection of index subsequences $\{\{n_{m,j}\}_{j=1}^{\infty}\}_{m=1}^{\infty}$, and choose $m = N$. Then $w_{n_{N,1}} \in A'_{n_{N,1}}$ by the definition of $\{w_n\}_{n=1}^{\infty}$, and $\pi_k(w_{n_{N,1}}) = y'_k$ for $k \leq N$. Further, $A'_{n_{N,1}} \subset A'_n$

by construction since $n_{N,1} \geq N \geq n$, and so $w_{n_{N,1}} \in A'_n$. But then $\pi_k(w_{n_{N,1}}) = \pi_k(w)$ for $k \leq N$, and this proves that $w \in A'_n$ since A'_n is defined by restrictions on at most these N coordinates.

Thus $w \in \bigcap_{n=1}^{\infty} A'_n$, and the proof is complete. ∎

9.5 Countable Additivity on $\mathcal{A}^+$

As in the finite dimensional case of Proposition 7.18, the proof of countable additivity of $\mu_{\mathcal{A}}$ on an algebra will proceed by demonstrating that this set function is continuous from above. Here, we will not need to assume that the probability spaces $\{(X_i, \sigma(X_i), \mu_i)\}_{i=1}^{\infty}$ are σ-finite to apply Proposition 6.19, since we have already restricted such spaces to be probability spaces. But for this proof, the probability spaces investigated will be further restricted to be probability spaces on $\mathbb{R}$. This latter restriction can be relaxed to include probability spaces on $\mathbb{R}^n$ with little more than added notational complexity. As this generalization requires no new ideas, we present the 1-dimensional result and leave it to the reader to generalize.

To appreciate the complexity of the countable additivity result in the current context, consider the following. Given are countably many disjoint sets $A_j \in \mathcal{A}$, each of which is a finite disjoint union of measurable rectangles from $\mathcal{A}'$:

$$A_j = \left\{ x \in X | (x_1, x_2, ..., x_{n(j)}) \in \bigcup_{k \leq m(j)} \prod_{i=1}^{n(j)} A_{j,i}^{(k)} \right\}.$$

Thus for each j, the set $A_j \in \mathcal{A}$ is a disjoint union of $m(j)$ measurable rectangles, each of which is defined in terms of an $n(j)$-product of measurable sets $A_{j,i}^{(k)} \in \sigma(X_i)$.

This notation is simplified by defining each of these rectangles in terms of sequential indexes, $1 \leq i \leq n(j)$, taking $n(j)$ to be the maximum index used for these finitely many rectangles, and for the finitely many rectangles in A_k-sets for $k \leq j$, allowing $A_{j,i}^{(k)} = X_i$ as necessary. In addition to having each measurable rectangle use sequential indexes, this also allows the convenience of assuming that all rectangles included in the A_j-unions use the same sequential index set.

For a proof of countable additivity, we will assume that $\bigcup_{j=1}^{\infty} A_j = A \in \mathcal{A}$, and hence using the same index convention:

$$A = \left\{ x \in X | (x_1, x_2, ..., x_n) \in \bigcup_{k \leq m} \prod_{i=1}^{n} A_i^{(k)} \right\}.$$

Note that A and all A_j-sets can be assumed to use the same notational convention by identifying n first, and then adding as necessary dimensions for $n(1)$, then $n(2)$, and so forth.

While it is easy to appreciate that $\{m(j)\}_{j=1}^{\infty}$ need not be bounded, as sets in $\mathcal{A}$ can be finite unions of any number of rectangles, it is tempting to assume that the set $\{n(j)\}_{j=1}^{\infty}$ is bounded and that this is assured by the requirement that $\bigcup_{j=1}^{\infty} A_j \in \mathcal{A}$. If this were true, the question of countable additivity could then be addressed using the Chapter 7 machinery of finite dimensional spaces. This approach was seen in proof of Proposition 9.8 for finite additivity.

Unfortunately, the set $\{n(j)\}_{j=1}^{\infty}$ need not be bounded despite the assumption that the union of these sets reflects restrictions on only n spaces. In fact, the same can be said within $\mathcal{A}'$, that a countable union of disjoint rectangles defined by $\{\prod_{i=1}^{n(j)} A_{ji}\}_{j=1}^{\infty}$ can be a disjoint rectangle defined by $\{\prod_{i=1}^{n} A_i\}$, even if the set $\{n(j)\}_{j=1}^{\infty}$ is unbounded.

Exercise 9.12 (On unbounded $\{n(j)\}_{j=1}^{\infty}$) *In an infinite dimensional product space, give examples in the semi-algebra $\mathcal{A}'$ of a countable collection of the disjoint rectangles, $\{A_j\}_{j=1}^{\infty} \equiv \{\prod_{i=1}^{n(j)} A_{ji}\}_{j=1}^{\infty}$, so that:*

1. *$\bigcup_j A_j \in \mathcal{A}'$, yet the set $\{n(j)\}_{j=1}^{\infty}$ is unbounded.*
2. *$A_{j+1} \subset A_j$ for all j and $\bigcap_j A_j = \emptyset$, yet again the set $\{n(j)\}_{j=1}^{\infty}$ is unbounded.*

Hint: Work with the infinite product of Lebesgue measure spaces for which $(X_i, \sigma(X_i), \mu_i) = (\mathbb{R}, \mathcal{M}_L, m)$, since these are not results which require that these spaces be probability spaces. For item 1 try an example for which $A_{j1} = A$ is fixed, and every rectangle restricts only three product factors. For item 2 the assumption that $A_{j+1} \subset A_j$ allows that the number of restrictions increases with j, so let $n(j) = j$.

9.5.1 Outline of Proof and Need for $\mathcal{A}^+$

The proof of countable additivity will utilize the approach of continuity from above where it will be given that $\mu_{\mathcal{A}}(A_1) < \infty$, $A_{j+1} \subset A_j$, and $\bigcap_{j=1}^{\infty} A_j = \emptyset$. This does not imply that the set $\{n(j)\}_{j=1}^{\infty}$ is bounded as noted in Exercise 9.12, but it is only the case of unbounded $\{n(j)\}_{j=1}^{\infty}$ that needs additional proof. Otherwise, the proof reduces to the finite dimensional case as for Proposition 9.8. Consequently, it will be explicitly assumed below that the set $\{n(j)\}_{j=1}^{\infty}$ is unbounded.

Recall that each $A_j \in \mathcal{A}$ is a finite disjoint union of cylinder sets as in the previous section. As noted in Remark 9.5, finitely many additional X_i-constraints can always be added to the definition of any cylinder set or the associated A_j-union. Thus we will assume for notational simplicity that each A_j is given by $J_j = (1, 2, ..., n(j))$ and a finite disjoint union of measurable rectangles in $\prod_{i=1}^{n(j)} X_i$, with $\{n(j)\}_{j=1}^{\infty}$ unbounded. Further, $n(j)$ can be assumed to be increasing in j by replacing each with $n'(j) \equiv \max(n(i)|i \leq j)$.

Following is an outline of the forthcoming proof and associated considerations.

1. Statement of desired result:

If $\{(X_i, \sigma(X_i), \mu_i)\}_{i=1}^{\infty}$ are probability spaces, then $\mu_{\mathcal{A}}$ is countably additive on the algebra $\mathcal{A}$ of finite disjoint unions of cylinder sets.

2. The continuity from the above approach:

To prove countable additivity, we use Proposition 6.19 and show that given $\{A_j\}_{j=1}^{\infty} \subset \mathcal{A}$ with $\mu_0(A_1) < \infty$ by assumption, where $A_{j+1} \subset A_j$ for all j, and $\bigcap_j A_j = \emptyset$, then $\lim_{j \to \infty} \mu_{\mathcal{A}}(A_j) = 0$. By monotonicity of $\mu_{\mathcal{A}}$, which is a corollary of finite additivity, it follows from $A_{j+1} \subset A_j$ that $\mu_{\mathcal{A}}(A_{j+1}) \leq \mu_{\mathcal{A}}(A_j)$. Thus $\lim_{j \to \infty} \mu_{\mathcal{A}}(A_j)$ always exists since these values are bounded from below by 0. Our goal is to prove that this limit is 0.

3. Proof by contradiction:

Given $\{A_j\}_{j=1}^{\infty} \subset \mathcal{A}$ with $\mu_{\mathcal{A}}(A_1) < \infty$ and $A_{j+1} \subset A_j$ for all j, we prove that if $\mu_{\mathcal{A}}(A_j) \geq \epsilon > 0$ for all j, then $\bigcap_{j=1}^{\infty} A_j \neq \emptyset$. By the comment in item 2, the assumption that $\mu_A(A_j) \geq \epsilon > 0$ for all j is equivalent to the assumption that $\lim_{j\to\infty} \mu_{\mathcal{A}}(A_j) \geq \epsilon$.

4. Initial insight to proof:

When does $\mu_{\mathcal{A}}(A_1) < \infty$, $A_{j+1} \subset A_j$ and $\mu_{\mathcal{A}}(A_j) \geq \epsilon$ for all j imply that $\bigcap_{j=1}^{\infty} A_j \neq \emptyset$? Rather than work in $\prod_{i=1}^{\infty} X_i$, it seems compelling that this investigation would be simplified if it could be mapped into a problem in the finite dimensional spaces $\{\prod_{i=1}^{n(j)} X_i\}$. Then since each A_j is a finite disjoint union of cylinder sets, it follows that each A_j is given by $J_j = (1, 2, ..., n(j))$ as noted above, and $H_j \subset \prod_{i=1}^{n(j)} X_i$, a finite disjoint union of measurable rectangles. Further, $\mu_{\mathcal{A}}(A_j) = \mu_{J_j}(H_j)$, where μ_{J_j} denotes the product measure of Proposition 7.20, with the set function of that result defined on $\mathcal{A}'$, the semi-algebra of measurable rectangles, as in (9.3) with $J_j \equiv (1, 2, ..., n(j))$.

The collection $\{H_j\}$ exists in different and increasing dimensional product spaces since $\{n(j)\}_{j=1}^{\infty}$ is increasing and unbounded, but the nesting of the sets $\pi_{J_j}^{-1}(H_j) = A_j$ tells us something useful.

If $\pi_1 : \prod_{i=1}^{n(j)} X_i \to X_1$ is the projection onto the first component, then $A_{j+1} \subset A_j$ assures that $\pi_1(H_{j+1}) \subset \pi_1(H_j)$. This follows since $\pi_1(H_j)$ obtains the first component of A_j by $\pi_{J_j}^{-1}$ (recall Notation 9.4). Thus, the collection $\{\pi_1(H_j)\}_{j=1}^{\infty} \subset X_1$ is a nested collection of sets. It is tempting to posit that there exists $x_1 \in \bigcap_{j=1}^{\infty} \pi_1(H_j)$, that x_1 is the first component of the sought after $x \in \bigcap_{j=1}^{\infty} A_j$, and that by repeating this argument for the other components we obtain x, the point in this intersection set.

For a general probability space $(X_i, \sigma(X_i), \mu_i)$, it is difficult to predict that nested sets have nonempty intersection. Indeed, even for $X_i = \mathbb{R}$ it is not difficult to develop examples of nonempty nested sets $\{B_j\} \subset \mathbb{R}$ which have empty intersection. For example the bounded open intervals $\{(0, 1/n)\}$ have empty intersection, as do the unbounded intervals $\{(n, \infty)\}$ or $\{[n, \infty)\}$. For a general metric space, such as the rationals with the standard distance function, $\{[\sqrt{2}, \sqrt{2}+1/n]\}$ is a nested collection of closed and bounded intervals with empty intersection.

5. The role of compactness:

It turns out that nested **compact** sets in $\mathbb{R}$ have nonempty intersection. This result is attributed to **Georg Cantor (1845–1918)** and generalizes to $\mathbb{R}^n$ and to complete metric spaces with some restrictions. Recall that on $\mathbb{R}$ (and $\mathbb{R}^n$), the **Heine-Borel theorem** of Proposition 2.27 states that a set is compact if and only if it is both closed and bounded.

Proposition 9.13 (Cantor's Intersection Theorem) *Let* $\{A_j\} \subset \mathbb{R}$ *be a collection of nonempty compact sets with* $A_{j+1} \subset A_j$ *for all* j. *Then* $\bigcap_{j=1}^{\infty} A_j \neq \emptyset$.

Proof. *Let* $x_j = \inf\{x | x \in A_j\}$. *Then* x_j *is finite since* A_j *is bounded, and* $x_j \in A_j$ *since* A_j *is closed. By the nested assumption,* $x_j \leq x_{j+1}$ *for all* j, *and so the sequence* $\{x_j\}$ *is increasing and also bounded since* $\{x_j\} \subset A_1$. *Consequently,* $x \equiv \lim x_j$ *exists and* $x \in A_1$ *since closed. Now for any* k, $x_j \in A_k$ *for all* $j \geq k$ *by nesting, and so also* $x \in A_k$ *since closed. As this is true for all* k, *it follows that* $x \in \bigcap_{j=1}^{\infty} A_j$. ∎

6. An application of inner regularity:

While item 5 provides an insight to the question in item 4, it is clear that we cannot assume that the sets $H_j \subset \prod_{i=1}^{n(j)} X_i$ are compact. Indeed, all that can be stated is that they are finite disjoint unions of measurable rectangles. But since measurable, if the probability measure on the product space $\left(\prod_{i=1}^{n(j)} X_i, \sigma(\prod_{i=1}^{n(j)} X_i), \mu_{J_j} \right)$ is **inner regular**, then

$$\mu_{J_j}(H_j) = \sup_{K_j \subset H_j} \mu_{J_j}(K_j), \ K_j \text{ compact.}$$

In other words, H_j contains compact subsets with μ_{J_j}-measure arbitrarily close to $\mu_{J_j}(H_j)$. Here as above, μ_{J_j} denotes the product measure of Proposition 7.20, with the set function of that result defined on $\mathcal{A}'$, the semi-algebra of measurable rectangles, as in (9.3) with $J_j \equiv (1, 2, ..., n(j))$.

In order to apply this insight, the result below will require that the probability spaces be defined on $\mathbb{R}$:

$$\{(X_i, \sigma(X_i), \mu_i)\}_{i=1}^\infty = \{(\mathbb{R}, \mathcal{B}(\mathbb{R}), \mu_i)\}_{i=1}^\infty,$$

where $\mathcal{B}(\mathbb{R})$ denotes the Borel sigma algebra of Definition 2.13. This assures the desired regularity result by Proposition 8.20. Then given nonempty measurable sets $H_j \subset \mathbb{R}^{n(j)} \equiv \prod_{i=1}^{n(j)} X_i$, there are compact sets $K_j \subset H_j$, the measure of which can be taken as close as desired to the measure of H_j.

7. The final construction:
From the compact collection $\{K_j\}$, each of which exists in a different finite dimensional space $\mathbb{R}^{n(j)}$, we will define a collection of cylinder sets $\{B_j\} \subset \mathbb{R}^\mathbb{N} \equiv \prod_{i=1}^\infty X_i$, with each B_j given by K_j and J_j:

$$B_j = \{x \in \mathbb{R}^\mathbb{N} | (x_1, x_2, ..., x_{n(j)}) \in K_j\}.$$

Defining $C_j \subset X$ by $C_j = \bigcap_{i \leq j} B_j$ obtains $C_j \subset B_j \subset A_j$. As $\mu_{\mathcal{A}}(A_j) = \mu_{J_j}(H_j) \geq \epsilon$ for all j, we will select $\{K_j\}$ so that $\mu_{\mathcal{A}}(C_j) \geq \epsilon/2$. The final step is to construct $y \in \bigcap_{j=1}^\infty C_j$, which then proves that $\bigcap_{j=1}^\infty A_j \neq \emptyset$.

8. A Technicality and Need for a new algebra $\mathcal{A}^+$:
The set B_j in step 7 need not be in the algebra $\mathcal{A}$ since the compact set K_j need not be a finite union of measurable rectangles in $\mathbb{R}^{n(j)}$. So we will need to expand the algebra $\mathcal{A}$ to an algebra $\mathcal{A}^+$ which contains these more general sets. Once the proof of countable additivity of $\mu_{\mathcal{A}}$ on $\mathcal{A}^+$ is complete, this naturally proves the same for $\mathcal{A}$. We will then prove that $\mathcal{A}^+$ generates the same sigma algebra $\sigma(X)$ of **Carathéodory measurable** sets using the outer measure $\mu_{\mathcal{A}^+}^*$, as would have been generated using analogously defined $\mu_{\mathcal{A}}^*$. Thus, this final result will provide the same measure and extension to $\sigma(X)$ as would have been achieved with $\mu_{\mathcal{A}}$ and $\mathcal{A}$.

As the proof of countable additivity typically requires manipulations that use finite additivity, basing this proof on $\mathcal{A}^+$ will require a demonstration that $\mu_{\mathcal{A}}$ is in fact finitely additive on $\mathcal{A}^+$. The next section introduces $\mathcal{A}^+$ and develops its properties.

9.5.2 Algebra $\mathcal{A}^+$ and Finite Additivity of $\mu_{\mathcal{A}}$

We begin with a definition which introduces the needed collection of sets. Although all measure spaces are defined with $X_j \equiv \mathbb{R}$, it is sometimes convenient to express product spaces such as $\prod_{i=1}^n \mathbb{R}_{j(i)}$ to identify which component measure space sigma algebras and probability measures are reflected.

Definition 9.14 (Product space; general cylinder sets: $\mathcal{A}^+$) *Given probability spaces* $\{(\mathbb{R}, \mathcal{B}(\mathbb{R}), \mu_i)\}_{i=1}^\infty$, *where $\mathcal{B}(\mathbb{R})$ denotes the Borel sigma algebra of Definition 2.13, the **product space** $\mathbb{R}^{\mathbb{N}} = \prod_{i=1}^\infty \mathbb{R}_i$ is defined by:*

$$\mathbb{R}^{\mathbb{N}} = \{(x_1, x_2, ...) | x_i \in \mathbb{R}\}.$$

*A **general finite dimensional measurable rectangle** or **general cylinder set** $A \subset \mathbb{R}^{\mathbb{N}}$ is defined for any n-tuple of positive integers $J = (j(1), j(2), ..., j(n))$ and $H \in \mathcal{B}\left(\prod_{i=1}^n \mathbb{R}_{j(i)}\right)$ by:*

$$A = \{x \in \mathbb{R}^{\mathbb{N}} | (x_{j(1)}, x_{j(2)}, ... x_{j(n)}) \in H\}. \tag{9.2}$$

Here $\mathcal{B}\left(\prod_{i=1}^n \mathbb{R}_{j(i)}\right)$ denotes the finite dimensional product space Borel sigma algebra of Proposition 7.20 associated with $\{(\mathbb{R}, \mathcal{B}(\mathbb{R}), \mu_{j(i)})\}_{i=1}^n$, and there denoted $\sigma(\mathcal{A})$. Recall Proposition 8.1.

*The cylinder set in (9.2) will be said **to be defined by** H and J. The collection of general cylinder sets in X is denoted by $\mathcal{A}^+$.*

The cylinder set A above can also be characterized in terms of the projection mapping:

$$\pi_J = \prod_{i=1}^n \pi_{j(i)} : \mathbb{R}^{\mathbb{N}} \to \prod_{i=1}^n \mathbb{R}_{j(i)},$$

by:

$$A = \pi_J^{-1}(H). \tag{9.3}$$

*For $A \in \mathcal{A}^+$ defined by H and J, the **product set function** μ_0 is defined by:*

$$\mu_0(A) = \mu_J(H), \tag{9.4}$$

where μ_J denotes the finite dimensional product space measure of Proposition 7.20 associated with $\{(\mathbb{R}, B(\mathbb{R}), \mu_{j(i)})\}_{i=1}^n$.

Remark 9.15 (On $\mathbb{R}^{\mathbb{N}}$; Nonuniqueness in $\mathcal{A}^+$)

1. *As noted above, each $\mathbb{R}_{j(i)} = \mathbb{R}$. The importance of this subscript notation is in the definition of $\mu_J(H)$, since this finite dimensional product measure reflects the component space measures $\{\mu_{j(i)}(\mathbb{R})\}_{i=1}^n$.*

2. *With $\{(X_i, \sigma(X_i), \mu_i)\}_{i=1}^\infty = \{(\mathbb{R}, \mathcal{B}(\mathbb{R}), \mu_i)\}_{i=1}^\infty$, the notational convention $\prod_{i=1}^\infty X_i \equiv \mathbb{R}^{\mathbb{N}}$, where $\mathbb{N}$ denotes the natural numbers $\{1, 2, 3, ...\}$, generalizes the finite dimensional*

notation of $\mathbb{R}^n$. While $x \in \mathbb{R}^n$ is identified with an n-tuple of reals $(x_1, ..., x_n)$, $x \in \mathbb{R}^N$ is identified with a countable sequence of reals $(x_1, x_2, ...)$.

3. *The set function μ_0 defined in (9.4) is a generalization of μ_0 as defined in (9.3). This is because if $H = \prod_{i=1}^n A_{j(i)}$, then by Proposition 7.20:*

$$\mu_J(H) = \prod_{i=1}^n \mu_{j(i)}(A_{j(i)}).$$

4. *Any such A in (9.4) can be defined by infinitely many (H, J)-pairs as was the case for cylinder sets defined in (9.2). This will often be used to simplify proofs. For example, if A is given by $J = (j(1), j(2), ..., j(n))$ and $H \in \mathcal{B}\left(\prod_{i=1}^n \mathbb{R}_{j(i)}\right)$, we can assume that A is also given by $J' = (1, 2, ..., j(n))$ and $H' \in \mathcal{B}\left(\prod_{i=1}^{j(n)} \mathbb{R}_i\right)$.*

To see this, define projection mappings π_J and $\pi_{J'}$ on $\mathbb{R}^N$ as above, and also define $\pi_J^{j(n)}$ on $\prod_{i=1}^{j(n)} \mathbb{R}_i$ by $\pi_J^{j(n)} : \prod_{i=1}^{j(n)} \mathbb{R}_i \to \prod_{i=1}^n \mathbb{R}_{j(i)}$. Then define $H' = \left(\pi_J^{j(n)}\right)^{-1}(H) \subset \prod_{i=1}^{j(n)} \mathbb{R}_i$. This set is a Borel set, since $\left(\pi_J^{j(n)}\right)^{-1}$ takes measurable rectangles in $\mathcal{A}'_J$ to measurable rectangles in $\mathcal{A}'_{J'}$, and so

$$\left(\pi_J^{j(n)}\right)^{-1}\left(\mathcal{A}'_J\right) \subset \mathcal{A}'_{J'}.$$

This applies to the associated algebras, since:

$$\left(\pi_J^{j(n)}\right)^{-1}\left(\bigcup_{k=1}^N A_k\right) = \bigcup_{k=1}^N \left(\pi_J^{j(n)}\right)^{-1}(A_k),$$

and then to the associated Borel sigma algebras by Proposition 8.1.

To show that A is given by J' and H', the key insight is that $\pi_J = \pi_J^{j(n)}\pi_{J'}$, and so $\left(\pi_J^{j(n)}\pi_{J'}\right)^{-1} = \pi_{J'}^{-1}\left(\pi_J^{j(n)}\right)^{-1}$. Thus:

$$\pi_{J'}^{-1}(H') = \pi_{J'}^{-1}\left(\left(\pi_J^{j(n)}\right)^{-1}(H)\right) = \pi_J^{-1}(H) = A.$$

5. *As any such A in (9.4) can be defined by infinitely many (H, J)-pairs, it will be necessary to verify that μ_0 is well defined. This fact will be demonstrated in (9.5) of the following proposition. But this demonstration is more subtle than that for (9.6) for simpler cylinder sets, since it is more challenging to tease out the redundant index components. That is, if A is given by both (H, J) and (K, I), the challenge will be to identify the components of J and I which are nonconstraining.*

This was relatively easy in the context of (9.2), while for the current demonstration this insight will follow from the judicious use of projection mappings.

We now turn to well-definedness of μ_0 on $\mathcal{A}^+$.

Proposition 9.16 (μ_0 is well defined on $\mathcal{A}^+$) *The product set function μ_0 in (9.4) is well defined on $\mathcal{A}^+$. Specifically, assume that $A \subset \mathbb{R}^\mathbb{N}$ is given by both (H, J) as in Definition 9.14, and (K, I), where $I = (i(1), i(2), ..., i(m))$ and $K \in \mathcal{B}\left(\prod_{k=1}^m \mathbb{R}_{i(k)}\right)$.*

Then $\mu_0(A)$ is uniquely defined:

$$\mu_J(H) = \mu_I(K). \tag{9.5}$$

Proof. *Consider the set $J \bigcap I$. If $J \bigcap I = \emptyset$, then by Definition 9.14 it must be the case that $H = \prod_{i=1}^n \mathbb{R}_{j(i)}$ and $K = \prod_{k=1}^m \mathbb{R}_{i(k)}$. So $H = K = \mathbb{R}^\mathbb{N}$, $\mu_J(H) = \mu_I(K) = 1$, and $\mu_0(A)$ is well defined.*

If $J \bigcap I \equiv L \neq \emptyset$, say $L \equiv (l(1), l(2), ..., l(p))$, then A is not constrained by the indexes in $J - L$ or $I - L$. Define π_L on $\mathbb{R}^\mathbb{N}$ by:

$$\pi_L : \mathbb{R}^\mathbb{N} \to \prod_{i=1}^p \mathbb{R}_{l(i)},$$

and similarly define $\pi_L^{(n)}$ on $\prod_{i=1}^n \mathbb{R}_{j(i)}$ and $\pi_L^{(m)}$ on $\prod_{k=1}^m \mathbb{R}_{i(k)}$. These latter projections are well defined since $L \subset J$ and $L \subset I$.

Now let $H' = \pi_L^{(n)}(H)$ and $K' = \pi_L^{(m)}(K)$. The proof is completed in three steps.

1. $H' = K'$:

We show that A is also given by (H', L) and (K', L). To this end, we first prove that the projections defining H' and K' are invertible. Regarding H', invertibility means that $H = \left(\pi_L^{(n)}\right)^{-1} H'$, and so:

$$\left(\pi_L^{(n)}\right)^{-1} \pi_L^{(n)}(H) = H. \tag{1}$$

This is certainly true if $L = J$, so assume that L is a proper subset of J.

Then $H \subset \left(\pi_L^{(n)}\right)^{-1} \pi_L^{(n)}(H)$ since if $x \in H$, and reordering indexes for notational convenience:

$$x = (x'_{l(1)}, ..., x'_{l(p)}, x_{l'(1)}, ..., x_{l'(n-p)}),$$

where $\left(x'_{l(1)}, ..., x'_{l(p)}\right) \in H'$. Thus:

$$\left(\pi_L^{(n)}\right)^{-1} \pi_L^{(n)}(x) = \{(x'_{l(1)}, ..., x'_{l(p)}, \cdot, ..., \cdot)\},$$

where the last $(n-p)$-coordinates are unconstrained. Hence in particular, $x \in \left(\pi_L^{(n)}\right)^{-1} \pi_L^{(n)}(H)$.

Conversely, let $x' \in \left(\pi_L^{(n)}\right)^{-1} \pi_L^{(n)}(H)$:

$$x' = (x'_{l(1)}, ..., x'_{l(p)}, x'_{l'(1)}, ..., x'_{l'(n-p)}),$$

where again $\left(x'_{l(1)}, ..., x'_{l(p)}\right) \in H'$. *If the above* $x \in H$ *yet* $x' \notin H$, *then* $\pi_J^{-1}(x) \in \pi_J^{-1}(H)$ *yet* $\pi_J^{-1}(x') \notin \pi_J^{-1}(H) = A$. *But this then contradicts the observation above that A is not defined by the indexes in* $J - L$ *or* $I - L$. *Thus* $\left(\pi_L^{(n)}\right)^{-1} \pi_L^{(n)}(H) \subset H$ *and (1) is proved. The same statement holds for* $\pi_L^{(m)}$ *relative to K.*

With π_J *and* π_K *defined analogously with* π_L, *we have that as defined on* $\mathbb{R}^{\mathbb{N}}$:

$$\pi_L^{(n)} \pi_J = \pi_L, \quad \pi_L^{(m)} \pi_K = \pi_L. \tag{2}$$

Now $(\pi_a \pi_b)^{-1} = \pi_b^{-1} \pi_a^{-1}$ *for compositions of projections, so from (9.3) and (2):*

$$A = \pi_J^{-1}(H) = \pi_J^{-1}\left(\left(\pi_L^{(n)}\right)^{-1} H'\right) = \left(\pi_L^{(n)} \pi_J\right)^{-1}(H') = \pi_L^{-1}(H').$$

Thus A is given by (H', L), *and also by* (K', L) *by the same argument. It then follows by definition that* $H' = K'$.

2. $H', K' \in \boldsymbol{\mathcal{B}}\left(\prod_{i=1}^p \mathbb{R}_{l(i)}\right)$ **and thus** $\mu_L(H') = \mu_L(K')$:

As these are identical proofs we prove that $H' \in \boldsymbol{\mathcal{B}}\left(\prod_{i=1}^p \mathbb{R}_{l(i)}\right)$. *If* $H = \prod_{i=1}^n H_{j(i)} \in \boldsymbol{\mathcal{B}}\left(\prod_{i=1}^n \mathbb{R}_{j(i)}\right)$ *with all* $H_{j(i)} \in \boldsymbol{\mathcal{B}}(\mathbb{R})$, *then:*

$$H' \equiv \pi_L^{(n)}(H) = \prod_{i=1}^p H_{l(i)} \in \boldsymbol{\mathcal{B}}\left(\prod_{i=1}^p \mathbb{R}_{l(i)}\right).$$

Thus, for the associated semi-algebras, $\pi_L^{(n)}\left(\boldsymbol{\mathcal{A}}_J'\right) = \boldsymbol{\mathcal{A}}_L'$, *and so too for the algebras,* $\pi_L^{(n)}\left(\boldsymbol{\mathcal{A}}_J\right) = \boldsymbol{\mathcal{A}}_L$. *But for any projection,* $\pi\left(A \bigcup B\right) = \pi(A) \bigcup \pi(B)$, *and similarly for intersections, while* $\pi\left(\widetilde{A}\right) = \widetilde{\pi(A)}$ *for complements. Thus by Proposition 8.1,* $\pi_L^{(n)}\left(\boldsymbol{\mathcal{A}}_J\right) = \boldsymbol{\mathcal{A}}_L$ *generalizes to the smallest sigma algebras that contain these algebras:*

$$\pi_L^{(n)}\left(\boldsymbol{\mathcal{B}}\left(\prod_{i=1}^n \mathbb{R}_{j(i)}\right)\right) = \boldsymbol{\mathcal{B}}\left(\prod_{i=1}^p \mathbb{R}_{l(i)}\right),$$

and so $H' \in \boldsymbol{\mathcal{B}}\left(\prod_{i=1}^p \mathbb{R}_{l(i)}\right)$.

The same argument obtains $K' \in \boldsymbol{\mathcal{B}}\left(\prod_{i=1}^p \mathbb{R}_{l(i)}\right)$, *and so* $\mu_L(H') = \mu_L(K')$ *by item 1.*

3. $\mu_J(H) = \mu_L(H')$ **and** $\mu_I(K) = \mu_L(K')$:

As these are identical proofs, we prove that $\mu_J(H) = \mu_L(H')$. *Stated equivalently, we prove that if* $H' \in \boldsymbol{\mathcal{B}}\left(\prod_{i=1}^p \mathbb{R}_{l(i)}\right)$ *and* $H \equiv \left(\pi_L^{(n)}\right)^{-1} H'$, *then* $H \in \boldsymbol{\mathcal{B}}\left(\prod_{i=1}^n \mathbb{R}_{j(i)}\right)$ *and:*

$$\mu_J(H) = \mu_L(H'). \tag{3}$$

That $H \in \boldsymbol{\mathcal{B}}\left(\prod_{i=1}^n \mathbb{R}_{j(i)}\right)$ *follows from the observation that* $\left(\pi_L^{(n)}\right)^{-1}\left(\boldsymbol{\mathcal{A}}_L'\right) \subset \boldsymbol{\mathcal{A}}_J'$ *in the notation of item 2. The same is then true of the associated Borel sigma algebras, recalling Proposition 8.1.*

To prove (3), define two set functions on $\mathcal{B}\left(\prod_{i=1}^{p} \mathbb{R}_{l(i)}\right)$ as follows. If $H' \in \mathcal{B}\left(\prod_{i=1}^{p} \mathbb{R}_{l(i)}\right)$, let:

$$\mu_1(H') \equiv \mu_L(H'), \quad \mu_2(H') \equiv \mu_J\left(\left(\pi_L^{(n)}\right)^{-1}(H')\right).$$

If $H' \in \mathcal{A}_L'$ is a measurable rectangle with $H' = \prod_{i=1}^{p} H'_{l(i)}$ and all $H_{j(i)} \in \mathcal{B}(\mathbb{R})$, then $\left(\pi_L^{(n)}\right)^{-1} H' \in \mathcal{A}_J'$ and by direct calculation:

$$\mu_1(H') = \mu_2(H').$$

Finite additivity of μ_L and μ_J obtains that $\mu_1 = \mu_2$ on the algebra $\mathcal{A}_L$, and indeed μ_1 and μ_2 are measures on $\mathcal{B}\left(\prod_{i=1}^{p} \mathbb{R}_{l(i)}\right)$. Thus by Proposition 6.14, $\mu_1 = \mu_2$ on $\sigma\left(\mathcal{A}_L\right) = \mathcal{B}\left(\prod_{i=1}^{p} \mathbb{R}_{l(i)}\right)$. By (1), $H = \left(\pi_L^{(n)}\right)^{-1}(H')$ and so (3) is confirmed and the proof is complete. ∎

With well-definedness confirmed, we now turn to the structure of $\mathcal{A}^+$, that it is an algebra, and then prove finite additivity of μ_0 on $\mathcal{A}^+$.

We leave it as an exercise for the reader to prove that μ_0 is **finitely subadditive** on $\mathcal{A}^+$, meaning that if $\{A_j\}_{j=1}^{m} \subset \mathcal{A}^+$, not necessarily disjoint, then:

$$\mu_0\left(\bigcup_{j=1}^{m} A_j\right) \leq \sum_{j=1}^{m} \mu_0\left(A_j\right).$$

Proposition 9.17 **(μ_0 is finitely additive on the algebra $\mathcal{A}^+$)** *The collection $\mathcal{A}^+$ of general cylinder sets in X is an algebra, with $\mathcal{A} \subset \mathcal{A}^+$. Further, the product set function μ_0 in (9.4) is finitely additive on $\mathcal{A}^+$.*

Proof. *Any set $A \in \mathcal{A}'$ of Definition 9.3 with all Borel sigma algebras, is given by $J = (j(1), ..., j(n))$ and a measurable rectangle $H = \prod_{i=1}^{n} A_{j(i)} \in \mathcal{B}\left(\prod_{i=1}^{n} \mathbb{R}_{j(i)}\right)$, and so $\mathcal{A}' \subset \mathcal{A}^+$ since the definitions of A in (9.5) and (9.3) agree. That $\mathcal{A} \subset \mathcal{A}^+$ will follow by showing $\mathcal{A}^+$ to be an algebra.*

To this end, let $A \in \mathcal{A}^+$ be given by J and $H \in \mathcal{B}\left(\prod_{i=1}^{n} \mathbb{R}_{j(i)}\right)$. Then $\widetilde{A}$ is given by J and $\widetilde{H} \in \mathcal{B}\left(\prod_{i=1}^{n} \mathbb{R}_{j(i)}\right)$, since by (9.2):

$$\widetilde{A} = \{x \in \mathbb{R}^{\mathbb{N}} | (x_{j(1)}, x_{j(2)}, ...x_{j(n)}) \notin H\},$$

so $\widetilde{A} \in \mathcal{A}^+$.

In addition, if A is given as above, and B is given by $K \in \mathcal{B}\left(\prod_{k=1}^{m} \mathbb{R}_{i(k)}\right)$ and $I = (i(1), i(2), ..., i(m))$, then the index set $J \bigcup I$ contains $p \leq n + m$ distinct indexes which can be ordered to produce a combined index set $L = (l(1), l(2), ..., l(p))$. Define $\pi_J^{(n)}$ and $\pi_I^{(m)}$ on $\prod_{i=1}^{p} \mathbb{R}_{l(i)}$ by:

$$\pi_J^{(n)} : \prod_{i=1}^{p} \mathbb{R}_{l(i)} \to \prod_{i=1}^{n} \mathbb{R}_{j(i)},$$

$$\pi_I^{(m)} : \prod_{i=1}^{p} \mathbb{R}_{l(i)} \to \prod_{k=1}^{m} \mathbb{R}_{i(k)}.$$

These are well-defined since $J \subset L$ and $I \subset L$.

The set A can now be expressed in terms of $H' \equiv \left(\pi_J^{(n)}\right)^{-1}(H) \in \mathcal{B}\left(\prod_{i=1}^{p} \mathbb{R}_{l(i)}\right)$ and L, and similarly B can be expressed in terms of $K' \equiv \left(\pi_I^{(m)}\right)^{-1}(K) \in \mathcal{B}\left(\prod_{i=1}^{p} \mathbb{R}_{l(i)}\right)$ and L. For example, define the projection:

$$\pi_L : \mathbb{R}^{\mathbb{N}} \to \prod_{i=1}^{p} \mathbb{R}_{l(i)},$$

and define π_J analogously. Recalling (9.3) and step 1 of the prior proof:

$$\pi_L^{-1}(H') = \pi_L^{-1}\left(\pi_J^{(n)}\right)^{-1}(H)$$

$$= \left(\pi_J^{(n)}\pi_L\right)^{-1}(H)$$

$$= \pi_J^{-1}(H) = A.$$

It then follows that $A \bigcup B$ is given by $H' \bigcup K' \in \mathcal{B}\left(\prod_{i=1}^{p} \mathbb{R}_{l(i)}\right)$ and L. Hence $A \bigcup B \in \mathcal{A}^+$, and $\mathcal{A}^+$ is an algebra. Then also $\mathcal{A} \subset \mathcal{A}^+$ follows from $\mathcal{A}' \subset \mathcal{A}^+$.

For finite additivity of μ_0, let A and B be given as above and assume these sets are disjoint, so $A \bigcap B = \emptyset$. With the same notation it follows that $H' \bigcap K' = \emptyset$. Hence since μ_L is a measure:

$$\mu_0(A \bigcup B) \equiv \mu_L(H' \bigcup K')$$

$$= \mu_L(H') + \mu_L(K')$$

$$\equiv \mu_0(A) + \mu_0(B),$$

where the last step follows from Proposition 9.16. Finite additivity is now obtained by induction. ∎

Remark 9.18 ($\mathcal{A}^+$ vs. $\mathcal{A}$) *In the next sections, we demonstrate that μ_0 is countably additive on the algebra $\mathcal{A}^+$, and thus that the Hahn-Kolmogorov result of Proposition 6.4 can be applied to obtain a measure $\mu_{\mathbb{N}}$ and complete sigma algebra $\sigma(\mathbb{R}^{\mathbb{N}})$ on $\mathbb{R}^{\mathbb{N}}$. As always, $\mathcal{A}^+ \subset \sigma(\mathbb{R}^{\mathbb{N}})$ and $\mu_{\mathbb{N}} = \mu_0$ on $\mathcal{A}^+$. Indirectly, μ_0 will then also have been proved to be countably additive on $\mathcal{A} \subset \mathcal{A}^+$, and hence the same extension theory could again be applied to develop a measure $\mu'_{\mathbb{N}}$ and complete sigma algebra $\sigma'(\mathbb{R}^{\mathbb{N}})$ where $\mathcal{A} \subset \sigma'(\mathbb{R}^{\mathbb{N}})$ and $\mu'_{\mathbb{N}} = \mu_0$ on $\mathcal{A}$.*

Then $\mathcal{A} \subset \mathcal{A}^+$ obtains $\sigma'(\mathbb{R}^{\mathbb{N}}) \subset \sigma(\mathbb{R}^{\mathbb{N}})$ and $\mu'_{\mathbb{N}} = \mu_{\mathbb{N}}$ on $\mathcal{A}$.

It will then be shown in Corollary 9.21 that $\mathcal{A}^+ \subset \sigma'(\mathbb{R}^{\mathbb{N}})$, and hence $\sigma(\mathcal{A}^+) \subset \sigma'(\mathbb{R}^{\mathbb{N}})$ where $\sigma(\mathcal{A}^+)$ denotes the sigma algebra generated by $\mathcal{A}^+$. But as $\sigma'(\mathbb{R}^{\mathbb{N}})$ is complete and $\sigma(\mathbb{R}^{\mathbb{N}})$ is the smallest completion of $\sigma(\mathcal{A}^+)$ by Proposition 6.24, it follows that $\sigma(\mathbb{R}^{\mathbb{N}}) \subset \sigma'(\mathbb{R}^{\mathbb{N}})$ and thus $\sigma(\mathbb{R}^{\mathbb{N}}) = \sigma'(\mathbb{R}^{\mathbb{N}})$.

It will thus finally be shown that starting with the algebra $\mathcal{A}^+$, the same complete measure space $(\mathbb{R}^{\mathbb{N}}, \sigma(\mathbb{R}^{\mathbb{N}}), \mu_{\mathbb{N}})$ is obtained as would have been obtained starting with algebra $\mathcal{A}$.

9.5.3 Countable Additivity of $\mu_{\mathcal{A}}$ on $\mathcal{A}^+$

In this section, we fill in the details of the proof as outlined in Section 9.5.1. To simplify notation and as justified in Remark 9.15, general cylinder sets $A_j \in \mathcal{A}^+$ will be given by $H_j \subset \mathbb{R}^{n(j)}$ and $J_j = (1, 2, ..., n(j))$, where H_j is a Borel measurable set in $\mathbb{R}^{n(j)}$, and μ_{J_j} will denote the finite dimensional Borel (probability) measure induced by $\prod_{i=1}^{n(j)} \mu_i$ on $\mathbb{R}^{n(j)}$. Also, as noted in Section 9.5.1, it is only the case of unbounded $\{n(j)\}$ that needs additional proof and so we explicitly assume that the set $\{n(j)\}$ is unbounded. In addition, we assume that the collection $\{n(j)\}$ is increasing by replacing each $n(j)$ with $m(j) \equiv \max(n(i)|i \leq j)$, and by replacing $H_j \subset \mathbb{R}^{n(j)}$ with $H'_j \subset \mathbb{R}^{m(j)}$ as in Remark 9.15.

Proposition 9.19 (μ_0 is countably additive on $\mathcal{A}^+$) *Given probability spaces $\{(\mathbb{R}, \mathcal{B}(\mathbb{R}), \mu_i)\}_{i=1}^{\infty}$, where $\mathcal{B}(\mathbb{R})$ denotes the Borel sigma algebra of Definition 2.13, let $\mathbb{R}^{\mathbb{N}}$ be given as in Definition 9.15, and $\mathcal{A}^+$ denote the algebra of general cylinder sets as defined in (9.2).*

Then the product set function μ_0 defined in (9.4) on $\mathcal{A}^+$ is countably additive and hence a probability measure $\mu_{\mathcal{A}}$ on this algebra.

Proof. *We will prove continuity from above and demonstrate this using a proof by contradiction. To this end, let $\{A_j\}_{j=1}^{\infty} \subset \mathcal{A}^+$ be a collection of nonempty nested cylinder sets, $A_{j+1} \subset A_j$. Note that since $\{(\mathbb{R}, \mathcal{B}(\mathbb{R}), \mu_i)\}_{i=1}^{\infty}$ are probability spaces, $\mu_0(A_1) < \infty$. We show that if $\mu_0(A_j) \geq \epsilon > 0$ for all j, then $\bigcap_{j=1}^{\infty} A_j \neq \emptyset$. Equivalently, if $\bigcap_{j=1}^{\infty} A_j = \emptyset$, then of necessity $\mu_0(A_j) \to 0$, since these measures are decreasing. Then since μ_0 is continuous from above, it is countably additive by Proposition 6.19.*

That μ_0 is a probability measure on $\mathcal{A}^+$ follows from the observation that $\mathbb{R}^{\mathbb{N}} = \pi_1^{-1}(\mathbb{R}) \in \mathcal{A}^+$, where π_1 is the projection to the first component, and thus:

$$\mu_0\left(\mathbb{R}^{\mathbb{N}}\right) = \mu_1(\mathbb{R}) = 1.$$

There are four main steps. As noted above, we assume that A_j is given by $H_j \subset \mathbb{R}^{n(j)}$ and $J_j = (1, 2, ..., n(j))$, where $\{n(j)\}$ is increasing and unbounded. In this notation, (9.4) is stated:

$$\mu_0(A_j) = \mu_{J_j}(H_j).$$

1. Replacing $\{A_j\}_{j=1}^{\infty}$ by Special Subsets $\{C_j\}_{j=1}^{\infty}$: As each A_j is given by Borel measurable $H_j \subset \mathbb{R}^{n(j)}$ and $\mu_{J_j}(H_j) = \mu_0(A_j)$, it follows that $\mu_{J_j}(H_j) \geq \epsilon > 0$ for all j. Each Borel measure

μ_{J_j} is inner regular by Proposition 8.20, so there exists compact $K_j \subset H_j$ such that $\mu_{J_j}(H_j - K_j) < \epsilon/2^{j+1}$. Define $B_j \in \mathcal{A}^+$ to be given by $K_j \subset \mathbb{R}^{n(j)}$ and $J_j = (1, 2, ..., n(j))$, so $B_j = \pi_{J_j}^{-1}(K_j)$, and define $C_j \subset \mathbb{R}^{\mathbb{N}}$ by $C_j = \bigcap_{i \leq j} B_i$. Note that $C_j \in \mathcal{A}^+$ since $\mathcal{A}^+$ is an algebra.

Then $C_j \subset B_j \subset A_j$, and since $A_j - B_j \equiv A_j \bigcap \tilde{B}_j \in \mathcal{A}^+$ is given by $H_j - K_j = H_j \bigcap \tilde{K}_j$ and J_j:

$$\mu_0(A_j - B_j) = \mu_{J_j}(H_j - K_j) < \epsilon/2^{j+1}. \tag{1}$$

Then by De Morgan's laws:

$$A_j - C_j = A_j \bigcap \left(\bigcup_{i \leq j} \tilde{B}_i \right)$$
$$= \bigcup_{i \leq j} \left(A_j \bigcap \tilde{B}_i \right)$$
$$= \bigcup_{i \leq j} (A_j - B_i).$$

As $\{A_j - B_i\}_{i \leq j}$ need not be disjoint, we use finite subadditivity for $\mu_0(A_j - C_j)$, noting that $\mu_0(A_j - C_j)$ is defined since $A_j - C_j \in \mathcal{A}^+$. Now $\mu_0(A_j - B_i) \leq \mu_0(A_i - B_i)$ for $i \leq j$ by the nesting property, and this obtains with (1) that for all j:

$$\mu_0(A_j - C_j) \leq \sum_{i \leq j} \mu_0(A_i - B_i) < \epsilon/2.$$

Since $\mu_0(A_j) \geq \epsilon$, it follows that $\mu_0(C_j) > \epsilon/2$ for all j. Hence, $\{C_j\}_{j=1}^{\infty}$ are nonempty sets in $\mathbb{R}^{\mathbb{N}}$.

Summarizing, $\{C_j\}_{j=1}^{\infty} \subset \mathbb{R}^{\mathbb{N}}$ is a nested sequence of nonempty subsets of $\{A_j\}_{j=1}^{\infty}$ with $C_{j+1} \subset C_j$ for all j.

The goal of the next steps is to use this sequence of nested sets to construct $y \in \bigcap_{j=1}^{\infty} B_j$, and it will then follow from $B_j \subset A_j$ that $\bigcap_{j=1}^{\infty} A_j \neq \emptyset$, completing the proof.

2. An Infinite Point Sequence: To prove $\bigcap_{j=1}^{\infty} B_j \neq \emptyset$, first choose $x^{(j)} \in C_j$ for all j, which is possible since all C_j are nonempty. Then, each $x^{(j)} \in A_j$, and since A_j is defined by $J_j = (1, 2, ..., n(j))$ with $\{n(j)\}$ increasing, it follows that for every j:

$$x^{(j)} = (x_1^{(j)}, ..., x_{n(1)}^{(j)}, x_{n(1)+1}^{(j)}, ..., x_{n(2)}^{(j)}, ..., x_{n(j-1)+1}^{(j)}, ..., x_{n(j)}^{(j)}, ...).$$

Further, since $H_j = \pi_{J_j}(A_j)$:

$$(x_1^{(j)}, ..., x_{n(1)}^{(j)}, x_{n(1)+1}^{(j)}, ..., x_{n(2)}^{(j)}, ..., x_{n(j-1)+1}^{(j)}, ..., x_{n(j)}^{(j)}) \in H_j.$$

As $x^{(j)} \in C_j = \bigcap_{i \leq j} B_i \equiv \bigcap_{i \leq j} \pi_{J_i}^{-1}(K_i)$, we can similarly conclude that:

$$(x_1^{(j)}, ..., x_{n(1)}^{(j)}) \in K_1, \text{ all } j;$$
$$(x_1^{(j)}, ..., x_{n(1)}^{(j)}, x_{n(1)+1}^{(j)}, ..., x_{n(2)}^{(j)}) \in K_2, \text{ all } j \geq 2;$$

and in general:

$$(x_1^{(j)}, ..., x_{n(1)}^{(j)}, x_{n(1)+1}^{(j)}, ..., x_{n(2)}^{(j)}, ..., x_{n(i-1)+1}^{(j)}, ..., x_{n(i)}^{(j)}) \in K_i, \text{ all } j \geq i.$$

3. A Convergent Subsequence: *Recall each $K_i \subset \mathbb{R}^{n(i)}$, so let $\pi_k : \mathbb{R}^{n(i)} \to \mathbb{R}$ denote the kth coordinate projection mapping for $k \leq n(i)$. As K_i is compact and hence closed and bounded, it follows that $\pi_k(K_i)$ is compact for all k. Specifically π_k is continuous by Proposition 3.12, and such mappings preserve compactness by an application of this proposition. Now consider $\left\{x_1^{(j)}\right\}_{j=1}^{\infty} \in \pi_1(K_1)$. Any infinite sequence of points in compact $\pi_1(K_1)$ has an accumulation point in $\pi_k(K_1)$ and a subsequence that converges to this accumulation point. Thus there exists $y_1 \in \pi_1(K_1)$ and subsequence $\left\{x_1^{(n_{1,m})}\right\}_{m=1}^{\infty}$ which converges to y_1.*

Next consider $\left\{x_2^{(n_{1,m})}\right\}_{m=1}^{\infty} \in \pi_2(K_1)$, defined to be a subsequence, using the original $\left\{x_2^{(j)}\right\}_{j=1}^{\infty} \in \pi_2(K_1)$. Then again, there exists $y_2 \in \pi_2(K_1)$ and subsequence $\left\{x_2^{(n_{2,m})}\right\}_{m=1}^{\infty}$ which converges to y_2. By construction, $\{n_{2,m}\}_{m=1}^{\infty} \subset \{n_{1,m}\}_{m=1}^{\infty}$. We continue this construction through the $x_{n(1)}$ components, then turn to compact K_2, which by part 2 contains all the sequences $\left\{x_k^{(j)}\right\}_{j=2}^{\infty}$ for $n(1) + 1 \leq k \leq n(2)$.

Each step constructs a subsequence of the prior index subsequence, which converges to a point $y_k \in \pi_k(K_2)$. And this process continues for each component in turn, using the compactness of each $\pi_k(K_i)$ and identifying an accumulation point for each component, say, y_k, as well as a subsequence of the original collection of k-components which converge to this point.

We now have a collection of subsequences, $\{\{n_{k,m}\}_{m=1}^{\infty}\}_{k=1}^{\infty}$ so that:

$$\{n_{k+1,m}\}_{m=1}^{\infty} \subset \{n_{k,m}\}_{m=1}^{\infty},$$

and $\left\{x_k^{(n_{k,m})}\right\}_{m=1}^{\infty}$ converges to y_k for each k.

4. The Nonempty Intersection: *Finally, consider the diagonal index sequence $\{n_{i,i}\}_{i=1}^{\infty}$. It is then true by construction that this index sequence is a subsequence of all the constructed index sequences, and hence with this single sequence, $\left\{x_k^{(n_{i,i})}\right\}_{i=1}^{\infty}$ converges to y_k for each k. We now prove that:*

$$y \equiv (y_1, y_2, y_3, ...) \in B_j \text{ for all } j.$$

To this end, note that $(y_1, y_2, y_3, ...y_{n(1)}) \in K_1$ by construction and so $y \in \pi_{J_1}^{-1}(K_1) = B_1$. But also $(y_1, y_2, y_3, ...y_{n(2)}) \in K_2$, since by construction this is a limit of a subsequence of $(x_1^{(j)}, ..., x_{n(1)}^{(j)}, x_{n(1)+1}^{(j)}, ..., x_{n(2)}^{(j)}) \in K_2$, and so $y \in \pi_{J_2}^{-1}(K_2) = B_2$, and so forth, completing the proof.

Hence $y \in A_j$ for all j, proving $\bigcap_{j=1}^{\infty} A_j \neq \emptyset$. ∎

9.6 Extension to a Probability Measure on $\mathbb{R}^{\mathbb{N}}$

The following result is a special case of **Kolmogorov's Existence theorem**, named for **Andrey Kolmogorov** (1903–1987). In Book VII, which considers the probability space $(\mathbb{R}^T, \sigma(\mathbb{R}^T), \mu)$ of Brownian Motion, a more general version will be developed. There $\mathbb{R}^T$ is the space of all functions $f : [0, T] \to \mathbb{R}$ defined on a real interval $[0, T]$ which could be $[0, \infty)$. This generalizes the notation $\mathbb{R}^{\mathbb{N}}$, which can be identified with the space of all functions $f : \mathbb{N} \to \mathbb{R}$, thereby producing all possible real sequences.

From the perspective of the more general formulation, the component spaces of Kolmogorov's result, here $\{(\mathbb{R}, \sigma_i(\mathbb{R}), \mu_i)\}_{i=1}^{\infty}$, are suppressed. Instead, one is given $\mathbb{R}^{\mathbb{N}}$ and a collection of Borel measures $\{\mu_J\}$ defined on $\mathbb{R}^n$ for all n, and the requirement that these measures satisfy certain consistency conditions. An example of such a consistency condition was noted in item 3 of the proof of Proposition 9.16.

Recall the set-up, that:

$$L \equiv (l(1), l(2), ..., l(p)) \subset J = (j(1), ..., j(n)),$$

and $\pi_L^{(n)}$ is the projection:

$$\pi_L^{(n)} : \prod_{i=1}^{n} \mathbb{R}_{j(i)} \to \prod_{i=1}^{p} \mathbb{R}_{l(i)}.$$

It was then proved that if $H' \in \mathcal{B}\left(\prod_{i=1}^{p} \mathbb{R}_{l(i)}\right)$ and $H \equiv \left(\pi_L^{(n)}\right)^{-1} H'$, then $H \in \mathcal{B}\left(\prod_{i=1}^{n} \mathbb{R}_{j(i)}\right)$ and:

$$\mu_J(H) = \mu_L(H').$$

The point is that H with J, and H' with L, convey the same information in terms of defining a set $A \subset \mathbb{R}^{\mathbb{N}}$. The consistency requirement of Kolmogorov's result demands that the measures of these "finite dimensional sets," here called cylinders sets, must therefore agree.

Proposition 9.20 (Kolmogorov's Existence Theorem) *Given probability spaces* $\{(\mathbb{R}, \mathcal{B}(\mathbb{R}), \mu_i)\}_{i=1}^{\infty}$, *let* $\mathcal{A}^+$ *denote the algebra of general cylinder sets defined in (9.2), and* $\mu_{\mathcal{A}}$ *the measure on* $\mathcal{A}^+$ *of Proposition 9.19 defined by* μ_0 *in (9.4).*

Then $\mu_{\mathcal{A}}$ *can be extended to a probability measure* $\mu_{\mathbb{N}}$ *on a complete sigma algebra* $\sigma(\mathbb{R}^{\mathbb{N}})$. *By extended it is meant that* $\mathcal{A}^+ \subset \sigma(\mathbb{R}^{\mathbb{N}})$ *and* $\mu_{\mathbb{N}}(A) = \mu_{\mathcal{A}}(A)$ *for all* $A \in \mathcal{A}^+$. *Thus,* $(\mathbb{R}^{\mathbb{N}}, \sigma(\mathbb{R}^{\mathbb{N}}), \mu_{\mathbb{N}})$ *is a complete probability space.*

Further, $\mu_{\mathbb{N}}$ *is the unique extension of* $\mu_{\mathcal{A}}$ *to* $\sigma(\mathcal{A}^+)$, *the smallest sigma algebra containing* $\mathcal{A}^+$, *and to* $\sigma(\mathbb{R}^{\mathbb{N}})$.

Proof. *By the Hahn-Kolmogorov extension theorem of Proposition 6.4, the measure* $\mu_{\mathcal{A}}$ *and algebra* $\mathcal{A}^+$ *define an outer measure* $\mu_{\mathcal{A}}^*$ *on* $\sigma(P(\mathbb{R}^{\mathbb{N}}))$, *the power sigma algebra on* $\mathbb{R}^{\mathbb{N}}$.

In addition, the collection of Carathéodory measurable sets is a sigma algebra $\sigma(\mathbb{R}^{\mathbb{N}})$ with $\mathcal{A}^+ \subset \sigma(\mathbb{R}^{\mathbb{N}})$, and $\mu^*_{\mathcal{A}}$ restricted to $\sigma(\mathbb{R}^{\mathbb{N}})$ is a measure on this sigma algebra. With $\mu_{\mathbb{N}}$ denoting the restriction of $\mu^*_{\mathcal{A}}$ to $\sigma(\mathbb{R}^{\mathbb{N}})$, it follows that $(\mathbb{R}^{\mathbb{N}}, \sigma(\mathbb{R}^{\mathbb{N}}), \mu_{\mathbb{N}})$ is a complete measure space and $\mu_{\mathbb{N}}(A) = \mu_{\mathcal{A}}(A)$ for all $A \in \mathcal{A}^+$.

The extension of $\mu_{\mathcal{A}}$ on $\mathcal{A}^+$ to $\mu_{\mathbb{N}}$ on $\sigma(\mathbb{R}^{\mathbb{N}})$ is uniquely defined on $\sigma(\mathcal{A}^+)$ by Proposition 6.14, and to $\sigma(\mathbb{R}^{\mathbb{N}})$ by Proposition 6.24.   ∎

The following corollary was promised in Remark 9.18. Since $\mathcal{A} \subset \mathcal{A}^+$, $\mu_{\mathcal{A}}$ is also a measure on $\mathcal{A}$ and thus can be extended by the same process as in the prior proof. We now show that this extension would coincide with that produced in Proposition 9.20, as given by Proposition 6.4.

Corollary 9.21 (Extension of $\mu_{\mathcal{A}}$ on $\mathcal{A}$) *Let $\sigma'(\mathbb{R}^{\mathbb{N}})$ denote the complete sigma algebra of Carathéodory measurable sets defined by $\mu_{\mathcal{A}}$ and $\mathcal{A}$, and $\mu'_{\mathbb{N}}$ the associated probability measure. Then in the notation of Proposition 9.20:*

$$\sigma'(\mathbb{R}^{\mathbb{N}}) = \sigma(\mathbb{R}^{\mathbb{N}}), \quad \mu'_{\mathbb{N}} = \mu_{\mathbb{N}}.$$

In other words:

$$(\mathbb{R}^{\mathbb{N}}, \sigma'(\mathbb{R}^{\mathbb{N}}), \mu'_{\mathbb{N}}) = (\mathbb{R}^{\mathbb{N}}, \sigma(\mathbb{R}^{\mathbb{N}}), \mu_{\mathbb{N}}).$$

Proof. *By Corollary 6.25, this result will follow if it can be proved that $\mathcal{A}^+ \subset \sigma'(\mathbb{R}^{\mathbb{N}})$.*

To this end, let $A \in \mathcal{A}^+$ be given by $J = (j(1), ..., j(n))$ and Borel measurable $H \in \mathcal{B}$ $\left(\prod_{i=1}^n \mathbb{R}_{j(i)}\right)$, and recall Corollary 7.23 on approximating measurable sets in finite dimensional product spaces. If $\mathcal{A}\left(\prod_{i=1}^n \mathbb{R}_{j(i)}\right)$ denotes the algebra of finite disjoint unions of measurable rectangles on $\prod_{i=1}^n \mathbb{R}_{j(i)}$, then there are sets $H^{(1)} \in \mathcal{A}_{\delta\sigma}\left(\prod_{i=1}^n \mathbb{R}_{j(i)}\right)$ and $H^{(2)} \in \mathcal{A}_{\sigma\delta}\left(\prod_{i=1}^n \mathbb{R}_{j(i)}\right)$ so that $H^{(1)} \subset H \subset H^{(2)}$ and $\mu_J(H^{(2)} - H^{(1)}) = 0$. Thus:

$$H^{(1)} = \bigcup_i \left[\bigcap_k B_{ik}^{(1)}\right], \quad H^{(2)} = \bigcap_i \left[\bigcup_k B_{ik}^{(2)}\right], \tag{1}$$

where each $B_{ik}^{(\cdot)} \in \mathcal{A}\left(\prod_{i=1}^n \mathbb{R}_{j(i)}\right)$.

Define $A_{ik}^{(\cdot)} \subset \mathbb{R}^{\mathbb{N}}$ to be given by $B_{ik}^{(\cdot)}$ and index set J. As each $B_{ik}^{(\cdot)}$ is a finite disjoint union of measurable rectangles from the associated semi-algebra $\mathcal{A}'\left(\prod_{i=1}^n \mathbb{R}_{j(i)}\right)$:

$$B_{ik}^{(\cdot)} = \bigcup_{j=1}^{N_{ik}^{(\cdot)}} B_{ikj}^{(\cdot)},$$

it follows that:

$$A_{ik}^{(\cdot)} = \pi_J^{-1}\left(B_{ik}^{(\cdot)}\right) = \bigcup_{j=1}^{N_{ik}^{(\cdot)}} \pi_J^{-1}\left(B_{ikj}^{(\cdot)}\right).$$

Thus since $\pi_J^{-1}\left(B_{ikj}^{(\cdot)}\right) \in \mathcal{A}'$ for all j by Definition 9.3, and are disjoint by construction, we have $A_{ik}^{(\cdot)} \in \mathcal{A}$ for all i, k.

Next, define $A^{(\cdot)} \subset \mathbb{R}^{\mathbb{N}}$ to be given by $H^{(\cdot)}$ in (1) and index set J. Then $A^{(1)} \in \sigma'(\mathbb{R}^{\mathbb{N}})$ since by the same calculation:

$$A^{(1)} \equiv \bigcup_i \left[\bigcap_k A_{ik}^{(1)} \right],$$

with $A_{ik}^{(1)} \in \mathcal{A} \subset \sigma'(\mathbb{R}^{\mathbb{N}})$. Similarly, $A^{(2)} \in \sigma'(\mathbb{R}^{\mathbb{N}})$.

Thus if $A \in \mathcal{A}^+$, there is $A^{(1)}, A^{(2)} \in \sigma'(\mathbb{R}^{\mathbb{N}})$ with $A^{(1)} \subset A \subset A^{(2)}$. Further, since $A^{(2)} - A^{(1)}$ is given by J and $H^{(2)} - H^{(1)}$:

$$\mu_{\mathbb{N}}(A^{(2)} - A^{(1)}) = \mu_0(A^{(2)} - A^{(1)}) = \mu_J(H^{(2)} - H^{(1)}) = 0.$$

As $\sigma'(\mathbb{R}^{\mathbb{N}})$ is complete with respect to $\mu_{\mathbb{N}}$, this proves that $A \in \sigma'(\mathbb{R}^{\mathbb{N}})$ and hence $\mathcal{A}^+ \subset \sigma'(\mathbb{R}^{\mathbb{N}})$. ∎

As noted prior to Corollary 7.23, one of the advantages of developing a measure space by the Carathéodory approach is that we immediately obtain the approximations of Proposition 6.5, which we state here for completeness, in two versions.

Corollary 9.22 (Approximating Carathéodory Measurable Sets 1) *Let $\mathcal{A}^+$ denote the algebra of general cylinder sets as in Definition 9.14, and $(\mathbb{R}^{\mathbb{N}}, \sigma(\mathbb{R}^{\mathbb{N}}), \mu_{\mathbb{N}})$ the associated complete infinite product measure space given in Proposition 9.20. If $B \in \sigma(\mathbb{R}^{\mathbb{N}})$, then given $\epsilon > 0$:*

1. *There is a set $A \in \mathcal{A}_\sigma^+$, the collection of countable unions of sets in the algebra $\mathcal{A}^+$, so that $B \subset A$ and:*

$$\mu_{\mathbb{N}}(A) \leq \mu_{\mathbb{N}}(B) + \epsilon, \qquad \mu_{\mathbb{N}}(A - B) < \epsilon. \tag{9.6}$$

2. *There is a set $C \in \mathcal{A}_\delta^+$, the collection of countable intersections of sets in the algebra $\mathcal{A}^+$, so that $C \subset B$ and:*

$$\mu_{\mathbb{N}}(B) \leq \mu_{\mathbb{N}}(C) + \epsilon, \qquad \mu_{\mathbb{N}}(B - C) < \epsilon. \tag{9.7}$$

3. *There is a set $A' \in \mathcal{A}_{\sigma\delta}^+$, the collection of countable intersections of sets in $\mathcal{A}_\sigma^+$, and a set $C' \subset \mathcal{A}_{\delta\sigma}^+$, the collection of countable unions of sets in $\mathcal{A}_\delta^+$, so that $C' \subset B \subset A'$ and:*

$$\mu_{\mathbb{N}}(A' - B) = \mu_{\mathbb{N}}(B - C') = 0. \tag{9.8}$$

Proof. *Propositions 6.5 and 9.20.* ∎

Corollary 9.23 (Approximating Carathéodory Measurable Sets 2) *Let $\mathcal{A}$ denote the algebra generated by the semi-algebra of cylinder sets $\mathcal{A}'$ of Definition 9.3, and $(\mathbb{R}^{\mathbb{N}}, \sigma(\mathbb{R}^{\mathbb{N}}), \mu_{\mathbb{N}})$ the associated complete infinite product measure space given in Corollary 9.21. If $B \in \sigma(\mathbb{R}^{\mathbb{N}})$, then given $\epsilon > 0$:*

1. *There is a set $A \in \mathcal{A}_\sigma$, the collection of countable unions of sets in the algebra $\mathcal{A}$, so that $B \subset A$ and:*

$$\mu_{\mathbb{N}}(A) \leq \mu_{\mathbb{N}}(B) + \epsilon, \qquad \mu_{\mathbb{N}}(A - B) < \epsilon.$$

2. *There is a set $C \in \mathcal{A}_\delta$, the collection of countable intersections of sets in the algebra $\mathcal{A}$, so that $C \subset B$ and:*

$$\mu_{\mathbb{N}}(B) \leq \mu_{\mathbb{N}}(C) + \epsilon, \qquad \mu_{\mathbb{N}}(B - C) < \epsilon.$$

3. *There is a set $A' \in \mathcal{A}_{\sigma\delta}$, the collection of countable intersections of sets in $\mathcal{A}_\sigma$, and a set $C' \subset \mathcal{A}_{\delta\sigma}$, the collection of countable unions of sets in $\mathcal{A}_\delta$, so that $C' \subset B \subset A'$ and:*

$$\mu_{\mathbb{N}}(A' - B) = \mu_{\mathbb{N}}(B - C') = 0.$$

Proof. *Proposition 6.5 and Corollary 9.21.* ∎

9.7 Probability of General Rectangles

We began this chapter with a naive attempt to define the measure of a general product set in an infinite dimensional product space defined on $\{(X_i, \sigma(X_i), \mu_i)\}_{i=1}^{\infty}$. Specifically, if:

$$A = \prod_{i=1}^{\infty} A_i, \text{ with } A_i \in \sigma(X_i),$$

we attempted to define a **product measure** denoted μ_0 by:

$$\mu_0(A) = \prod_{i=1}^{\infty} \mu_i(A_i).$$

This definition required the ability to work with infinite products, the basic results of which were addressed in Exercise 9.1. More importantly, defining the measure of such a set without the necessary framework of a measure theory created open questions on the properties of such a "measure," and indeed whether such a definition could ultimately be proved to be consistent and welldefined on a sigma algebra of interest.

As the necessary framework has now been developed for the infinite dimensional complete probability space $(\mathbb{R}^{\mathbb{N}}, \sigma(\mathbb{R}^{\mathbb{N}}), \mu_{\mathbb{N}})$, we now return to a consideration of the question of the probabilities of such infinite dimensional rectangles. Of course, we must also prove that such rectangles are measurable.

Proposition 9.24 (Probability of general rectangles) *Let $A \subset \mathbb{R}^{\mathbb{N}}$ be defined by $\mathbb{N}$ and $\prod_{i=1}^{\infty} A_i$, where $A_i \in \mathcal{B}(\mathbb{R})$. In other words,*

$$A = \{x \in \mathbb{R}^{\mathbb{N}} | x_i \in A_i\}.$$

Then $A \in \sigma(\mathbb{R}^{\mathbb{N}})$, and:

$$\mu_{\mathbb{N}}(A) = \prod_{i=1}^{\infty} \mu_i(A_i). \tag{9.9}$$

Proof. *Define $B_j = \{x \in \mathbb{R}^{\mathbb{N}} | x_i \in A_i \text{ for } i \leq j\}$. Then B_j is a finite dimensional measurable rectangle for all j, so $B_j \in \mathcal{A}' \subset \sigma(\mathbb{R}^{\mathbb{N}})$, and hence $\bigcap_j B_j \in \sigma(\mathbb{R}^{\mathbb{N}})$. Since $A \subset B_j$ for all j, it follows that $A \subset \bigcap_j B_j$. But if $x \in \bigcap_j B_j$, then by definition $x_i \in A_i$ for all i, and so $x \in A$. Thus $A = \bigcap_j B_j$ and consequently $A \in \sigma(\mathbb{R}^{\mathbb{N}})$.*

Now since $\mu_{\mathbb{N}}(B_1) = \mu_1(A_1) < \infty$ and $B_{j+1} \subset B_j$ for all j, it follows by continuity from above of the measure $\mu_{\mathbb{N}}$ that:

$$\mu_{\mathbb{N}}(A) = \lim_{j \to \infty} \mu_{\mathbb{N}}(B_j)$$
$$= \lim_{j \to \infty} \prod_{i=1}^{j} \mu_i(A_i). \qquad \blacksquare$$

Corollary 9.25 (Probability 0 general rectangles) *Let $A \subset \mathbb{R}^{\mathbb{N}}$ be defined by $\mathbb{N}$ and $\prod_{i=1}^{\infty} A_i$, where $A_i \in \mathcal{B}(\mathbb{R})$. If there is an $\epsilon > 0$ so that $0 \leq \mu_i(A_i) \leq 1 - \epsilon$ for all i, then $\mu_{\mathbb{N}}(A) = 0$.*

Proof. *From the above proof:*

$$\mu_{\mathbb{N}}(A) = \lim_{j \to \infty} \prod_{i=1}^{j} \mu_i(A_i)$$
$$\leq \lim_{j \to \infty} (1 - \epsilon)^j$$
$$= 0. \qquad \blacksquare$$

Remark 9.26 (On nonzero probability general rectangles) *In the infinite dimensional probability space $(\mathbb{R}^{\mathbb{N}}, \sigma(\mathbb{R}^{\mathbb{N}}), \mu_{\mathbb{N}})$, an infinite dimensional rectangle $A = \prod_{i=1}^{\infty} A_i$ can have nonzero probability only if with $\mu_i(A_i) \equiv 1 - \epsilon_i > 0$, that $\epsilon_i \to 0$ fast enough that $\sum_i \ln(1 - \epsilon_i)$ converges. By a Taylor series analysis and Exercise 9.1, $\sum_i \ln(1 - \epsilon_i)$ converges if and only if $\sum_i \epsilon_i$ converges.*

For an obvious example, this occurs when $\mu_i(A_i) = 1$ for all but finitely many i, which is to say that A is then a finite dimensional measurable rectangle. More generally, every convergent series $\sum_i \epsilon_i$ with $0 < \epsilon_i < 1$ provides a criterion for an infinite dimensional rectangle to have positive probability.

References

I have listed below a number of textbook references for the mathematics and finance presented in this series of books. Many provide both theoretical and applied materials in their respective areas that are beyond those developed here, and they would be worth pursuing by those interested in gaining a greater depth or breadth of knowledge in the given subjects. This list is by no means complete and is intended only as a guide to further study. In addition, a limited number of research papers will be identified in each book if they are referenced therein. A more complete guide to published papers can be found in the references below.

The reader will no doubt observe that the mathematics references are somewhat older than the finance references and upon web searching will find that several of the older texts in each category have been updated to newer editions, sometimes with additional authors. Since I own and use the editions below, I decided to present these editions rather than reference the newer editions which I have not reviewed. As many of these older texts are considered "classics," they are also likely to be found in university and other libraries.

That said, there are undoubtedly many very good new texts by both new and established authors with similar titles that are also worth investigating. One that I will, at the risk of immodesty, recommend for more introductory materials on Mathematics, Probability Theory and Finance is:

1. Reitano, Robert, R. *Introduction to Quantitative Finance: A Math Tool Kit.* Cambridge, MA: The MIT Press, 2010.

Topology, Measure, Integration, Linear Algebra

1. Doob, J. L. *Measure Theory.* New York, NY: Springer-Verlag, 1994.
2. Dugundji, James. *Topology.* Boston, MA: Allyn and Bacon, 1970.
3. Edwards, Jr., C. H. *Advanced Calculus of Several Variables.* New York, NY: Academic Press, 1973.
4. Gemignani, M. C. *Elementary Topology.* Reading, MA: Addison-Wesley Publishing, 1967.
5. Halmos, Paul R. *Measure Theory.* New York, NY: D. Van Nostrand, 1950.
6. Hewitt, Edwin, and Karl Stromberg. *Real and Abstract Analysis.* New York, NY: Springer-Verlag, 1965.
7. Royden, H. L. *Real Analysis*, 2nd Edition. New York, NY: The MacMillan Company, 1971.
8. Rudin, Walter. *Principals of Mathematical Analysis*, 3rd Edition. New York, NY: McGraw-Hill, 1976.
9. Rudin, Walter. *Real and Complex Analysis*, 2nd Edition. New York, NY: McGraw-Hill, 1974.
10. Shilov, G. E., and B. L. Gurevich. *Integral, Measure & Derivative: A Unified Approach.* New York, NY: Dover Publications, 1977.
11. Strang, Gilbert. *Introduction to Linear Algebra*, 4th Edition. Wellesley, MA: Cambridge Press, 2009.

Probability Theory, Stochastic Processes

1. Billingsley, Patrick. *Probability and Measure*, 3rd Edition. New York, NY: John Wiley & Sons, 1995.

2. Chung, K. L., and R. J. Williams. *Introduction to Stochastic Integration*. Boston, MA: Birkhäuser, 1983.

3. Davidson, James. *Stochastic Limit Theory*. New York, NY: Oxford University Press, 1997.

4. de Haan, Laurens, and Ana Ferreira. *Extreme Value Theory, An Introduction*. New York, NY: Springer Science, 2006.

5. Durrett, Richard. *Probability: Theory and Examples*, 2nd Edition. Belmont, CA: Wadsworth Publishing, 1996.

6. Durrett, Richard. *Stochastic Calculus, A Practical Introduction*. Boca Raton, FL: CRC Press, 1996.

7. Feller, William. *An Introduction to Probability Theory and Its Applications*, Volume I. New York, NY: John Wiley & Sons, 1968.

8. Feller, William. *An Introduction to Probability Theory and Its Applications*, Volume II, 2nd Edition. New York, NY: John Wiley & Sons, 1971.

9. Friedman, Avner. *Stochastic Differential Equations and Applications, Volumes 1 and 2*. New York, NY: Academic Press, 1975.

10. Ikeda, Nobuyuki, and Shinzo Watanabe. *Stochastic Differential Equations and Diffusion Processes*. Tokyo, Japan: Kodansha Scientific, 1981.

11. Karatzas, Ioannis, and Steven E. Shreve. *Brownian Motion and Stochastic Calculus*. New York, NY: Springer-Verlag, 1988.

12. Kloeden, Peter E., and Eckhard Platen. *Numerical Solution of Stochastic Differential Equations*. New York, NY: Springer-Verlag, 1992.

13. Lowther, George, *Almost Sure, A Maths Blog on Stochastic Calculus*, 2018–2000, https://almostsure.wordpress.com/stochastic-calculus/

14. Lukacs, Eugene. *Characteristic Functions*. New York, NY: Hafner Publishing, 1960.

15. Nelson, Roger B. *An Introduction to Copulas*, 2nd Edition. New York, NY: Springer Science, 2006.

16. ksendal, Bernt. *Stochastic Differential Equations, An Introduction with Applications*, 5th Edition. New York, NY: Springer-Verlag, 1998.

17. Protter, Phillip. *Stochastic Integration and Differential Equations, A New Approach*. New York, NY: Springer-Verlag, 1992.

18. Revuz, Daniel, and Marc Yor. *Continuous Martingales and Brownian Motion*, 3rd Edition. New York, NY: Springer-Verlag, 1991.

19. Rogers, L. C. G., and D. Williams. *Diffusions, Markov Processes and Martingales*, Volume 1, Foundations, 2nd Edition. Cambridge, UK: Cambridge University Press, 2000.

20. Rogers, L. C. G., and D. Williams. *Diffusions, Markov Processes and Martingales*, Volume 2, Itô Calculus, 2nd Edition. Cambridge, UK: Cambridge University Press, 2000.

21. Sato, Ken-Iti. *Lévy Processes and Infinitely Divisible Distributions*. Cambridge University Press, Cambridge, UK, 1999.

22. Schilling, René L. and Lothar Partzsch. *Brownian Motion: An Introduction to Stochastic Processes*, 2nd Edition. Berlin/Boston: Walter de Gruyter GmbH, 2014.

23. Schuss, Zeev, *Theory and Applications of Stochastic Differential Equations*. New York, NY: John Wiley & Sons, 1980.

Finance Applications

1. Etheridge, Alison. *A Course in Financial Calculus*. Cambridge, UK: Cambridge University Press, 2002.

2. Embrechts, Paul, Claudia Klüppelberg, and Thomas Mikosch. *Modelling Extremal Events for Insurance and Finance*. New York, NY: Springer-Verlag, 1997.

3. Hunt, P. J., and J. E. Kennedy. *Financial Derivatives in Theory and Practice*, Revised Edition. Chichester, UK: John Wiley & Sons, 2004.

4. McLeish, Don L. *Monte Carlo Simulation and Finance*. New York, NY: John Wiley, 2005.

5. McNeil, Alexander J., Rüdiger Frey, and Paul Embrechts. *Quantitative Risk Management: Concepts, Techniques, and Tools*. Princeton, NJ.: Princeton University Press, 2005.

Research Papers for Book I

1. Solovay, Robert M. "A model of set-theory in which every set of reals is Lebesgue measurable," Annals of Mathematics. Second Series, 92: 1–56, 1970.

Index

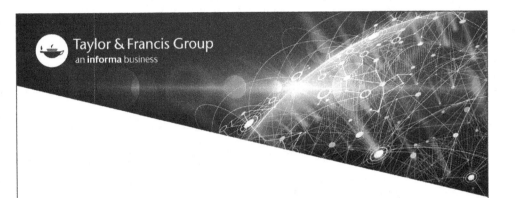

Taylor & Francis eBooks

www.taylorfrancis.com

A single destination for eBooks from Taylor & Francis
with increased functionality and an improved user
experience to meet the needs of our customers.

90,000+ eBooks of award-winning academic content in
Humanities, Social Science, Science, Technology, Engineering,
and Medical written by a global network of editors and authors.

TAYLOR & FRANCIS EBOOKS OFFERS:

A streamlined
experience for
our library
customers

A single point
of discovery
for all of our
eBook content

Improved
search and
discovery of
content at both
book and
chapter level

REQUEST A FREE TRIAL
support@taylorfrancis.com

 Routledge
Taylor & Francis Group

 CRC Press
Taylor & Francis Group

Printed in the United States
by Baker & Taylor Publisher Services

Printed in the United States
by Baker & Taylor Publisher Services